AQA Science
Science A

GCSE

Jim Breithaupt

Ann Fullick

Lawrie Ryan

Pauline Anning

Bev Cox

Niva Miles

Gavin Reeder

John Scottow

Editor
Lawrie Ryan

WITHDRAWN

D1335729

OXFORD
UNIVERSITY PRESS

Leeds City College
Technology Campus

TC07550

OXFORD
UNIVERSITY PRESS

Keighley Campus Library
Leeds City College
16|17   TC07550

Great Clarendon Street, Oxford, OX2 6DP, United Kingdom

Oxford University Press is a department of the University of Oxford. It furthers the University's objective of excellence in research, scholarship, and education by publishing worldwide. Oxford is a registered trade mark of Oxford University Press in the UK and in certain other countries

Text © Jim Breithaupt, Ann Fullick, Patrick Fullick and Lawrie Ryan 2011
Original illustrations © Oxford University Press 2014

The moral rights of the authors have been asserted

First published by Nelson Thornes Ltd in 2011
This edition published by Oxford University Press in 2014

All rights reserved. No part of this publication may be reproduced, stored in a retrieval system, or transmitted, in any form or by any means, without the prior permission in writing of Oxford University Press, or as expressly permitted by law, by licence or under terms agreed with the appropriate reprographics rights organization. Enquiries concerning reproduction outside the scope of the above should be sent to the Rights Department, Oxford University Press, at the address above.

You must not circulate this work in any other form and you must impose this same condition on any acquirer

British Library Cataloguing in Publication Data
Data available

978-1-4085-0822-0

10 9 8 7 6 5 4 3

Printed in China by Golden Cup Printing Co. Ltd

Acknowledgements

Cover photograph: Suzanne Laird/Getty Images (girls); Andrew Butterton/Alamy (background)
Illustrations: include artwork drawn by Wearset Ltd and David Russell Illustration
Page make-up: Wearset Ltd
Index: created by Indexing Specialists (UK) Ltd

AQA examination questions are reproduced by permission of the Assessment and Qualifications Alliance.

H1.1 Science Source/Science Photo Library; H1.2 iStockphoto; H2.1 Martyn F. Chillmaid/Science Photo Library; H2.2 Steve Taylor/Science Photo Library; H3.1 John Kaprielian/Science Photo Library; H3.2 iStockphoto; H3.3 Martyn F. Chillmaid; H4.1 iStockphoto; H4.2 Cordelia Molloy/Science Photo Library; H5.1 Patrick Eden/Alamy; H7.1 CNRI/Science Photo Library; H9.1 NASA; H10.1 iStockphoto; H10.SQ3 iStockphoto; H10.SQ4 Andrew Lambert Photography/Science Photo Library; H10.SQ8 Rubberball/Photolibrary; H10.SQ10 iStockphoto.

B1.1.1 iStockphoto; B1.1.2 Christian Forchner/epa/Corbis; B1.1.3 Flip Nicklin/Minden Pictures/FLPA; B1.2.1 Getty Images; B1.2.2 iStockphoto; B1.2.3 Rex Features; B1.3.1 iStockphoto; B1.3.2 iStockphoto; B1.4.1 Eric Erbe/Science Photo Library; B1.4.2 Dr Harold Fisher, Visuals Unlimited/Science Photo Library; B1.4.3 Science Photo Library; B1.5.1 Kallista Images/CDC/Getty Images; B1.5.2 Eye of Science/Science Photo Library; B1.6.1 Tek Image/Science Photo Library; B1.6.2 Cordelia Moloy/Science Photo Library; B1.6.3 St Mary's Hospital Medical School/Science Photo Library; B1.7.1 Geoff Tompkinson/Science Photo Library; B1.7.2 CDC/Science Photo Library; B1.9.1 Jenifer Harrington/Getty Images; B1.10.3 iStockphoto; B1.10.4 Martin Lee/Rex Features; B2.1.1 Philippe Lissac/Godong/Corbis; B2.1.2 iStockphoto; B2.2.3 Anthony Bradshaw/Getty Images; B2.4.1 Saturn Stills/Science Photo Library; B2.4.2 Mirrorpix; B2.5.1 Gavin Rodgers/Rex Features; B2.5.2 Copyright 2010 Photolibrary; B2.5.3 iStockphoto; B2.6.1 Martin Shields/Science Photo Library; B2.7.1 Barcroft Media/Getty Images; B2.EQ1 Ann Fullick; B2.EQ3 FLPA; B3.1.1 Colin Cuthbert/Science Photo Library; B3.1.2 Andrew Dunsmore/Rex Features; B3.2.1 iStockphoto; B3.3.1 AP/PA Photos; B3.4.1 NASA/Science Photo Library; B3.4.2 ©Handout/Reuters/Corbis; B3.5.3 iStockphoto; B3.6.1 AFP/Getty Images; B3.6.2 Rex Features; B3.6.3 Sipa Press/Rex Features; B4.1.1 iStockphoto; B4.1.2 Norbert Wu/Minden Pictures/FLPA; B4.1.3 Ralph White/Corbis; B4.2.2 iStockphoto; B4.2.3a LANDOV/Press Association Images; B4.2.3b iStockphoto; B4.3.2 iStockphoto; B4.3.3 iStockphoto; B4.4.1 iStockphoto; B4.4.2 iStockphoto; B4.4.3 iStockphoto; B4.4.4 iStockphoto; B4.5.1 Lizzie Harper/Science Photo Library; B4.5.2 iStockphoto; B4.5.3 iStockphoto; B4.6.1 Silvestre Silva/FLPA; B4.6.2 James Cook, University of Reading; B4.6.3 Dembinsky Photo Ass./FLPA; B4.6.4 iStockphoto; B4.7.1 Copyright 2010 Photolibrary; B4.7.2 William Mullins/Science Photo Library; B4.8.1 Mike Lane/FLPA; B4.8.3 iStockphoto; B4.EQ1 Natural Visions/Heather Angel; B5.1.1 iStockphoto; B5.2.1 iStockphoto; B5.2.2 iStockphoto; B5.2.3 iStockphoto; B5.3.1 iStockphoto; B5.3.2 The

Bridgeman Art Library/Getty Images; B5.4.1 Christopher Baines/Alamy; B5.4.3 iStockphoto; B5.5.1 Erica Olsen/FLPA; B5.5.2a iStockphoto; B5.5.2b ©Ian Wood/Alamy; B5.5.2c John Eveson/FLPA; B5.5.2d iStockphoto; B6.1.1 iStockphoto; B6.1.2 Steve Gschmeissner/Science Photo Library; B6.2.1 iStockphoto; B6.2.2 iStockphoto; B6.2.3 Allison Michael Orenstein/Getty Images; B6.3.1 Gary Roberts/Rex Features; B6.3.2 iStockphoto; B6.3.3 iStockphoto; B6.4.1 Copyright 2010 Photolibrary; B6.4.2 Nigel Cattlin/FLPA; B6.5.1 Press Association Images; B6.6.2 Nick Cobbing/Rex Features; B6.7.1a A & M University/Rex Features; B6.7.1b AP/EMPICS; B6.7.2 epa european pressphoto agency b.v.; B6.7.3 Courtesy Golden Rice Humanitarian Board. www.goldenrice.org; B7.1.2 Frans Lanting/FLPA; B7.1.3 English Heritage Photo Library; B7.2.2 Frans Lanting/FLPA; B7.2.3 Science Photo Library; B7.3.1 Scott Linstead/Minden Pictures/FLPA; B7.3.2 Brian Jackson/Fotolia; B7.4.1 iStockphoto; B7.4.2 Getty Images; B7.4.4a Pat & Tom Leeson/Science Photo Library; B7.4.4b Pat & Tom Leeson/Science Photo Library; B7.SQ4 Dr Jeremy Burgess/Science Photo Library; B7.EQ3 iStockphoto.

C1.1.1 Charles D. Winters/Science Photo Library; C1.3.3 Martyn F. Chillmaid/Science Photo Library; C1.4.1 iStockphoto; C2.1.1 iStockphoto; C2.1.2 Last Refuge/Getty Images; C2.1.3 Alison Bowden/Fotolia; C2.2.1 Peter Arnold Images/Photolibrary; C2.2.2 iStockphoto; C2.2.3 Andrew Lambert Photography/Science Photo Library; C2.4.1 Ru Baile/Fotolia; C2.4.2 iStockphoto; C2.4.3 Cordelia Molloy/Science Photo Library; C2.5.1 Mark Thomas/Science Photo Library; C2.5.2 Digital Light Source/Photolibrary; C2.5.3a Yuri Arcurs/Fotolia; C2.5.3b Absolut/Fotolia; C2.5.3c Joetex1/Fotolia; C2.5.3d Studio Vision1/Fotolia; C2.5.3e iStockphoto; C2.5.3f iStockphoto; C2.5.3g iStockphoto; C2.5.4 iStockphoto; C2.5.5 Thomas Sztanek/Fotolia; C3.1.1 Britain on View/Photolibrary; C3.1.3 iStockphoto; C3.2.1 Luis Veiga/Getty Images; C3.2.2 Ton Kinsbergen/Science Photo Library; C3.2.3 iStockphoto; C3.2.4 iStockphoto; C3.3.1 Dmitriy Ystuyjanin/Fotolia; C3.3.2 iStockphoto; C3.3.3 Image Source/Rex Features; C3.4.1 Lee Prince/Fotolia; C3.4.3 CSIRO; C3.5.2 Jeff Greenberg/Photolibrary; C3.5.3 iStockphoto; C3.5.4 iStockphoto; C3.5.5 Michellepix/Fotolia; C3.6.2 Eco Images/Universal Images Group/Getty Images; C3.6.3 iStockphoto; C4.1.1 Tim Graham/Getty Images; C4.2.3 iStockphoto; C4.3.1 John Millar/Getty Images; C4.3.3 Copyright 2010 Photolibrary; C4.5.1 USDA; C4.5.3 Bloomberg via Getty Images; C5.1.1 Paul Rapson/Science Photo Library; C5.2.1 Cordelia Molloy/Science Photo Library; C5.2.3 Charles D. Winters/Science Photo Library; C5.3.1 Image Source/Rex Features; C5.4.1 iStockphoto; C5.4.2 Ap Photo/Josh Reynolds; C5.4.4a Pixel Shepherd/Photolibrary; C5.4.4b Northscape/Alamy; C5.5.1 Martyn F. Chillmaid/Science Photo Library; C5.5.2 Scott Sinklier/Agstockusa/Science Photo Library; C6.1.1 iStockphoto; C6.1.2 Cordelia Molloy/Science Photo Library; C6.2.1 Morphy Richards Ltd; C6.2.2 Cordelia Molloy/Science Photo Library; C6.2.4 © Rnl – Fotolia; C6.3.1 iStockphoto; C6.3.2a iStockphoto; C6.3.2b Alain Pol Ism/Science Photo Library; C6.3.3 iStockphoto; C6.4.1 Cordelia Molloy/Science Photo Library; C6.4.2 Martyn F. Chillmaid; C6.4.3 iStockphoto; C6.4.4 Garo/Phanie/Rex Features; C7.1.2 Noaa/Science Photo Library; C7.2.1 John Cancalosi/Photolibrary; C7.2.3 Canadian Press/Rex Features; C7.3.1 iStockphoto; C7.3.2 Stocktrek RF/Getty Images; C7.3.3 Georgette Douwma/Science Photo Library; C7.3.4 Penn State University/Science Photo Library; C7.4.2 Science Source/Science Photo Library; C7.4.3b. Murton/Southampton Oceanography Centre/Science Photo Library; C7.5.1 Psamtik/Fotolia; C7.5.4 H. Raguet/Eurelios/Science Photo Library; C7.6.2 Copyright 2010 Photolibrary.

P1.1.1 Ted Kinsman/Getty Images; P1.1.3 Tony Craddock/Science Photo Library; P1.1.4 Photolibrary/Tsuneo Nakamura; P1.2.1 AP/PA Photos; P1.3.1 Charles D. Winters/Science Photo Library; P1.4.1 Fotolia; P1.4.3 Gary Ombler/Getty Images; P1.5.1 iStockphoto; P1.6.2 Spohn Matthieu/Getty Images; P1.6.5 iStockphoto; P1.7.1 Cordelia Molloy/Science Photo Library; P1.7.2 iStockphoto; P1.7.3 iStockphoto; P1.7.5a Fotolia; P1.7.5b iStockphoto; P1.8.2 G&D Images/Alamy; P1.SQ1 Fotolia; P2.1.1 SNCF; P2.1.4 Photolibrary/Superstock; P2.2.1 Fotolia; P2.2.3 iStockphoto; P2.3.1 iStockphoto; P2.3.2a iStockphoto; P3.1.1 Martyn F. Chillmaid/Science Photo Library; P3.1.2 Emmeline Watkins/Science Photo Library; P3.1.3 Getty Images; P3.2.1 www.powerstudies.com; P3.2.2 Getty Images; P3.2.3 iStockphoto; P3.3.1 Jim Breithaupt; P3.3.3 iStockphoto; P3.4.2 Ted Kinsman/Science Photo Library; P3.4.3a Fotolia, P3.4.3b iStockphoto, P3.4.3c Dimplex; P3.SQ1 iStockphoto; P4.1.2 Adam Gault/Science Photo Library; P4.1.3 PA Archive/Press Association Images; P4.2.1 Skyscan/Science Photo Library; P4.2.3 iStockphoto; P4.2.4 Canada Press/PA Photos; P4.3.2 G. Brad Lewis/Science Photo Library; P4.3.4 Peter Menzel/Science Photo Library; P4.4.3 Fotobank/Rex Features; P4.4.4 Photolibrary; P4.5.2 Fotolia; P4.5.4 Tony Gwynne/Alamy; P4.6.1 iStockphoto; P4.6.3a iStockphoto; P4.6.3b iStockphoto; P5.1.1 iStockphoto; P5.3.1 Photolibrary; P5.3.5 Shout/Rex Features; P5.4.1a Sciencephotos/Alamy; P5.4.3 Photolibrary/Peter Arnold Images; P5.4.5 iStockphoto; P5.5.2 NASA/ESA/STSCI/Hubble Heritage Team/Science Photo Library; P5.5.3 iStockphoto; P5.6.1 Fotolia; P5.7.1 Photolibrary/Imagebroker; P6.3.1 iStockphoto; P6.3.2 iStockphoto; P6.4.1 NASA/ESA/Getty Images; P6.5.1 Mark Garlick/Science Photo Library; P6.5.2 NASA/Science Photo Library.

Although we have made every effort to trace and contact all copyright holders before publication this has not been possible in all cases. If notified, the publisher will rectify any errors or omissions at the earliest opportunity.

Links to third party websites are provided by Oxford in good faith and for information only. Oxford disclaims any responsibility for the materials contained in any third party website referenced in this work.

Science A — Contents

Welcome to AQA GCSE Science!

This book has been written for you by very experienced teachers and subject experts. It covers the information you need to know for your exams and is packed full of features to help you achieve the very best that you can.

Questions in yellow boxes check that you understand what you are learning as you go along. The answers are all within the text so if you don't know the answer, you can go back and reread the relevant section.

Figure 1 Many diagrams are as important for you to learn as the text, so make sure you revise them carefully.

Key words are highlighted in the text. You can look them up in the glossary at the back of the book if you are not sure what they mean.

 Where you see this icon, you will know that this part of the topic involves How Science Works – a really important part of your GCSE and an interesting way to understand 'how science works' in real life.

Where you see this icon, there are supporting electronic resources in our Kerboodle online service.

Learning objectives

Each topic begins with key questions that you should be able to answer by the end of the lesson.

Study tip

Hints that provide important advice on things to remember and what to watch out for.

 Did you know ...?

There are lots of interesting and often strange facts about science. This feature tells you about many of them.

⚬⚬ links

Links will tell you where you can find more information about what you are learning.

Activity

An activity is linked to a main lesson and could be a discussion or task in pairs, groups or by yourself.

 Maths skills

This feature highlights the maths skills that you will need for your GCSE Science exams with short, visual explanations.

Practical

This feature helps you become familiar with key practicals. It may be a simple introduction, a reminder or the basis for a practical in the classroom.

Anything in the Higher Tier boxes must be learned by those sitting the Higher Tier exam. If you'll be taking the Foundation Tier exam, these boxes can be missed out.

The same is true for any other places which are marked Higher or [H].

Summary questions

These questions give you the chance to test whether you have learned and understood everything in the topic. If you get any wrong, go back and have another look.

And at the end of each chapter you will find …

Summary questions

These will test you on what you have learned throughout the whole chapter, helping you to work out what you have understood and where you need to go back and revise.

Practice questions

These questions are examples of the types of questions you will answer in your actual GCSE exam, so you can get lots of practice during your course.

Key points

At the end of the topic are the important points that you must remember. They can be used to help with revision and summarising your knowledge.

H1   How does science work? (k)

Learning objectives

- What is meant by 'How Science Works'?
- What is a hypothesis?
- What is a 'prediction' and why should you make one?
- How can we investigate a problem scientifically?

⊂⊃ links

You can find out more about your ISA by looking at H10 The ISA at the end of this chapter.

This first chapter looks at 'How Science Works'. It is an important part of your GCSE because the ideas introduced here will crop up throughout your course. You will be expected to collect scientific **evidence** and to understand how we use evidence. These concepts will be assessed as the major part of your internal school assessment.

You will take one or more 45-minute tests. These tests are based on **data** you have collected previously plus data supplied for you in the test. They are called '**Investigative Skills Assignments**' **(ISA)**. The ideas in 'How Science Works' will also be assessed in your examinations.

How science works for us

Science works for us all day, every day. You do not need to know how a mobile phone works to enjoy sending text messages. But, think about how you started to use your mobile phone or your television remote control. Did you work through pages of instructions? Probably not!

You knew that pressing the buttons would change something on the screen (**knowledge**). You played around with the buttons, to see what would happen (**observation**). You had a guess based on your knowledge and observations at what you thought might be happening (**prediction**) and then tested your idea (**experiment**).

Perhaps 'How Science Works' should really be called 'How Scientists Work'.

Science moves forward by slow steady steps. When a genius, such as Einstein, comes along then it takes a giant leap. Those small steps build on knowledge and experience that we already have.

The steps don't always lead in a straight line, starting with an observation and ending with a conclusion. More often than not you find yourself going round in circles, but each time you go around the loop you gain more knowledge and so can make better predictions.

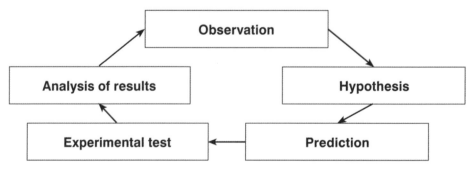

Each small step is important in its own way. It builds on the body of knowledge that we have. In 1675 a German chemist tried to extract gold from urine. He must have thought that there was a connection between the two colours. He was wrong. But after a while, with a terrible stench coming from his laboratory, the urine began to glow.

He had discovered phosphorus. Phosphorus catches fire easily. A Swedish scientist worked out how to manufacture phosphorus without the smell of urine. That is why most matches these days are manufactured in Sweden.

Figure 1 Albert Einstein was a genius, but he worked through scientific problems in the same way as you will in your GCSE

Activity

Investigating fireworks

Fireworks must be safe to light. Therefore you need a fuse that will last long enough to give you time to get well out of the way.

- Fuses can be made by dipping a special type of cotton into a mixture of two chemicals. One chemical (A) reacts by burning; the other (B) doesn't.
- The chemicals stick to the cotton. Once it is lit, the cotton will continue to burn, setting the firework off. The concentrations of the two chemicals will affect how quickly the fuse burns.
- In groups discuss how you could work out the correct concentrations of the chemicals to use. You want the fuse to last long enough to get out of the way. However, you don't want it to burn so long that we all get bored waiting for the firework to go off!

You can use the following headings to discuss your investigation. One person should be writing your ideas down, so that you can discuss them with the rest of the class.

- What prediction can you make about the concentration of the two chemicals (A and B) and the fuse?
- What would you vary in each test? This is called the 'independent variable'.
- What would you measure to judge the effect of varying the independent variable? This is called the 'dependent variable'.
- What would you need to keep unchanged to make this a fair test? These are called 'control variables'.
- Write a plan for your investigation.

Figure 2 Fireworks

Did you know ...?

The Greeks were arguably the first true scientists. They challenged traditional myths about life. They put forward ideas that they knew would be challenged. They were keen to argue the point and come to a reasoned conclusion.

Other cultures relied on long established myths, and argument was seen as heresy.

Key points

- **Observations** are often the starting point for an investigation.
- A **hypothesis** is a proposal intended to explain certain facts or observations.
- A **prediction** is an intelligent guess, based on some **knowledge.**
- An **experiment** is a way of testing your prediction.

Summary questions

1 Copy and complete the paragraph using the words below:

prediction   knowledge   experiment   observation   conclusion

You have learned before that a cup of tea loses heat if it is left standing. This is a piece of .... You make an .... that dark coloured cups will cool faster. So you make a .... that if you have a black cup, this will cool fastest of all. You carry out an .... to get some results, and from these you make a ....

H2

Fundamental ideas about how science works

Learning objectives

- How do you spot when an opinion is not based on good science?

- What is the importance of continuous and categoric variables?

- What does it mean to say that evidence is valid?

- What is the difference between a result being repeatable and a result being reproducible?

- How can two sets of data be linked?

Study tip

Read a newspaper article or watch the news on TV. Ask yourself if any research presented is valid. Ask yourself if you can trust that person's opinion and why.

Figure 1 Student recording a range of temperatures – an example of a continuous variable

Science is too important for us to get it wrong

Sometimes it is easy to spot when people try to use science poorly. Sometimes it can be funny. You might have seen adverts claiming to give your hair 'body' or sprays that give your feet 'lift'!

On the other hand, poor scientific practice can cost lives or have serious consequences.

Some years ago a company sold the drug thalidomide to people as a sleeping pill. Research was carried out on animals to see if it was safe. The research did not include work on pregnant animals. The **opinion** of the people in charge was that the animal research showed the drug could be used safely with humans.

Then the drug was also found to help ease morning sickness in pregnant women. Unfortunately, doctors prescribed it to many women, resulting in thousands of babies being born with deformed limbs. It was far from safe.

These are very difficult decisions to make. You need to be absolutely certain of what the science is telling you.

a Why was the opinion of the people in charge of developing thalidomide based on poor science?

Deciding on what to measure: variables

Variables are physical, chemical or biological quantities or characteristics.

In an investigation, you normally choose one thing to change or vary. This is called the **independent variable.**

When you change the independent variable, it may cause something else to change. This is called the **dependent variable**.

A **control variable** is one that is kept the same and is not changed during the investigation.

You need to know about two different types of these variables:

- A **categoric variable** is one that is best described by a label (usually a word). The 'colour of eyes' is a categoric variable, e.g. blue or brown eyes.
- A **continuous variable** is one that we measure, so its value could be any number. Temperature (as measured by a thermometer or temperature sensor) is a continuous variable, e.g. 37.6 °C, 45.2 °C. Continuous variables can have values (called a quantity) that can be found by making measurements (e.g. light intensity, flow rate, etc.).

b Imagine you were growing seedlings using different volumes of water. Would it be better to say that some were tall and some were short, or some were taller than others, or to measure the heights of all the seedlings?

Making your evidence repeatable, reproducible and valid

When you are designing an investigation you must make sure that other people can get the same results as you. This makes the evidence you collect reproducible. A repeatable measurement is one that consistently remains the same after several repeats.

A measurement is **repeatable** if the original experimenter repeats the investigation using the same method and equipment and obtains the same results.

A measurement is **reproducible** if the investigation is repeated by another person, or by using different equipment or techniques, and the same results are obtained.

You must also make sure you are measuring the actual thing you want to measure. If you don't, your data can't be used to answer your original question. This seems very obvious but it is not always quite so easy. You need to make sure that you have controlled as many other variables as you can, so that no one can say that your investigation is not **valid**. A measurement is valid if it measures what it is supposed to be measuring with an appropriate level of performance.

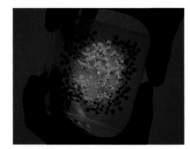
Figure 2 Cress seedlings growing in a Petri dish

c State one way in which you can show that your results are repeatable.

How might an independent variable be linked to a dependent variable?

Looking for a link between your independent and dependent variable is very important. The pattern in your graph or bar chart can often help you to see if there is a link.

But beware! There may not be a link! If your results seem to show that there is no link, don't be afraid to say so. Look at Figure 3.

The points on the top graph show a clear pattern, but the bottom graph shows random scatter.

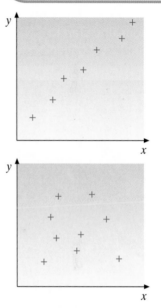
Figure 3 Which graph shows that there might be a link between x and y?

Study tip

When designing your investigation you should always try to measure continuous data whenever you can. This is not always possible, so then you have to use a label (categoric variable). You might still be able to put the variables in an order so that you can spot a pattern. For example, you could describe flow rate as 'fast flowing', 'steady flow' or 'slow flowing'.

??? Did you know ... ?

Aristotle, a brilliant Greek scientist, once proclaimed that men had more teeth than women! Do you think that his data collection was reproducible?

Key points

● Be on the lookout for non-scientific opinions.

● Continuous data give more information than other types of data.

● Check that evidence is reproducible and valid.

Summary questions

1 Copy and complete the paragraph using the words below:

categoric   independent   dependent   continuous

Stefan wanted to find out which was the strongest supermarket plastic carrier bag. He tested five different bags by adding weight to them until they broke. The type of bag he used was the .... variable and the weight that it took to break it was the .... variable. The 'type of bag' is called a .... variable and the 'weight needed to break' it was a .... variable.

2 A researcher claimed that the metal tungsten 'alters the growth of leukaemia cells' in laboratory tests. A newspaper wrote that they would 'wait until other scientists had reviewed the research before giving their opinion.' Why is this a good idea?

H3 Starting an investigation

Learning objectives

- How can you use your scientific knowledge to observe the world around you?

- How can you use your observations to make a hypothesis?

- How can you make predictions and start to design an investigation?

Figure 1 A plant showing positive phototropism

Observation

As humans we are sensitive to the world around us. We can use our senses to detect what is happening. As scientists we use observations to ask questions. We can only ask useful questions if we know something about the observed event. We will not have all of the answers, but we know enough to start asking relevant questions.

If we observe that the weather has been hot today, we would not ask if it was due to global warming. If the weather was hotter than normal for several years then we could ask that question. We know that global warming takes many years to show its effect.

When you are designing an investigation you have to observe carefully which variables are likely to have an effect.

> **a** Would it be reasonable to ask if the plant in Figure 1 is 'growing towards the glass'? Explain your answer.

A farmer noticed that her corn was much smaller at the edge of the field than in the middle (observation). She noticed that the trees were quite large on that side of the field. She came up with the following ideas that might explain why this is happening:

1 The trees at the edge of the field were blocking out the light.
2 The trees were taking too many nutrients out of the soil.
3 The leaves from the tree had covered the young corn plants in the spring.
4 The trees had taken too much water out of the soil.
5 The seeds at the edge of the field were genetically small plants.
6 They had planted fewer seeds on that side of the field.
7 The fertiliser spray had not reached the side of the field.
8 The wind had been too strong over winter and had moved the roots of the plants.
9 The plants at the edge of the field had a disease.

> **b** Discuss each of these ideas and use your knowledge of science to decide which four are the most likely to have caused the poor growth of the corn.

Observations, backed up by really creative thinking and good scientific knowledge can lead to a **hypothesis**.

Testing scientific ideas

Scientists always try to think of ways to explain how things work or why they behave in the way that they do.

After their observations, they use their understanding of science to come up with an idea that could explain what is going on. This idea is sometimes called a hypothesis. They use this idea to make a prediction. A prediction is like a guess, but it is not just a wild guess – it is based on previous understanding.

A scientist will say, 'If it works the way I think it does, I should be able to change **this** (the independent variable) and **that** will happen (the dependent variable).'

?? Did you know ...?

Some biologists think that we still have about one hundred millions species of insects to discover – plenty to go for then! Of course, observing one is the easy part – knowing that it is undiscovered is the difficult bit!

Predictions are what make science so powerful. They mean that we can work out rules that tell us what will happen in the future. For example, a weather forecaster can use knowledge and understanding to predict **wind** speeds. Knowing this, sailors and windsurfers can decide if it would be a good day to enjoy their sport.

Knowledge of energy transfer could lead to an idea that the inside of chips cook by energy being conducted from the outside. You might predict that small, thinly sliced chips will cook faster than large, fat chips.

Figure 2 Which cook faster? Small, thinly sliced chips or larger, fat chips?

c Look at the photograph in Figure 2. How could you test your prediction about how fast chips cook?

Not all predictions are correct. If scientists find that the prediction doesn't work, then it's back to the drawing board! They either amend their original idea or think of a completely new one.

Starting to design a valid investigation

observation + knowledge ➡ hypothesis ➡ prediction ➡ investigation

We can test a prediction by carrying out an **investigation**. You, as the scientist, predict that there is a relationship between two variables.

The independent variable is one that is selected and changed by you, the investigator. The dependent variable is measured for each change in your independent variable. Then all other variables become control variables, kept constant so that your investigation is a fair test.

If your measurements are going to be accepted by other people then they must be valid. Part of this is making sure that you are really measuring the effect of changing your chosen variable. For example, if other variables aren't controlled properly, they might be affecting the data collected.

Figure 3 Measuring a pulse

d Look at Figure 3. When investigating his heart rate before and after exercise, Darren got his girlfriend to measure his pulse. Would Darren's investigation be valid? Explain your answer.

Summary questions

1 Copy and complete the paragraph using the words below:

controlled dependent independent knowledge prediction hypothesis

An observation linked with scientific .... can be used to make a .... A .... links an .... variable to a .... variable. All other variables need to be ....

2 What is the difference between a prediction and a guess?

3 Imagine you were testing the rate of a chemical reaction by using different concentrations of the reactants. The chemical reaction you are investigating might release energy which would alter the temperature of the solution.

   a How could you monitor the temperature?

   b What other control variables can you think of that might affect the results?

Key points

- Observation is often the starting point for an investigation.
- Testing predictions can lead to new scientific understanding.
- You must design investigations that produce valid results if you are to be believed.

H4

Planning an investigation

Learning objectives

- How do you design a fair test?
- How do you set up a survey?
- How do you set up a control group or control experiment?
- How do you reduce risks in hazardous situations?

Study tip

If you are asked about why it is important to keep control variables constant, you need to give a detailed explanation. Don't just answer 'To make it a fair test'.

When you are asked to write a plan for your investigation, make sure that you give all the details. Ask yourself 'Would someone else be able to follow my written plan and use it to do the investigation?'

Fair testing

A **fair test** is one in which only the independent variable affects the dependent variable. All other variables called control variables should be kept the same. If the test is not fair, then the results of your investigation will not be valid.

Sometimes it is very difficult to keep control variables the same. However, at least you can **monitor** them, so that you know whether they have changed or not.

Surveys

Not all scientific investigations involve deliberately changing the independent variable.

If you were investigating the effect that using a mobile phone may have on health you wouldn't put a group of people in a room and make them use their mobile phones to see if they developed brain cancer!

Instead, you might conduct a **survey.** You might study the health of a large number of people who regularly use a mobile phone. You could then compare their health with those who never use a mobile phone.

You would have to choose people of the same age and same family history to test. The larger the sample size you test, the more valid your results will be.

Figure 1 Investigating the effect of using a mobile phone on health would involve using data from surveys as well as laboratory studies

Control group

Control groups are used in investigations to try to make sure that you are measuring the variable that you intend to measure. When investigating the effects of a new drug, the control group will be given a **placebo**. This is a 'pretend' drug that actually has no effect on the patient at all. The control group think they are taking a drug but the placebo does not contain the drug. This way you can control the variable of 'thinking that the drug is working' and separate out the effect of the actual drug.

Usually neither the patient nor the doctor knows until after the trials have been completed which of the patients were given the placebo. This is known as a **double-blind trial**.

Risks and hazards

One of the first things you must do is to think about any potential **hazards** and then assess the **risk**.

Everything you do in life presents a hazard. What you have to do is to identify the hazard and then decide the degree of risk that it gives. If the risk is very high, you must do something to reduce it.

For example, if you decide to go out in the pouring rain, lightning could be a possible hazard. However, you decide that the risk is so small that you decide to ignore it and go out anyway.

If you decide to cross a busy road, the cars travelling along it at high speed represent a hazard. You decide to reduce the risk by crossing at a pedestrian crossing.

Figure 2 The hazard is the busy road. We reduce the risk by using a pedestrian crossing.

Activity

Burning alcohols

Imagine you were testing alcohols to see how much energy they release when burned.

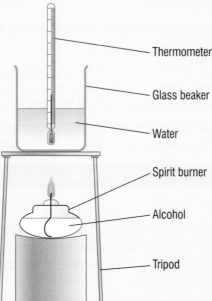

- Thermometer
- Glass beaker
- Water
- Spirit burner
- Alcohol
- Tripod

- What are the **hazards** that are present?
- What could you do to reduce the **risk** from these hazards?

Summary questions

1 Copy and complete the paragraph using the words below:

investigation hazards assessment risks

Before you carry out any practical ...., you need to carry out a risk .... You can do this by looking for any potential .... and making sure that the .... are as small as possible.

2 Explain the difference between a control group and a control variable.

3 Briefly describe how you would go about setting up a fair test in a laboratory investigation. Give your answer as general advice.

Study tip

Before you start your practical work you must make sure that it is safe. What are the likely hazards? How could you reduce the risk caused by these hazards? This is known as a **risk assessment**. You may well be asked questions like this on your ISA paper.

Key points

- Care must be taken to ensure fair testing – as far as is possible.
- Control variables must be kept the same during an investigation.
- Surveys are often used when it is impossible to carry out an experiment in which the independent variable is changed.
- Control groups allow you to make a comparison.
- A risk assessment must be made when planning a practical investigation.

H5 | Designing an investigation

Learning objectives

- How do you make sure that you choose the best values for your variables?
- How do you decide on a suitable range?
- How do you decide on a suitable interval?
- How do you ensure accuracy and precision?

Choosing values of a variable

Trial runs will tell you a lot about how your early thoughts are going to work out.

Do you have the correct conditions?

A photosynthesis investigation that produces tiny amounts of oxygen might not have enough light, pondweed or carbon dioxide. Alternatively, the temperature might not be high enough.

Have you chosen a sensible range?

Range means the maximum and minimum values of the independent or dependent variables. It is important to choose a suitable range for the independent variable, otherwise you may not be able to see any change in the dependent variable.

For example, if results are all very similar, you might not have chosen a wide enough range of light intensities.

Have you got enough readings that are close together?

The gap between the readings is known as the **interval**.

For example, you might alter the light intensity by moving a lamp to different distances from the pondweed. A set of 11 readings equally spaced over a distance of 1 metre would give an interval of 10 centimetres.

If the results are very different from each other, you might not see a pattern if you have large gaps between readings over the important part of the range.

Figure 1 Measuring the extension of a spring

Practical

Springs

In this experiment you hang weights from the spring using a weight hanger and measure the extension of the spring. Once you have done some preliminary testing, suggest a suitable range of weights to test and a suitable interval for the weights.

Accuracy

Accurate measurements are very close to the **true value**.

Your investigation should provide data that are accurate enough to answer your original question.

However, it is not always possible to know what that true value is.

How do you get accurate data?

- You can repeat your measurements and your mean is more likely to be accurate.
- Try repeating your measurements with a different instrument and see if you get the same readings.
- Use high quality instruments that measure accurately.
- The more carefully you use the measuring instruments, the more accuracy you will get.

Precision, resolution, repeatability and reproducibility

A **precise** measurement is one in which there is very little spread about the mean value.

If your repeated measurements are closely grouped together then you have precision. Your measurements must be made with an instrument that has a suitable **resolution**. Resolution of a measuring instrument is the smallest change in the quantity being measured (input) that gives a perceptible change in the reading.

It's no use measuring the time for a fast reaction to finish using the seconds hand on a clock! If there are big differences within sets of repeat readings, you will not be able to make a valid conclusion. You won't be able to trust your data!

How do you get precise data?

- You have to use measuring instruments with sufficiently small scale divisions.
- You have to repeat your tests as often as necessary.
- You have to repeat your tests in exactly the same way each time.

If you repeat your investigation using the same method and equipment and obtain the same results, your results are said to be **repeatable.**

If someone else repeats your investigation in the same way, or if you repeat it by using different equipment or techniques, and the same results are obtained, it is said to be **reproducible.**

You may be asked to compare your results with those of others in your group, or with data from other scientists. Research like this is a good way of checking your results.

A word of caution!

Precision depends only on the extent of random errors – it gives no indication of how close results are to the true value. Just because your results show precision does not mean your results are accurate.

   a Draw a thermometer scale reading 49.5 °C, showing four results that are both accurate and precise.

Study tip

You must know the difference between accurate and precise results.

Imagine measuring the temperature after a set time when a fuel is used to heat a fixed volume of water. Two students repeated this experiment, four times each. Their results are marked on the thermometer scales below:

- A **precise** set of repeat readings will be grouped closely together.
- An **accurate** set of repeat readings will have a mean (average) close to the true value.

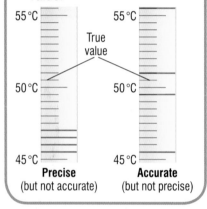

Summary questions

1 Copy and complete the paragraph using the words below:

  range repeat conditions readings

  Trial runs give you a good idea of whether you have the correct .... to collect any data, whether you have chosen the correct .... for the independent variable, whether you have enough ...., and if you need to do .... readings.

2 Use an example to explain how a set of repeat measurements could be accurate, but not precise.

3 Explain the difference between a set of results that are reproducible and a set of results that are repeatable.

Key points

- You can use a trial run to make sure that you choose the best values for your variables.

- The range states the maximum and the minimum values of a variable.

- The interval is the gap between the values of a variable.

- Careful use of the correct equipment can improve accuracy and precision.

- You should try to reproduce your results carefully.

H6 Making measurements

Learning objectives

- Why do results always vary?

- How do you choose instruments that will give you accurate results?

- What do we mean by the 'resolution' of an instrument?

- What is the difference between a systematic error and a random error?

- How does human error affect results and what do you do with anomalies?

Using instruments

Try measuring the temperature of a beaker of water using a digital thermometer. Do you always get the same result? Probably not. So can we say that any measurement is absolutely correct?

In any experiment there will be doubts about actual measurements.

When you choose an instrument you need to know that it will give you the accuracy that you want. You need to be confident that it is giving a true reading.

If you have used an electric water bath, would you trust the temperature on the dial? How do you know it is the true temperature? You could use a very expensive thermometer to calibrate your water bath. The expensive thermometer is more likely to show the true temperature. But can you really be sure it is accurate?

Instruments that measure the same thing can have different sensitivities. The **resolution** of an instrument refers to the smallest change in a value that can be detected. This is one factor that determines the precision of your measurements.

Errors

Even when an instrument is used correctly, the results can still show differences.

Results may differ because of **random error**. This is most likely to be due to a poor measurement being made. It could be due to not carrying out the method consistently.

If you repeat your measurements several times and then calculate a mean, you will reduce the effect of random errors.

The **error** might be a **systematic error**. This means that the method was carried out consistently but an error was being repeated. A systematic error will make your readings be spread about some value other than the true value. This is because your results will differ from the true value by a consistent amount each time a measurement is made.

No amount of repeats can do anything about systematic errors. If you think that you have a systematic error, you need to repeat using a different set of equipment or a different technique. Then compare your results and spot the difference!

A **zero error** is one kind of systematic error. Suppose that you were trying to measure the length of your desk with a metre rule, but you hadn't noticed that someone had sawn off half a centimetre from the end of the ruler. It wouldn't matter how many times you repeated the measurement, you would never get any nearer to the true value.

Study tip

If you are asked what may have caused an error, never answer simply '**human error**' – you won't get any marks for this.

You need to say what the experimenter may have done to cause the error, or give more detail, e.g. 'Human reaction time might have caused an error in the timing when using a stopwatch'.

Check out these two sets of data that were taken from the investigation that Matt did. He tested five different oils. The bottom row is the time calculated from knowing the viscosity of the different oils:

Type of oil used	A	B	C	D	E
Time taken to flow down tile (seconds)	23.2	45.9	49.5	62.7	75.9
	24.1	36.4	48.7	61.5	76.1
Calculated time (seconds)	18.2	30.4	42.5	55.6	70.7

a Discuss whether there is any evidence for random error in these results.
b Discuss whether there is any evidence for systematic error in these results.

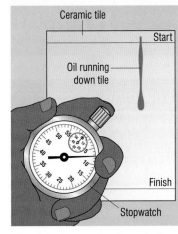

Figure 1 Matt timing the flow of oil

Anomalies

Anomalous results are clearly out of line. They are not those that are due to the natural variation you get from any measurement. These should be looked at carefully. There might be a very interesting reason why they are so different. You should always look for anomalous results and discard them before you calculate a mean, if necessary.

● If anomalies can be identified while you are doing an investigation, then it is best to repeat that part of the investigation.

● If you find anomalies after you have finished collecting data for an investigation, then they must be discarded.

?? Did you know ... ?

Sir Alexander Fleming had grown bacteria on agar plates. He noticed an anomaly. There was some mould growing on one of the plates and around it there were no bacteria. He decided to investigate further and grew more of the mould. Only because Fleming checked out his anomaly did it lead to the discovery of penicillin.

⚭ links

B1 1.6 Using drugs to treat disease.

Key points

● Results will nearly always vary.

● Better quality instruments give more accurate results.

● The resolution of an instrument refers to the smallest change that it can detect.

● Human error can produce random and/or systematic errors.

● We examine anomalies as they might give us some interesting ideas. If they are due to a random error, we repeat the measurements. If there is no time to repeat them, we discard them.

Summary questions

1 Copy and complete the paragraph using the words below:

accurate   discarded   random   resolution   systematic use   variation

There will always be some .... in results. You should always choose the best instruments that you can in order to get the most .... results. You must know how to .... the instrument properly. The .... of an instrument refers to the smallest change that can be detected. There are two types of error: .... and .... Anomalies due to random error should be ....

2 What kind of error will most likely occur in the following situations?
   a Asking everyone in the class to measure the length of the bench.
   b Using a ruler that has a piece missing from the zero end.

H7

Presenting data

- How do we calculate the mean from a set of data?
- How do you use tables of results?
- What is the range of the data?
- How do you display your data?

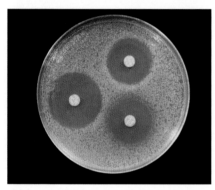

Figure 1 Petri dish with discs showing growth inhibition of bacteria

For this section you will be working with data from this investigation:

Mel spread some bacteria onto a dish containing nutrient jelly. She also placed some discs onto the jelly. The discs contained different concentrations of an antibiotic. The dish was taped and then left for a couple of days.

Then she measured the diameter of the clear part around each disc. The clear part is where the bacteria have not been able to grow. The bacteria grew all over the rest of the dish.

Tables

Tables are really good for getting your results down quickly and clearly. You should design your table **before** you start your investigation.

Your table should be constructed to fit in all the data to be collected. It should be fully labelled, including units.

You may want to have extra columns for repeats, calculations of means or calculated values.

Checking for anomalies

While filling in your table of results you should be constantly looking for anomalies.

- Check to see if any reading in a set of repeat readings is significantly different from others.
- Check to see if the pattern you are getting as you change the independent variable is what you expected.

Remember, a result that looks anomalous should be checked out to see if it really is a poor reading.

Planning your table

Mel had decided on the values for her independent variable. We always put these in the first column of a table. The dependent variable goes in the second column. Mel will find its values as she carries out the investigation.

So she could plan a table like this:

Concentration of antibiotic (µg/ml)	Size of clear zone (mm)
4	
8	
16	
32	
64	

Or like this:

Concentration of antibiotic (µg/ml)	4	8	16	32	64
Size of clear zone (mm)					

All she had to do in the investigation was to write the correct numbers in the second column to complete the top table.

Mel's results are shown in the alternative format in the table below.

Concentration of antibiotic (µg/ml)	4	8	16	32	64
Size of clear zone (mm)	4	16	22	26	28

The range of the data

Pick out the maximum and the minimum values and you have the range of a variable. You should always quote these two numbers when asked for a range. For example, 'the range of the dependent variables is between 4 mm (the lowest value) and 28 mm (the highest value)' – and don't forget to include the units!

a What is the range for the independent variable in Mel's set of data?

 Maths skills

The mean of the data

Often you have to find the **mean** of each repeated set of measurements. The first thing you should do is to look for any anomalous results. If you find any, miss these out of the calculation. Then add together the remaining measurements and divide by how many there are. For example:

● Mel takes four readings 15 mm, 18 mm, 29 mm, 15 mm

● 29 mm is an anomalous result and so is missed out. So:
15 + 18 + 15 = 48

● 48 divided by three (the number of valid results) = **16 mm**

The repeat values and mean can be recorded as shown below:

Concentration of antibiotic (µg/ml)	Size of clear zone (mm)			
	1st test	2nd test	3rd test	Mean
8	15	18	15	16

Displaying your results

Bar charts

If one of your variables is categoric then you should use a **bar chart**.

Line graphs

If you have a continuous independent and a continuous dependent variable then a **line graph** should be used. Plot the points as small 'plus' signs (+).

Summary questions

1 Copy and complete the paragraph using the words below:

categoric   continuous   mean   range

The maximum and minimum values show the .... of the data. The sum of the values in a set of repeat readings divided by the total number of these repeat values gives the .... Bar charts are used when you have a .... independent variable and a continuous dependent variable. Line graphs are used when you have .... independent and dependent variables.

2 Draw a graph of Mel's results from the top of this page.

Study tip

When you make a table for your results remember to include:

● headings, including the units

● a title.

When you draw a line graph or bar chart remember to:

● use a sensible scale that is easy to work out

● use as much of the graph paper as possible; your data should occupy at least one-third of each axis

● label both axes

● draw a line of best fit if it is a line graph

● label each bar if it is a bar chart.

Study tip

Marks are often dropped in the ISA by candidates plotting points incorrectly. Also use **a line of best fit** where appropriate – don't just join the points 'dot-to-dot'!

Key points

● The **range** states the maximum and the minimum values.

● The **mean** is the sum of the values divided by how many values there are.

● Tables are best used during an investigation to record results.

● Bar charts are used when you have a **categoric** variable.

● Line graphs are used to display data that are **continuous**.

H8

Using data to draw conclusions

Learning objectives

- How do we best use charts and graphs to identify patterns?

- What are the possible relationships we can identify from charts and graphs?

- How do we draw conclusions from relationships?

- How can we decide if our results are good and our conclusions are valid?

Identifying patterns and relationships

Now that you have a bar chart or a line graph of your results you can begin to look for patterns. You must have an open mind at this point.

Firstly, there could still be some anomalous results. You might not have picked these out earlier. How do you spot an anomaly? It must be a significant distance away from the pattern, not just within normal variation. If you do have any anomalous results plotted on your graph, circle these and ignore them when drawing the **line of best fit**.

Now look at your graph. Is there a pattern that you can see? When you have decided, draw a line of best fit that shows this pattern.

A line of best fit is a kind of visual averaging process. You should draw the line so that it leaves as many points slightly above the line as there are points below. In other words it is a line that steers a middle course through the field of points.

The vast majority of results that you get from continuous data require a line of best fit.

Remember that a line of best fit can be a straight line or it can be a curve – you have to decide from your results.

You need to consider whether your graph shows a linear **relationship**. This simply means, can you be confident about drawing a straight line of best fit on your graph? If the answer is yes – then is this line positive or negative?

a Say whether graphs **i** and **ii** in Figure 1 show a positive or a negative linear relationship.

Look at the graph in Figure 2. It shows a positive linear relationship. It also goes through the origin (0, 0). We call this a **directly proportional** relationship.

Your results might also show a curved line of best fit. These can be predictable, complex or very complex! Look at Figure 3 below.

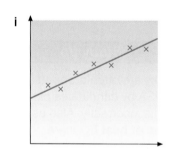

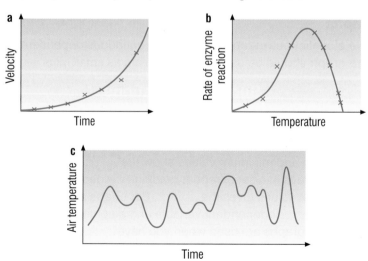

Figure 1 Graphs showing linear relationships

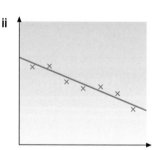

Figure 2 Graph showing a directly proportional relationship

Figure 3 a Graph showing predictable results **b** Graph showing complex results **c** Graph showing very complex results

Drawing conclusions

If there is a pattern to be seen (for example, as one variable gets bigger the other also gets bigger), it may be that:

- changing one has caused the other to change
- the two are related, but one is not necessarily the cause of the other.

Your conclusion must go no further than the evidence that you have.

Activity

Looking at relationships

Some people think that watching too much television can cause an increase in violence.

The table shows the number of television sets in the UK for four different years, and the number of murders committed in those years.

Year	Number of televisions (millions)	Number of murders
1970	15	310
1980	25	500
1990	42	550
2000	60	750

Plot a graph to show the relationship. Do you think this proves that watching television causes violence? Explain your answer.

Poor science can often happen if a wrong decision is made here. Newspapers have said that living near electricity substations can cause cancer. All that scientists would say is that there is possibly an association.

Evaluation

You will often be asked to evaluate either the method of the investigation or the conclusion that has been reached. Ask yourself: Could the method have been improved? Is the conclusion that has been made a valid one?

Summary questions

1 Copy and complete the paragraph using the words below:

anomalous   complex   directly   negative   positive

Lines of best fit can be used to identify .... results. Linear relationships can be .... or .... If a straight line goes through the origin of a graph then the relationship is .... proportional. Often a line of best fit is a curve which can be predictable or ....

2 Nasma knew about the possible link between cancer and living near to electricity substations. She found a quote from a National Grid Company survey of substations:

Measurements of the magnetic field were taken at 0.5 metre above ground level within 1 metre of fences and revealed 1.9 microteslas. After 5 metres this dropped to the normal levels measured in any house.

Discuss the type of experiment and the data you would expect to see to support a conclusion that it is safe to build houses over 5 metres from an electricity substation.

Study tip

When you read scientific claims, think carefully about the evidence that should be there to back up the claim.

Key points

- Drawing lines of best fit help us to study the relationship between variables.

- The possible relationships are linear, positive and negative, directly proportional, predictable and complex curves.

- Conclusions must go no further than the data available.

- The reproducibility of data can be checked by looking at other similar work done by others, perhaps on the internet. It can also be checked by using a different method or by others checking your method.

H9 Scientific evidence and society

Learning objectives

- How can science encourage people to trust its research?

- How might bias affect people's judgement of science?

- Can politics influence judgements about science?

- Do you have to be a professor to be believed?

STAR IN SCANDAL SHOCK

We Find Out What They Don't Want You To Know... And WE TELL YOU!

MOBILE PHONE TUMOUR RISK?

Swedish researchers found that the risk of developing an ear tumour increased if you used a mobile phone. The study was of 750 people. This type of tumour affects one in 100 000 people and the risk increased four times if you used the phone for more than 10 years.

A spoke...
that...

?? Did you know ... ?

A scientist who rejected the idea of a causal link between smoking and lung cancer was later found to be being paid by a tobacco company.

Study tip

If you are asked about bias in scientific evidence, there are two types:

- The measuring instruments may have introduced a bias because they were not calibrated correctly.

- The scientists themselves may have a biased opinion (e.g. if they are paid by a company to promote their product).

Now you have reached a conclusion about a piece of scientific research. So what is next? If it is pure research then your fellow scientists will want to look at it very carefully. If it affects the lives of ordinary people then society will also want to examine it closely.

You can help your cause by giving a balanced account of what you have found out. It is much the same as any argument you might have. If you make ridiculous claims then nobody will believe anything you have to say.

Be open and honest. If you only tell part of the story then someone will want to know why! Equally, if somebody is only telling you part of the truth, you cannot be confident with anything they say.

a 'X-rays are safe, but should be limited' is the headline in an American newspaper. What information is missing? Is it important?

You must be on the lookout for people who might be biased when representing scientific evidence. Some scientists are paid by companies to do research. When you are told that a certain product is harmless, just check out who is telling you.

b Suppose you wanted to know about safe levels of noise at work. Would you ask the scientist who helped to develop the machinery or a scientist working in the local university? What questions would you ask, so that you could make a valid judgement?

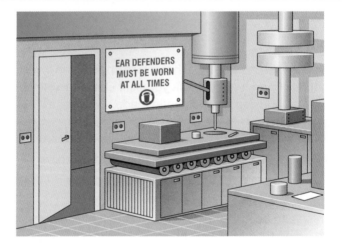

We also have to be very careful in reaching judgements according to who is presenting scientific evidence to us. For example, if the evidence might provoke public or political problems, then it might be played down.

Equally others might want to exaggerate the findings. They might make more of the results than the evidence suggests. Take as an example the siting of mobile phone masts. Local people may well present the same data in a totally different way from those with a wider view of the need for mobile phones.

c Check out some websites on mobile phone masts. Get the opinions of people who think they are dangerous and those who believe they are safe. Try to identify any political bias there might be in their opinions.

Science can often lead to the development of new materials or techniques. Sometimes these cause a problem for society where hard choices have to be made.

Scientists can give us the answers to many questions, but not to every question. Scientists have a contribution to make to a debate, but so do others like environmentalists, economists, and politicians.

The limitations of science

Science can help us in many ways but it cannot supply all the answers. We are still finding out about things and developing our scientific knowledge. For example, the Hubble telescope has helped us to revise our ideas about the beginnings of the universe.

There are some questions that we cannot answer, maybe because we do not have enough reproducible, repeatable and valid evidence. For example, research into the causes of cancer still needs much work to be done to provide data.

Figure 1 The Hubble space telescope can look deep into space and tell us things about the Universe's beginning from the formations of early galaxies

There are some questions that science cannot answer at all. These tend to be questions where beliefs, opinions and ethics are important. For example, science can suggest what the universe was like when it was first formed, but cannot answer the question of why it was formed.

BIODIESEL
The Fuel of the Future?

The demand for palm oil has grown tremendously in the last few years. It is used in many food products like margarine and chocolate, in cosmetics, and increasingly for making biodiesel.

Some scientists say that it is the answer to the dwindling supplies of crude oil because palm oil is a renewable resource.

Other people say that planting millions of acres of palm trees is destroying natural habitats such as rainforests and peat bogs.

Summary questions

1 Copy and complete the paragraph using the words below:

   status balanced bias political

   Evidence from scientific investigations should be given in a .... way. It must be checked for any .... from the experimenter. Evidence can be given too little or too much weight if it is of .... significance. The .... of the experimenter is likely to influence people in their judgement of the evidence.

2 Collect some newspaper articles to show how scientific evidence is used. Discuss in groups whether these articles are honest and fair representations of the science. Consider whether they carry any bias.

3 Extract from a newspaper report about Sizewell nuclear power station:

   A radioactive leak can have devastating results but one small pill could protect you. Our reporter reveals how for the first time these life-saving pills will be available to families living close to the Sizewell nuclear power station.

   Suppose you were living near Sizewell power station. Who would you trust to tell you whether these pills would protect you from radiation?

Key points

- Scientific evidence must be presented in a balanced way that points out clearly how valid the evidence is.

- The evidence must not contain any bias from the experimenter.

- The evidence must be checked to appreciate whether there has been any political influence.

- The status of the experimenter can influence the weight placed on the evidence.

H10

The ISA

There are several different stages to the ISA (Investigate Skills Assignment) that you will complete for your Controlled Assessment. This will make up 25% of your GCSE marks.

Learning objectives

- How do you write a plan?
- How do you make a risk assessment?
- What is a hypothesis?
- How do you make a conclusion?

Stage 1

Your teacher will tell you the problem that you are going to investigate, and will give you a hypothesis. They will also set the problem in a context – in other words, where in real life your investigation could be useful. You should have a discussion about it, and talk about different ways that you might solve the problem. Your teacher should show you the equipment that you can use, and you should research one or two possible methods for carrying out an experiment to test the hypothesis. You should also research the context and do a risk assessment for your practical work. You will be allowed to make one side of notes on this research, which you can take into the written part of the ISA.

Figure 1 Doing practical work allows you to develop the skills needed to do well in the ISA

You should be allowed to handle the equipment and you may be allowed to carry out a preliminary experiment.

Make sure that you understand what you have to do – now is the time to ask questions if you are not sure.

Study tip

When you are making a blank table or drawing a graph or bar chart, make sure that you use full headings, e.g.

- 'the length of the leaf', **not** just 'length'
- 'the time taken for the reaction', **not** just 'time'
- 'the height from which the ball was dropped', **not** just 'height'
- and don't forget to include any units.

How Science Works

Section 1 of the ISA

At the end of this stage, you will answer Section 1 of the ISA. Among other things you will need to:

- identify one or more variables that you need to control
- describe how you would carry out the main experiment
- identify possible hazards and say what you would do to reduce any risk
- make a blank table ready for your results.

a What features should you include in your written plan?
b What should you include in your blank table?

Stage 2

This is where you carry out the experiment and get some results. Don't worry too much about spending a long time getting fantastically accurate results – it is more important to get some results that you can analyse. You will have to draw a graph or a bar chart.

c How do you decide whether you should draw a bar chart or a line graph?

Stage 3

This is where you answer Section 2 of the ISA. Section 2 of the ISA is all about your own results, so make sure that you look at your table and graph when you are answering this section. To get the best marks you will need to quote some data from your results.

How Science Works

Section 2 of the ISA

In this section you will need to:

- say what you were trying to find out
- analyse data that is given in the paper. This data will be in the same topic area as your investigation
- use ideas from your own investigation to answer questions about this data
- write a conclusion
- compare your conclusion with the hypothesis you have tested.

Study tip

When you are comparing your conclusion with the hypothesis, make sure that you also talk about the **extent** to which your results support the hypothesis.

Which of these answers do you think would score the most marks?

- My results support the hypothesis.
- In my results, as x got bigger, y got bigger, the same as stated in the hypothesis.
- In my results, as x got bigger, y got bigger, the same as stated in the hypothesis, but unlike the hypothesis, y stopped increasing after a while.

Key points

- When you are writing the plan make sure that you include details about:
  – the range and interval of the independent variable
  – the control variables
  – the number of repeats.

- Try to put down at least two possible hazards, and say how you are going to minimise the risk from them.

- Look carefully at the hypothesis that you are given – this should give you a good clue about how to do the experiment.

- Always refer back to the hypothesis when you are writing your conclusion.

Summary questions

1 Copy and complete the sentences using the words below:

independent   dependent   control

When writing a plan, you need to state the variable that you are deliberately going to change, called the .... variable. You also need to say what you expect will change because of this; this is called the .... variable. You must also say what .... variables you will keep the same in order to make it a fair test.

Summary questions

1 a Put these words into order. They should be in the order that you might use them in an investigation.

design, prediction, conclusion, method, repeat, controls, graph, results, table, improve, safety, hypothesis

2 a How would you tell the difference between an opinion that was scientific and a biased or prejudiced opinion?

b Suppose you were investigating the amount of gas produced in a reaction. Would you choose to investigate a categoric or a continuous variable? Explain why.

3 You might have seen that marble statues weather badly where there is air pollution. You want to find out why.

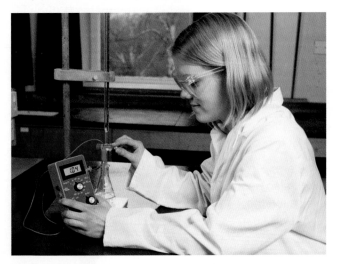

a You know that sulfur dioxide in the air forms an acid. How could this knowledge help you to develop a hypothesis about the effect of sulfur dioxide on marble statues?

b Develop a hypothesis about the effect of sulfur dioxide on marble statues.

c What experiment could you do to test your hypothesis?

d Suppose you are not able to carry out an experiment. How else could you test your hypothesis?

4 a What do you understand by a 'fair test'?

b Suppose you were carrying out an investigation into what effect diluting acid had on its pH. You would need to carry out a trial. Describe what a trial would tell you about how to plan your method.

c How could you decide if your results were repeatable?

d It is possible to calculate the effect of dilution on the pH of an acid. How could you use this to check on the accuracy of your results?

5 Suppose you were watching a friend carry out an investigation using the equipment shown on page 13. You have to mark your friend on how accurately he is making his measurements. Make a list of points that you would be looking for.

6 a How do you decide on the range of a set of data?

b How do you calculate the mean?

c When should you use a bar chart?

d When should you use a line graph?

7 a What should happen to anomalous results?

b What does a line of best fit allow you to do?

c When making a conclusion, what must you take into consideration?

d How can you check on the repeatability and reproducibility of your results?

8 a Why is it important when reporting science to 'tell the truth, the whole truth and nothing but the truth'?

b Why might some people be tempted not to be completely fair when reporting their opinions on scientific data?

9 a 'Science can advance technology and technology can advance science.' What do you think is meant by this statement?

b Who answers the questions that start with 'Should we .... '?

10 Wind turbines are an increasingly popular way of generating electricity. It is very important that they are sited in the best place to maximise energy output. Clearly they need to be where there is plenty of wind. Energy companies have to be confident that they get value for money. Therefore they must consider the most economic height to build them. Put them too high and they might not get enough extra energy to justify the extra cost of the turbine. Before deciding finally on a site they will carry out an investigation to decide the best height.

The prediction is that increasing the height will increase the power output of the wind turbine. A test platform was erected and the turbine placed on it. The lowest height that would allow the turbines to move was 32 m. The correct weather conditions were waited for and the turbine began turning and the power output was measured in kilowatts.

The results are in the table.

Height of turbine (m)	Power output Test 1 (kW)	Power output Test 2 (kW)
32	162	139
40	192	195
50	223	219
60	248	245
70	278	270
80	302	304
85	315	312

a What was the prediction for this test?

b What was the independent variable?

c What was the dependent variable?

d What is the range of the heights for the turbine?

e Suggest a control variable that should have been used.

f This is a fieldwork investigation. Is it possible to control all of the variables? If not, say what you think the scientist should have done to produce more accurate results.

g Is there any evidence for an anomalous result in this investigation? Explain your answer.

h Was the resolution of the power output measurement satisfactory? Provide some evidence for your answer from the data in the table.

i Draw a graph of the results for the second test.

j Draw a line of best fit.

k Describe the pattern in these results.

l What conclusion can you make?

m How might these data be of use to people who might want to stop a wind farm being built?

n Who should carry out these tests for those who might object to a wind farm being built?

B1 1.1

Diet and exercise

Learning objectives

- What does a healthy diet contain?

- Why can some people eat lots of food without getting fat?

- How does an athlete's diet differ from yours?

Did you know ...?

Whether you prefer sushi, dahl, or roast chicken, you need to eat a varied diet that includes everything you need to keep your body healthy.

What makes a healthy diet?

A balanced diet contains the correct amounts of:

- carbohydrates
- proteins
- fats
- vitamins
- minerals
- fibre
- water.

Your body uses carbohydrates, proteins and fats to release the energy you need to live and to build new cells. You need small amounts of vitamins and minerals for your body to work healthily. Without them you will suffer deficiency diseases. If you don't have a balanced diet then you will end up **malnourished**.

Figure 1 A balanced diet provides everything you need to survive, including plenty of energy

Fortunately, in countries like the UK, most of us take in all the minerals and vitamins we need from the food we eat. However, our diet can easily be unbalanced in terms of the amount of energy we take in. If we take in too much energy we put on weight. If we don't take in enough we become underweight.

It isn't always easy to get it right because different people need different amounts of energy. Even if you eat a lot, you can still lack vitamins and minerals if you don't eat the right food.

a Why do you need to eat food?

How much energy do you need?

The amount of energy you need to live depends on lots of different things. Some of these things you can change and some you can't.

Males need to take in more energy than a female of the same age – unless she is pregnant.

If you are a teenager, you will need more energy than if you are in your 70s.

b Why does a pregnant woman need more energy than a woman who isn't pregnant?

Your food supplies energy to your muscles as they work. So the amount of exercise you do affects the amount of energy you need. If you do very little exercise, then you don't need as much food. The more you exercise the more food you need to take in.

Figure 2 Athletes have a great deal of muscle tissue so they have to eat a lot of food to supply the energy they need

People who exercise regularly are usually much fitter than people who take little exercise. They make bigger muscles – up to 40% of their body mass. Muscle tissue transfers much more energy than fat. But exercise doesn't always mean time spent training or 'working out' in the gym. Walking to school, running around the house looking after small children or doing a physically active job all count as exercise too.

> **c** Why do athletes need to eat more food than the average person?

The temperature where you live affects how much energy you need as well. In warmer countries you need to eat less food. This is because you use less energy keeping your body temperature at a steady level.

The metabolic rate

Think of a friend who is very similar in age, gender and size to you. Despite these similarities, you may need quite different amounts of energy in your diet. This is because the rate of chemical reactions in your cells (the **metabolic rate**) varies from person to person.

Men generally have a higher metabolic rate than women. The proportion of muscle to fat in your body affects your metabolic rate. Men often have a higher proportion of muscle to fat than women. You can change the proportion of muscle to fat in your body by exercising. This will build up more muscle.

Your metabolic rate is also affected by the amount of activity you do. Exercise increases your metabolic rate for a time even after you stop exercising.

Scientists think that your basic metabolic rate may be affected by genetic factors you inherit from your parents. This is an example of how **inherited** factors can affect our health.

Did you know ... ?

Between 60–75% of your daily energy needs are used up in the basic reactions needed to keep you alive. About 10% is needed to digest your food – and only the final 15–30% is affected by your physical activity!

Study tip

'Metabolic rate' refers to the chemical reactions which take place in cells.

Figure 3 If you work somewhere really cold your metabolic rate will go up to keep you warm. You will need lots of fat in your diet to supply the energy you need.

Keighley Campus Library
Leeds City College

Key points

- Most people eat a varied diet, which includes everything needed to keep the body healthy.
- Different people need different amounts of energy.
- The metabolic rate varies from person to person.
- The more exercise you take, the more food you need.

Summary questions

1 What is 'a balanced diet'?

2 **a** Why do you need more energy in your diet when you are 18 than when you are 80?
  b Why does a top athlete need more energy in their diet than you do? Where does the energy in the diet come from?

3 **a** What is the 'metabolic rate'?
  b Explain why some people put on weight more easily than others.

B1 1.2

Weight problems

Learning objectives

- What health problems are linked to being overweight?

- Why is it unhealthy to be too thin?

- Why are people who do exercise usually healthier than those who do not?

Figure 1 In spite of some of the media hype, most people are not obese – but the amount of weight people carry varies a great deal!

Figure 2 Fitness instructors can help with improving health and fitness

Obesity

If you take in more energy than you use, the excess is stored as fat. You need some body fat to cushion your internal organs. Your fat also acts as an energy store for when you don't feel like eating. But if someone eats a lot more food than they need, this is a form of malnourishment. Over time they could become **overweight** or even **obese**.

Carrying too much weight is often inconvenient and uncomfortable. Obesity can also lead to serious health problems such as arthritis, type 2 diabetes (high blood sugar levels which are hard to control), high blood pressure and heart disease. Obese people are more likely to die at an earlier age than non-obese people.

> **a** What health problems are linked to obesity?

Losing weight

Many people want to be thinner. This might be for their health or just to look better. You gain fat by taking in more energy than you need. You lose **mass** when the energy content of your food is less than the energy you use in your daily life. There are three main ways you can lose mass.

- You can reduce the amount of energy you take in by cutting back the amount of food you eat. In particular, you can cut down on energy-rich foods like biscuits, crisps and chips.

- You can increase the amount of energy you use by doing more exercise.

- The best way to lose weight is to do both – reduce your energy intake and exercise more!

Scientists talk about 'mass', but most people talk about losing weight. Many people find it easier to lose weight by attending slimming groups. At these weekly meetings they get lots of advice and support from other slimmers. All slimming programmes involve eating fewer energy-rich foods and/or taking more exercise.

Exercise can make you healthier by helping to control your weight. It increases the amount of energy used by your body and increases the proportion of muscle to fat. It can make your heart healthier too. However, you need to take care. If you suddenly start taking vigorous exercise, you can cause other health problems.

Fitness instructors can measure the proportion of your body that is made up of fat. They can advise on the right food to eat and the exercise you need to become thinner, fitter, or both.

Different slimming programmes approach weight loss in different ways. Many simply give advice on healthy living. They advise lots of fruit and vegetables, avoiding too much fat or too many calories and plenty of exercise. Some are more extreme and suggest that you cut out almost all of the fat or the carbohydrates from your diet.

> **b** What must you do to lose weight?

How Science Works

You can find lots of slimming products in the supermarket. Used in the right way, they can help you to lose weight. Some people claim that 'slimming teas' or 'herbal pills' will enable you to eat what you like and still lose weight.

● What sort of evidence would you look for to decide which approaches to losing weight work best?

Figure 3 Slimming products can help you lose weight, but only if you control the total amount of energy you take in

?¿? Did you know … ?

The number of obese and overweight people is growing. The WHO (World Health Organisation) says over 1 billion adults worldwide are now overweight or obese.

Lack of food

In some parts of the world many people are underweight and malnourished because there is not enough food to eat. Civil wars, droughts and pests can all destroy local crops.

Deficiency diseases, due to lack of mineral **ions** and vitamins, are common in both children and adults when they never have enough food. Deficiency diseases can also occur if you do not have a balanced diet.

Study tip

The word 'malnourished' can be used to describe people who do not have a balanced diet. They may have too little food or too much food, or take in the wrong combination of foods.

Summary questions

1 Copy and complete using the words below:

energy fat less more obese

If you take in more .... than you use, the excess is stored as .... If you eat too much over a long period of time, you will eventually become .... To lose weight you need to eat .... and exercise ....

2 Why do people who are very thin, and some people who are obese, suffer from deficiency diseases?

3 One slimming programme controls your food intake. Another controls your food intake but also has an exercise programme. Which do you think would be the most effective? Explain your answer.

Key points

● If you take in more energy than you use, you will store the excess as fat.

● Obese people have more health problems than others.

● People who do not have enough to eat can develop serious health problems.

● Exercise helps reduce weight and maintain health.

B1 1.3   Inheritance, exercise and health

Learning objectives

- How can inherited factors affect your health?

- Why does your cholesterol level matter?

- Does exercise make you healthier?

⬡⬡ **links**

For information on metabolic rate, look back at B1 1.1 Diet and exercise.

Inheriting health

Inherited factors from your parents affect your appearance, such as the colour of your eyes. They also have a big effect on your health. They affect your metabolic rate, which affects how easily you lose and gain mass. Being overweight has a bad effect on your health. Inherited factors affect the proportion of muscle to fat in your body. They also affect your risk of heart disease, partly because they influence the levels of cholesterol in your blood.

Figure 1 Lots of things affect your health – your diet, how much exercise you take and what you inherit from your parents

Controlling cholesterol

The way your body balances cholesterol is an example of how an inherited factor can affect your health. You need cholesterol for your cell membranes and to make vital hormones. There are two forms of cholesterol carried around your body in your blood. One form is healthy but the other can cause health problems. If the balance of your cholesterol levels is wrong, your risk of getting heart disease increases.

a Why do you need cholesterol in your body?

The way your liver deals with the fat in your diet and makes the different types of cholesterol is inherited from your parents. For most people, eating a balanced diet means your liver can keep the balance of cholesterol right.

Eating lots of high-fat food means you are likely to have raised levels of harmful cholesterol and an increased risk of heart disease. But 1 in every 500 people inherit factors which mean they will have high levels of harmful cholesterol and an increased risk of heart disease whatever they eat. This is an example of how an inherited factor can affect your health.

⁇⁇ Did you know …?

The maximum healthy blood cholestrol is given as 6 mmol/l, 5 mmol/l and 4 mmol/l on different medical websites.

Scientists don't always agree!

Figure 2 Next time you eat a burger and fries, think about all the fat you are taking in. Will your body be able to deal with it, or are your blood cholesterol levels about to go up?

Exercise and health

Scientists have collected lots of evidence about exercise and health. It shows that people who exercise regularly are generally healthier than people who don't do much exercise. The graph in Figure 3 shows the results of an American study published in the journal *Circulation.* 6213 men were studied. The least active men were 4.5 times more likely to die early than the fittest, most active men.

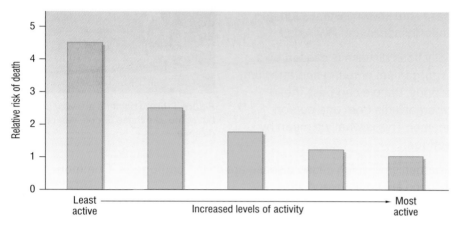

Figure 3 The effect of exercise on risk of death (Source: Jonathan Myers, *Circulation*, 2003)

These are some of the scientific explanations why exercise helps to keep you healthy.

- You are less likely to be overweight if you exercise regularly. This is partly because you will be using more energy.
- You will have more muscle tissue, which will increase your metabolic rate. If you can control your weight, you are less likely to be affected by problems such as arthritis, diabetes and high blood pressure.
- Your cholesterol levels are affected by exercise. Regular exercise lowers your blood cholesterol levels. It also helps the balance of the different types of cholesterol. When you exercise, your good cholesterol level goes up and the harmful cholesterol level goes down. This lowers your risk of heart disease and other health problems.

b How could you change your cholesterol levels?

Summary questions

1 Copy and complete using the words below:

heart metabolic inherited cholesterol balance

There are .... factors such as your .... rate that can affect your health. The way your liver makes .... is inherited and if the .... of cholesterol is wrong it can increase your risk of .... disease.

2 Why are people who exercise regularly usually healthier than people who take little exercise?

3 Using the data in Figure 3, which group of people do you think are most at risk of death? Why do you think this might be? What could they do to reduce the risk?

Key points

- Inherited factors affect our health. These include our metabolic rate and cholesterol level.

- People who exercise regularly are usually healthier than people who take little exercise.

B1 1.4 Pathogens and disease

Learning objectives

- What are pathogens?
- How do pathogens cause disease?
- How did Ignaz Semmelweis change the way we look at disease?

Infectious diseases are found all over the world, in every country. Some diseases are fairly mild ones, such as the common cold and tonsillitis. Other diseases are known killers, such as tetanus, influenza and HIV/Aids.

An infectious disease is caused by a **microorganism** entering and attacking your body. People can pass these microorganisms from one person to another. This is what we mean by **infectious**.

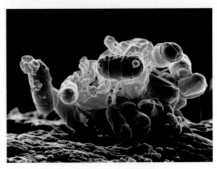

Figure 1 Many bacteria are very useful but some, like these *E. coli*, can cause disease

Microorganisms which cause disease are called **pathogens**. Common pathogens are bacteria and viruses.

a What causes infectious diseases?

The differences between bacteria and viruses

Bacteria are single-celled living organisms that are much smaller than animal and plant cells.

Although some bacteria cause disease, many are harmless and some are really useful to us. We use them to make food like yoghurt and cheese, to treat sewage and to make medicines.

Viruses are even smaller than bacteria. They usually have regular shapes. Viruses cause diseases in every type of living organism from people to bacteria.

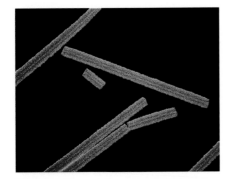

Figure 2 These tobacco mosaic viruses cause disease in plants

b How do viruses differ from bacteria?

How pathogens cause disease

Once bacteria and viruses are inside your body they reproduce rapidly. This is how they make you ill. Bacteria simply split in two – they often produce toxins (poisons) which affect your body. Sometimes they directly damage your cells. Viruses take over the cells of your body as they reproduce, damaging and destroying the cells. They very rarely produce toxins.

Common disease symptoms are a high temperature, headaches and rashes. These are caused by the damage and toxins produced by the pathogens. The symptoms also appear as a result of the way your body responds to the damage and toxins.

You catch an infectious disease when you pick up a pathogen from someone else who is infected with the disease.

⚭ links

For more information on bacteria that are resistant to antibiotics, see B1 1.8 Changing pathogens.

c How do pathogens make you feel ill?

How Science Works

The work of Ignaz Semmelweis

Ignaz Philipp Semmelweis was a doctor in the mid-1850s. At the time, many women in hospital died from childbed fever a few days after giving birth. However, no one knew what caused it.

Semmelweis noticed something about his medical students. They went straight from dissecting a dead body to delivering a baby without washing their hands. He wondered if they were carrying the cause of disease from the corpses to their patients.

Then another doctor cut himself while working on a body. He died from symptoms which were identical to childbed fever. Semmelweis was sure that the fever was caused by something that could be passed on – some kind of infectious agent.

He insisted that his medical students wash their hands before delivering babies. Immediately, fewer mothers died from the fever.

Getting his ideas accepted

Semmelweis talked to other doctors. He thought his evidence would prove to them that childbed fever was spread by doctors. But his ideas were mocked.

Many doctors thought that childbed fever was God's punishment to women. No one had ever seen bacteria or viruses. So it was hard to believe that disease was caused by something invisible passed from person to person. Doctors didn't like the idea that they might have been spreading disease. They were being told that their actions had killed patients instead of curing them.

In hospitals today, bacteria such as MRSA, which are resistant to antibiotics, are causing lots of problems. Getting doctors, nurses and visitors to wash their hands more often is part of the answer – just as it was in Semmelweis's time!

?? ? *Did you know … ?*

Semmelweis couldn't bear to think of the thousands of women who died because other doctors ignored his findings. By the 1860s he suffered a major breakdown and in 1865, aged only 47, he died – from an infection picked up from a patient during an operation.

Figure 3 Ignaz Semmelweis – his battle to get medical staff to wash their hands to prevent infections is still going on today

Key points

- Infectious diseases are caused by microorganisms called pathogens, such as bacteria and viruses.

- Bacteria and viruses reproduce rapidly inside your body. Bacteria can produce toxins which make you feel ill.

- Viruses damage your cells as they reproduce. This can also make you feel ill.

- Semmelweis recognised the importance of hand-washing in preventing the spread of infectious diseases in hospital.

Summary questions

1 Copy and complete using the words below:

toxins viruses microorganisms reproduce pathogens damage symptoms bacteria

The .... which cause infectious diseases are known as .... Once .... and .... get inside your body they .... rapidly. They .... your tissues and may produce .... which cause the .... of disease.

2 Give five examples of things we now know we can do to reduce the spread of pathogens to lower the risk of disease, e.g. hand-washing in hospitals.

3 Write a letter by Ignaz Semmelweis to a friend explaining how he formed his ideas and the struggle to get them accepted.

B1 1.5 — Defence mechanisms

Learning objectives

- How does your body stop pathogens getting in?
- How do white blood cells protect us from disease?

There are a number of ways in which pathogens spread from one person to another. The more pathogens that get into your body, the more likely it is that you will get an infectious disease.

Droplet infection: When you cough, sneeze or talk you expel tiny droplets full of pathogens from your breathing system. Other people breathe in the droplets, along with the pathogens they contain. So they pick up the infection, e.g. flu (influenza), tuberculosis or the common cold.

Figure 1 Droplets carrying millions of pathogens fly out of your mouth and nose at up to 100 miles an hour when you sneeze

Direct contact: Some diseases are spread by direct contact of the skin, e.g. impetigo and some sexually transmitted diseases like genital herpes.

Contaminated food and drink: Eating raw or undercooked food, or drinking water containing sewage can spread disease, e.g. diarrhoea or salmonellosis. You get these by taking large numbers of microorganisms straight into your gut.

Through a break in your skin: Pathogens can enter your body through cuts, scratches and needle punctures, e.g. HIV/Aids or hepatitis.

When people live in crowded conditions, with no sewage treatment, infectious diseases can spread very rapidly.

a What are the four main ways in which infectious diseases are spread?

Preventing microbes getting into your body

Each day you come across millions of disease-causing microorganisms. Fortunately your body has several ways of stopping these pathogens getting inside.

Your skin covers your body and acts as a barrier. It prevents bacteria and viruses from reaching the tissues beneath that can be infected.

If you damage or cut your skin you bleed. Your blood quickly forms a clot which dries into a scab. The scab forms a seal over the cut, stopping pathogens getting in through the wound.

Your breathing system could be a weak link in your body defences. Every time you breathe you draw air full of pathogens inside your body. However, your breathing system produces sticky liquid, called mucus. This mucus covers the lining of your lungs and tubes. It traps the pathogens. The mucus is then moved out of your body or swallowed down into your gut. Then the acid in your stomach destroys the microorganisms. In the same way, the stomach acid destroys most of the pathogens you take in through your mouth.

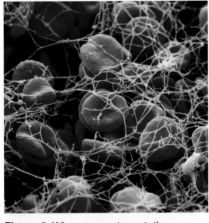

Figure 2 When you get a cut, the platelets in your blood set up a chain of events to form a clot that dries to a scab. This stops pathogens from getting into your body. It also stops you bleeding to death!

b What are the three main ways in which your body prevents pathogens from getting in?

How white blood cells protect you from disease

In spite of your body's defence mechanisms, some pathogens still get inside your body. Once there, they will meet your second line of defence – the **white blood cells** of your **immune system**.

The white blood cells help to defend your body against pathogens in several ways.

Table 1 Ways in which your white blood cells destroy pathogens and protect you against disease

Role of white blood cell	How it protects you against disease
Ingesting microorganisms	Some white blood cells ingest (take in) pathogens, destroying them so they can't make you ill.
Producing antibodies Antibody, Antigen, Bacterium, White blood cell, Antibody attached to antigen	Some white blood cells produce special chemicals called **antibodies**. These target particular bacteria or viruses and destroy them. You need a unique antibody for each type of pathogen. Once your white blood cells have produced antibodies once against a particular pathogen, they can be made very quickly if that pathogen gets into the body again.
Producing antitoxins Antitoxin molecule, Toxin and antitoxin joined together, Toxin molecule, Bacterium	Some white blood cells produce antitoxins. These counteract (cancel out) the toxins (poisons) released by pathogens.

Did you know ...?

Mucus produced from your nose turns green when you have a cold. This happens because some white blood cells contain green-coloured enzymes. These white blood cells destroy the cold viruses and any bacteria in the mucus of your nose when you have a cold. The dead white blood cells along with the dead bacteria and viruses are removed in the mucus, making it look green.

Summary questions

1 Explain how diseases are spread by:
 a droplet infection
 b direct contact
 c contaminated food and drink
 d through a cut in the skin.

2 Certain diseases mean you cannot fight infections very well. Explain why the following symptoms would make you less able to cope with pathogens.
 a Your blood won't clot properly.
 b The number of white cells in your blood falls.

3 Here are three common things we do. Explain carefully how each one helps to prevent the spread of disease.
 a Washing your hands before preparing a salad.
 b Throwing away tissues after you have blown your nose.
 c Making sure that sewage does not get into drinking water.

4 Explain in detail how the white blood cells in your body work.

Key points

- Your body has several methods of defending itself against the entry of pathogens using the skin, the mucus of the breathing system and the clotting of the blood.

- Your white blood cells help to defend you against pathogens by ingesting them, making antibodies and making antitoxins.

B1 1.6 | Using drugs to treat disease

Learning objectives

- What is a medicine?
- How do medicines work?
- Why can't we use antibiotics to treat diseases caused by viruses?

When you have an infectious disease, you generally take medicines which contain useful drugs. Often the medicine doesn't affect the pathogen that is causing the problems. It just eases the symptoms and makes you feel better.

Drugs like aspirin and paracetamol are very useful as painkillers. When you have a cold they will help relieve your headache and sore throat. On the other hand, they will have no effect on the viruses which have entered your tissues and made you feel ill.

Many of the medicines you can buy at a chemist's or supermarket are like this. They relieve your symptoms but do not kill the pathogens. They do not cure you any faster. You have to wait for your immune system to overcome the pathogens.

Figure 1 Taking paracetamol will make this child feel better, but she will not actually get well any faster as a result

a Why don't medicines like aspirin actually cure your illness?

Antibiotics

Drugs that make us feel better are useful but what we really need are drugs that can cure us. We use antiseptics and disinfectants to kill bacteria outside the body. But they are far too poisonous to use inside your body. They would kill you and your pathogens at the same time!

The drugs that have really changed the way we treat infectious diseases are **antibiotics**. These are medicines that can work inside your body to kill the bacteria that cause diseases.

b What is an antibiotic?

How antibiotics work

Antibiotics like penicillin work by killing the bacteria that cause disease while they are inside your body. They damage the bacterial cells without harming your own cells. They have had an enormous effect on our society. We can now cure bacterial diseases that killed millions of people in the past.

Unfortunately antibiotics are not the complete answer to the problem of infectious diseases. They have no effect on diseases caused by viruses.

The problem with viral pathogens is that they reproduce inside the cells of your body. It is extremely difficult to develop drugs that kill the viruses without damaging the cells and tissues of your body at the same time.

Figure 2 Penicillin was the first antibiotic. Now we have many different ones which kill different types of bacterium. Scientists are always on the look out for new antibiotics to keep us ahead in the battle against pathogens.

c How do antibiotics work?

How Science Works

Discovering penicillin

Alexander Fleming was a scientist who studied bacteria and wanted to find ways of killing them. In 1928, he was growing lots of bacteria on agar plates. Alexander was rather careless, and his lab was quite untidy. He often left the lids off his plates for a long time and forgot about experiments he had set up!

After one holiday, Fleming saw that lots of his culture plates had mould growing on them. He noticed a clear ring in the jelly around some of the spots of mould. Something had killed the bacteria covering the jelly.

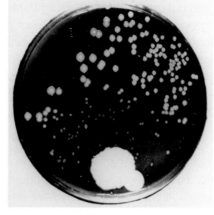

Figure 3 Alexander Fleming was on the lookout for something that would kill bacteria. As a result of him noticing the effect of this mould on his cultures, millions of lives have been saved around the world.

Fleming saw how important this was. He called the mould 'penicillin'. He worked hard to extract a juice from the mould. But he couldn't get much penicillin and he couldn't make it survive, even in a fridge. So Fleming couldn't prove it would actually kill bacteria and make people better. By 1934 he gave up on penicillin and went on to do different work.

About 10 years after penicillin was first discovered, Ernst Chain and Howard Florey set about trying to use it on people. They gave some penicillin they extracted to Albert Alexander, who was dying of a blood infection. The effect was amazing and Albert recovered. But then the penicillin ran out. Florey and Chain even tried to collect unused penicillin from Albert's urine, but it was no good. The infection came back and sadly Albert died.

They kept working and eventually they managed to make penicillin on an industrial scale. The process was able to produce enough penicillin to supply the demands of the Second World War. We have used it as a medicine ever since.

d Who was the first person to discover penicillin?

Summary questions

1 What is the main difference between drugs such as paracetamol and drugs such as penicillin?

2 **a** How did Alexander Fleming discover penicillin?
  b Why was it so difficult to make a medicine out of penicillin?
  c Who developed the industrial process which made it possible to mass-produce penicillin?

3 Explain why it is so much more difficult to develop medicines against viruses than it has been to develop antibacterial drugs.

Study tip

Remember:
- Antibiotics are drugs which kill bacteria.
- Antibodies are produced by white blood cells to kill bacteria.

Key points

- Some medicines relieve the symptoms of disease but do not kill the pathogens which cause it.

- Antibiotics cure bacterial diseases by killing the bacteria inside your body.

- Antibiotics do not destroy viruses because viruses reproduce inside the cells. It is difficult to develop drugs that can destroy viruses without damaging your body cells.

B1 1.7 Growing and investigating bacteria

Learning objectives

- How can we grow an uncontaminated culture of bacteria in the lab?

- Why do we need uncontaminated cultures?

- Why do we incubate bacteria at no more than 25 °C in schools and colleges?

To find out more about microorganisms we need to culture them. This means we grow very large numbers of them so that we can see all of the bacteria (the colony) as a whole. Many microorganisms can be grown in the laboratory. This helps us to learn more about them. We can find out what nutrients they need to grow and investigate which chemicals are best at killing them. Bacteria are the most commonly cultured microorganisms.

Growing microorganisms in the lab

To culture (grow) microorganisms you must provide them with everything they need. This means giving them a liquid or gel containing nutrients – a **culture medium**. It contains carbohydrate as an energy source along with various minerals and sometimes other chemicals. Most microorganisms also need warmth and oxygen to grow.

You usually provide the nutrients in **agar** jelly. Hot agar containing all the nutrients your bacteria will need is poured into a Petri dish. It is then left to cool and set before you add the microorganisms.

You must take great care when you are culturing microorganisms. The bacteria you want to grow may be harmless. However, there is always the risk that a **mutation** (a change in the DNA) will take place and produce a new and dangerous pathogen.

Figure 1 Culturing microorganisms like bacteria makes it possible for us to observe them and see how different chemicals affect them

You also want to keep the pure strains of bacteria you are culturing free from any other microorganisms. Such contamination might come from your skin, the air, the soil or the water around you. Investigations need uncontaminated cultures of microorganism. Whenever you are culturing microorganisms you must carry out strict health and safety procedures to protect yourself and others.

a What is agar jelly?

?? Did you know …?

You are surrounded by disease-causing bacteria all the time. If you cultured bacteria at 37 °C – human body temperature – there would be a very high risk of growing some dangerous pathogens.

Growing useful organisms

You can prepare an uncontaminated culture of microorganisms in the laboratory by following a number of steps.

The Petri dishes on which you will grow your microorganisms must be sterilised before using them. The nutrient agar, which will provide their food, must also be sterilised. This kills off any unwanted microorganisms. You can use heat to sterilise glass dishes. A special oven called an autoclave is often used. It sterilises by using steam at high pressure. Plastic Petri dishes are often bought ready-sterilised. UV light or gamma radiation is used to kill the bacteria.

Figure 2 When working with the most dangerous pathogens, scientists need to be very careful. Sensible safety precautions are needed when working with microorganisms.

b Why must everything be sterilised before you start a culture?

The next step is to **inoculate** the sterile agar with the microorganisms you want to grow.

Study tip

Make sure you understand why we sterilise. We boil solutions and heat-treat apparatus in an autoclave to **kill bacteria** already in them. This is sterilising.

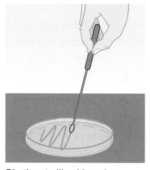

Sterilise the inoculating loop used to transfer micro-organisms to the agar by heating it until it is red hot in the flame of a Bunsen and then letting it cool. Do not put the loop down or blow on it as it cools.

Dip the sterilised loop in a suspension of the bacteria you want to grow and use it to make zigzag streaks across the surface of the agar. Replace the lid on the dish as quickly as possible to avoid contamination.

Seal the lid of the Petri dish with adhesive tape to prevent microorganisms from the air contaminating the culture – or microbes from the culture escaping. Do not seal all the way around the edge so oxygen can get into the dish and harmful anaerobic bacteria do not grow.

Figure 3 Culturing microorganisms safely in the laboratory

Once you have inoculated your plates, the sealed Petri dishes need to be incubated (kept warm) for several days so the microorganisms can grow. In school and college laboratories the maximum temperature at which cultures are incubated is 25 °C. This greatly reduces the likelihood that you will grow pathogens that might be harmful to people. In industrial conditions, bacterial cultures are often grown at higher temperatures, which allow the microorganisms to grow more rapidly.

Practical

Investigating the action of disinfectants and antibiotics

You can use cultures you set up yourself or pre-inoculated agar to investigate the effect of disinfectants and antibiotics on the growth of bacteria. An area of clear jelly indicates that the bacteria have been killed or cannot grow.

- What are the safety issues in this investigation and how will you manage any risks?

Summary questions

1 Why do we culture microorganisms in the laboratory?

2 Why don't we culture bacteria at 37 °C in the school lab?

3 When you set up a culture of bacteria in a Petri dish (see Figure 3) you give the bacteria everything they need to grow as fast as possible. However these ideal conditions do not last forever. What might limit the growth of the bacteria in a culture on a Petri dish?

Key points

- An uncontaminated culture of microorganisms can be grown using sterilised Petri dishes and agar. You sterilise the inoculating loop before use and seal the lid of the Petri dish to prevent unwanted microorganisms getting in. The culture is left at about 25 °C for a few days.

- Uncontaminated cultures are needed so we can investigate the effect of chemicals such as disinfectants and antibiotics on microorganisms.

- Cultures should be incubated at a maximum temperature of 25 °C in schools and colleges to reduce the likelihood of harmful pathogens growing.

Changing pathogens

Learning objectives

● What is antibiotic resistance?

● How can we prevent antibiotic resistance developing? **[H]**

● Why is mutation in bacteria and viruses such a problem?

If you are given an antibiotic and use it properly, the bacteria that have made you ill are killed off. However some bacteria develop resistance to antibiotics. They have a natural mutation (change in the genetic material) that means they are not affected by the antibiotic. These mutations happen by chance and they produce new strains of bacteria by **natural selection**.

More types of bacteria are becoming resistant to more antibiotics. Diseases caused by bacteria are becoming more difficult to treat. Over the years antibiotics have been overused and used when they are not really needed. This increases the rate at which antibiotic resistant strains have developed.

Antibiotic-resistant bacteria

Normally an antibiotic kills the bacteria of a non-resistant strain. However individual resistant bacteria survive and reproduce, so the population of **resistant** bacteria increases.

Antibiotics are no longer used to treat non-serious infections such as mild throat infections, which are often caused by viruses. Hopefully this will slow down the rate of development of resistant strains.

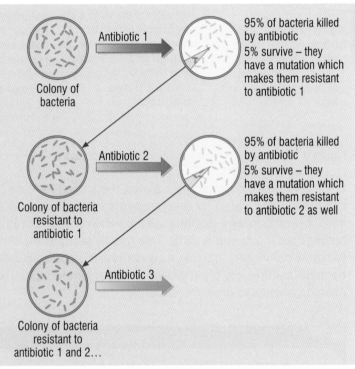

Colony of bacteria

Antibiotic 1

95% of bacteria killed by antibiotic
5% survive – they have a mutation which makes them resistant to antibiotic 1

Colony of bacteria resistant to antibiotic 1

Antibiotic 2

95% of bacteria killed by antibiotic
5% survive – they have a mutation which makes them resistant to antibiotic 2 as well

Colony of bacteria resistant to antibiotic 1 and 2…

Antibiotic 3

Figure 1 Bacteria can develop resistance to many different antibiotics in a process of natural selection as this simple model shows

To prevent more resistant strains of bacteria appearing it is important not to overuse antibiotics. It's best to only use them when you really need them. Antibiotics don't affect viruses so people should not demand antibiotics to treat an illness which the doctor thinks is viral.

Some antibiotics treat very specific bacteria. Others treat many different types of bacteria. The right type of antibiotic must be used to treat each bacterial infection to prevent further antibiotic resistance developing. It is also important that people finish their course of medicine every time.

a Why is it important not to use antibiotics too frequently?

Study tip

Washing hands removes the pathogens on them, but it may not kill the pathogens.

The MRSA story

Hospitals use a lot of antibiotics to treat infections. As a result of natural selection, some of the bacteria in hospitals are resistant to many antibiotics. This is what has happened with **MRSA** (the bacterium methicillin-resistant *Staphylococcus aureus*).

As doctors and nurses move from patient to patient, these antibiotic-resistant bacteria are spread easily. MRSA alone now contributes to around 1000 deaths every year in UK hospitals.

There are a number of simple steps which can reduce the spread of microorganisms such as MRSA. We have known some of them since the time of Semmelweis, but they sometimes get forgotten!

- Antibiotics should only be used when they are really needed.
- Specific bacteria should be treated with specific antibiotics.
- Medical staff should wash their hands with soap and water or alcohol gel between patients. They should wear disposable clothing or clothing that is regularly sterilised.
- Visitors should wash their hands as they enter and leave the hospital.
- Patients infected with antibiotic-resistant bacteria should be looked after in isolation from other patients.
- Hospitals should be kept clean – there should be high standards of hygiene.

b Is MRSA a bacterium or a virus?

Mutation and pandemics

Another problem caused by the mutation of pathogens is that new forms of diseases can appear. These new strains can spread quickly and cause widespread illness because no one is immune to them and there is no effective treatment. For example the flu virus mutates easily. Every year there are new strains of the virus that your immune system doesn't recognise. There is no effective treatment against viruses at all. The existing flu vaccine is not effective against new strains of the virus, and it takes time to develop a new vaccine.

There may be a flu **epidemic** (in one country) or even a **pandemic** (across several countries). In 1918–19, a new strain of flu virus killed over 40 million people around the world.

With modern international travel, a new strain of pathogen can spread very quickly. In 2009 there was a pandemic of a new strain of flu, known as swine flu, which spread very fast. Internationally, countries worked to stop it spreading and the death toll was kept relatively low.

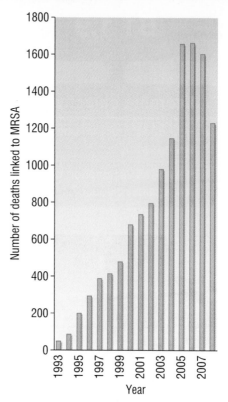

Figure 2 Data that show how the number of deaths in which MRSA played a part from 1993 (Source: National Statistics Office)

∞ links
For more information on the work of Semmelweis, look back at B1 1.4 Pathogens and disease.

Summary questions

1 Copy and complete using the words below:

antibiotics bacterium (virus) better disease mutation mutate resistant virus (bacterium)

If bacteria change or .... they may become .... to .... This means the medicine no longer makes you .... A .... in a .... or .... can also lead to a new form of ....

2 Make a flow chart to show how bacteria develop resistance to antibiotics.

3 Use Figure 2 to help you answer these questions.
   a How could you explain the increase in deaths linked to MRSA?
   b How could you explain the fall in deaths linked to MRSA, which still continues?

Key points

- Many types of bacterium have developed antibiotic resistance as a result of natural selection. To prevent the problem getting worse we must not overuse antibiotics.

- If bacteria or viruses mutate, new strains of the pathogen can appear causing disease.

- New strains of disease which spread rapidly can cause epidemics and pandemics. Antibiotics and vaccinations may not be effective against the new strain.

B1 1.9

Immunity

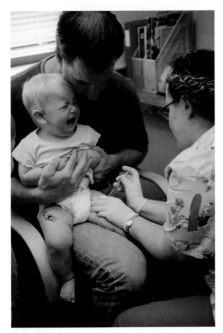

Figure 1 No one likes having a vaccination very much – but they save millions of lives!

Every cell has unique proteins on its surface called **antigens**. The antigens on the microorganisms that get into your body are different to the ones on your own cells. Your immune system recognises they are different.

Your white blood cells then make antibodies which join up with the antigens. This destroys the pathogens.

Your white blood cells 'remember' the right antibody needed to tackle a particular pathogen. If you meet that pathogen again, they can make the same antibody very quickly. So you become immune to that disease.

The first time you meet a new pathogen you get ill. That's because there is a delay while your body sorts out the right antibody needed. The next time, you completely destroy the invaders before they have time to make you feel unwell.

a What is an antigen?

Vaccination

Some pathogens can make you seriously ill very quickly. In fact you can die before your body manages to make the right antibodies. Fortunately, you can be protected against many of these serious diseases by **immunisation** (also known as **vaccination**).

Immunisation involves giving you a **vaccine**. A vaccine is usually made of a dead or weakened form of the disease-causing microorganism. It works by triggering your body's natural immune response to invading pathogens.

A small amount of dead or inactive pathogen is introduced into your body. This gives your white blood cells the chance to develop the right antibodies against the pathogen without you getting ill.

Then, if you meet the live pathogens, your white blood cells can respond rapidly. They can make the right antibodies just as if you had already had the disease, so you are protected against it.

b What is an antibody?

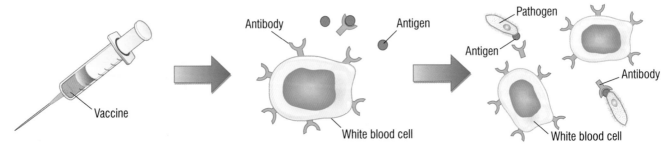

Small amounts of dead or inactive pathogen are put into your body, often by injection.

The antigens in the vaccine stimulate your white blood cells into making antibodies. The antibodies destroy the antigens without any risk of you getting the disease.

You are immune to future infections by the pathogen. That's because your body can respond rapidly and make the correct antibody as if you had already had the disease.

Figure 2 This is how vaccines protect you against dangerous infectious diseases

We use vaccines to protect us against both bacterial diseases (e.g. tetanus and diphtheria) and viral diseases (e.g. polio, measles and mumps). For example, the MMR vaccine protects against measles, mumps and rubella. Vaccines have saved millions of lives around the world. If a large proportion of the population is immune to a disease, the spread of the pathogen is very much reduced. One disease – smallpox – has been completely wiped out by vaccinations. Doctors hope polio will also disappear in the next few years.

c Give an example of one bacterial and one viral disease which you can be immunised against.

How Science Works

The vaccine debate

No medicine is completely risk free. Very rarely, a child will react badly to a vaccine with tragic results. Making the decision to have your baby immunised can be difficult.

Society needs as many people as possible to be immunised against as many diseases as possible. This keeps the pool of infection in the population very low. On the other hand, you know there is a remote chance that something may go wrong with a vaccination.

Because vaccines are so successful, we rarely see the terrible diseases they protect us against. A hundred years ago nearly 50% of all deaths of children and young people were caused by infectious diseases. The development of antibiotics and vaccines means that now only 0.5% of all deaths in the same age group are due to infectious disease. Many children were also left permanently damaged by serious infections. Parents today are often aware of the very small risks from vaccination – but sometimes forget about the terrible dangers of the diseases we vaccinate against.

If you are a parent it can be difficult to find unbiased advice to help you make a decision. The media highlight scare stories which make good headlines. The pharmaceutical companies want to sell vaccines. Doctors and health visitors can weigh up all the information, but they have vaccination targets set by the government.

Summary questions

1 Copy and complete using the words below:

antibodies   pathogen   immunised   dead   immune
inactive   white

People can be .... against a disease by introducing small quantities of .... or .... forms of a .... into your body. They stimulate the .... blood cells to produce .... to destroy the pathogen. This makes you .... to the disease in future.

2 Explain carefully, using diagrams if they help you:
   a how the immune system of your body works
   b how vaccines use your natural immune system to protect you against serious diseases.

3 Explain why vaccines can be used against both bacterial and viral diseases but antibiotics only work against bacteria.

Study tip

High levels of antibodies do not stay in your blood forever – immunity is the ability of your white blood cells to produce the right antibodies quickly if you are reinfected by a disease.

links

For more information on antibiotics, look back at B1 1.8 Changing pathogens.

Key points

- Your white blood cells produce antibodies to destroy the pathogens. Then your body will respond rapidly to future infections by the same pathogen, by making the correct antibody. You become immune to the disease.

- You can be immunised against a disease by introducing small amounts of dead or inactive pathogens into your body.

- We can use vaccinations to protect against both bacterial and viral pathogens.

B1 1.10 How do we deal with disease?

Learning objectives

- What are the advantages and disadvantages of being vaccinated?

- How has the treatment of disease changed over time?

The whooping cough story

In the 1970s, Dr John Wilson, a UK specialist in treating children, published a report suggesting that the pertussis (whooping cough) vaccine might cause brain damage in some children. The report was based on his study of a small group of 36 patients.

The media publicised the story and parents began to panic. The number of children being vaccinated against whooping cough fell from over 80% to around 30%. This was too low to protect the population from the disease.

People were so worried about the vaccine that they forgot that whooping cough itself can cause brain damage and death. In Scotland about 100 000 children suffered from whooping cough between 1977 and 1991. About 75 of them died. A similar pattern was seen across the whole of the UK.

An investigation into the original research discovered that it had serious flaws. Identical twin girls who were included in the study, and later died of a rare genetic disorder, had never actually had the whooping cough vaccine. It was a small study and only 12 of the children investigated had shown any symptoms close to the time of their whooping cough vaccination. Their parents were involved in claims for compensation from the vaccine manufacturers.

Activity

Design a webpage for parents that answers the sort of questions they might ask about their child having the normal vaccines. Make it user-friendly, i.e. the sort of thing a health worker could use to help reassure worried parents.

OR

Produce a PowerPoint presentation on the importance of responsible media reporting of science and medicine, using the whooping cough case as one of your main examples.

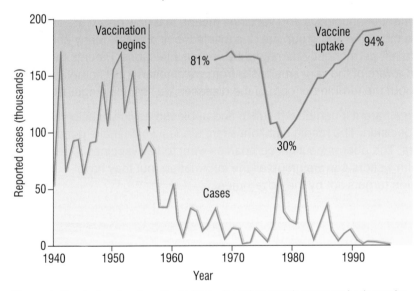

Figure 1 Graph showing the effect of the whooping cough scare on both uptake of the vaccine and the number of cases of the disease (Source: Open University)

No medical treatment (including vaccinations) is completely safe, but when the claims for compensation came to court, the whole study was questioned. After hearing all the evidence, the judge decided that the risks of whooping cough were far worse than any possible damage caused by the vaccine itself.

However, this judgement on the study got much less media coverage than the original scare story. Parents still felt there was 'no smoke without fire'. It was 20 years before vaccination levels, and the levels of whooping cough, returned to the levels before the scare. The number of people having vaccinations now is over 90%, and deaths from whooping cough are almost unknown in the UK.

Medicines for the future

Overuse of antibiotics has lead to spreading antibiotic resistance in many different bacteria. In recent years doctors have found strains of bacteria that are resistant to even the strongest antibiotics. When that happens, there is nothing more that antibiotics can do for a patient and he or she may well die.

The development of antibiotic resistant strains of bacteria means scientitsts are constantly looking for new antibiotics. It isn't easy to find chemicals which kill bacteria without damaging human cells.

Penicillin and several other antibiotics are made by moulds. Scientists are collecting soil samples from all over the world to try and find another mould to produce a new antibiotic that will kill antibiotic-resistant bacteria such as MRSA.

Crocodiles have teeth full of rotting meat. They live in dirty water and fight a lot. But scientists noticed that although crocodiles often give each other terrible bites, the bites do not become infected. They have extracted a chemical known as 'crocodillin' from crocodile blood and it seems to be a powerful antibiotic. Now the race is on to try and turn these amazing chemicals into antibiotics we can use.

Fish such as this plaice are covered with a slime which helps to protect them from damage and infection. Scientists have analysed this slime and found it contains proteins which have antibiotic properties. The proteins have been isolated from the slime and they still kill bacteria. So maybe fish will provide us with an antibiotic for the future.

Honey has been used since the time of the Ancient Egyptians to help heal wounds. Scientists in Germany and Australia have found that certain types of honey have antibiotic properties. They kill many bacteria, including MRSA. Doctors are using manuka honey dressings more and more to treat infected wounds.

Figure 2 Where will the next antibiotic be found?

Activity

Produce a poster on antibiotic resistance in bacteria and the search for new antibiotics. Make sure you explain how antibiotic resistance has developed and why we need more antibiotics. Use the ideas given here and, if possible, look for more examples of possible sources of new antibiotics.

Summary questions

1 Give one advantage and one disadvantage of being vaccinated.
2 List three examples of bad science from the story of the whooping cough vaccine and explain why the story should never have been published.

Key points

- Vaccination protects individuals and society from the effects of a disease.
- The treatment of disease has changed as our understanding of how antibiotics and immunity has increased.

Summary questions (k)

1 **a** Define the term 'balanced diet'.

b A top athlete needs to eat a lot of food each day. This includes protein and carbohydrate. Explain how they can eat so much without putting on weight.

2 Two young people have written to a lifestyle magazine problem page for advice about their diet and lifestyle. Produce an 'answer page' for the next edition of the magazine.

a Melanie: *I'm 16 and I worry about my weight a lot. I'm not really overweight but I want to be thinner. I've tried to diet but I just feel so tired when I do – and then I buy chocolate bars on the way home from school when my friends can't see me! What can I do?*

b Jaz: *I'm nearly 17 and I've grown so fast in the last year that I look like a stick! So my clothes look pretty silly. I'm also really good at football, but I don't seem as strong as I was and my legs get really tired by the end of a match. I want to build up a bit more muscle and stamina – but I don't just want to eat so much I end up getting really heavy. What can I do about it?*

3 **a** What factors affect the cholesterol levels in your blood?

b What can you do to help reduce your blood cholesterol levels?

c Cholesterol is one inherited factor which affects your health. Give one other example of an inherited factor which affects your health and explain how it does this.

4 How do tiny organisms like bacteria and viruses make a person ill?

5 There is going to be a campaign to try and stop the spread of colds in Year 7 of your school. There is going to be a poster and a simple PowerPoint presentation. Make a list of all the important things that the Year 7 children need to know about how diseases are spread. Also cover how the spread of infectious diseases from one person to another can be reduced.

6 **a** Vancomycin is an antibiotic which doctors used for patients infected with MRSA and other antibiotic-resistant bacteria. Now they are finding some infections are resistant to vancomycin. Explain how this may have happened. [H]

b What can we do to prevent the problem of antibiotic resistance getting worse? [H]

7 **a** How would you set up a culture of bacteria in a school lab?

b Describe how you would test to find out the right strength of disinfectant to use to wash the school floors.

Practice questions *k*

1 It is possible to grow microorganisms in the laboratory.
List A shows some temperatures.
List B shows situations for which these temperatures might be suitable.
Match each temperature to the correct situation.

List A	List B
25°C	Used in industrial laboratories to grow microorganisms quickly
35°C	Used in school laboratory to grow microorganisms safely
100°C	Used to stop microorganisms growing without killing them
	Used to kill microorganisms (3)

2 *In this question you will be assessed on using good English, organising information clearly and using specialist terms where appropriate.*

We need a balanced diet to keep us healthy. Explain the ways in which an unbalanced diet can affect the body. (6)

3 A person's metabolic rate varies with the amount of activity they do.

a Metabolic rate is ....

Choose one answer.
the breathing rate
the rate of chemical reactions in cells
the heart rate (1)

b Suggest **one** other factor which can change a person's metabolic rate. (1)

4 Polio is a disease caused by a virus. In the UK, children are given polio vaccine to protect them against the disease.

a Choose the correct words from each list to complete the sentences below.

i It is difficult to kill the polio virus inside the body because the virus .... (1)

is not affected by drugs   lives inside cells
produces antitoxins

ii The vaccine contains an .... form of the polio virus. (1)

active   infective   inactive

iii The vaccine stimulates the white blood cells to produce .... which destroy the virus. (1)

antibiotics   antibodies   drugs

b The graph shows the number of cases of polio in the UK between 1948 and 1968.

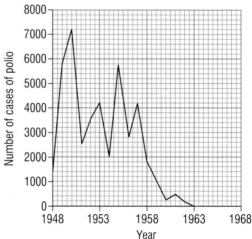

i In which year was the number of cases of polio highest? (1)

ii Polio vaccination was first used in the UK in 1955. How many years did it take for the number of cases of polio to fall to zero? (1)

iii There have been no cases of polio in the UK for many years. But children are still vaccinated against the disease.

Suggest **one** reason for this. (1)

AQA, 2006

5 Controlling infections in hospitals has become much more difficult in recent years.

a Suggest **two** reasons why MRSA is causing problems in many hospitals. (2)

b The pioneer in methods of treating infections in hospitals was Ignaz Semmelweis. He observed that women whose babies were delivered by doctors in hospital had a death rate of 18% from infections caught in the hospital. Women whose babies were delivered by midwives in the hospital had a death rate of 2%. He observed that doctors often came straight from examining dead bodies to the delivery ward.

i In a controlled experiment, Semmelweis made doctors wash their hands in chloride of lime solution before delivering the babies. The death rate fell to about 2% – down to the same level as the death rate in mothers whose babies were delivered by midwives.
Explain why the death rate fell. (1)

ii Explain how Semmelweis's results could be used to reduce the spread of MRSA in a modern hospital. (2)

AQA, 2005

B1 2.1

Responding to change

Learning objectives

- Why do you need a nervous system?

- What is a receptor?

- How do you respond to changes in your surroundings?

You need to know what is going on in the world around you. Your **nervous system** makes this possible. It enables you to react to your surroundings and coordinate your behaviour.

Your nervous system carries electrical signals (**impulses**) that travel fast – from 1 to 120 metres per second. This means you can react to changes in your surroundings very quickly.

a What is the main job of the nervous system?

The nervous system

Like all living things, you need to avoid danger, find food and, eventually, find a mate! This is where your nervous system comes into its own. Your body is particularly sensitive to changes in the world around you. Any changes (known as **stimuli**) are picked up by cells called **receptors**.

Receptor cells (e.g. the light receptor cells in your eyes) are like most animal cells. They have a nucleus, cytoplasm and a cell membrane. These receptors are usually found clustered together in special **sense organs**, such as your eyes and your skin. You have many different types of sensory receptor (see Figure 2).

b Where would you find receptors that respond to:
  i a loud noise
  ii touching a hot oven
  iii a strong perfume?

Figure 1 Your body is made up of millions of cells which have to work together. Whatever you do with your body – whether it's walking to school or playing on the computer – your movements need to be coordinated.

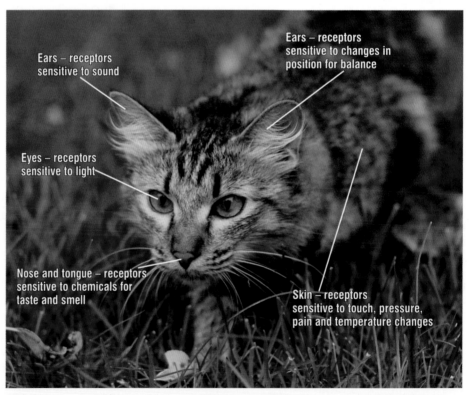

Ears – receptors sensitive to changes in position for balance

Ears – receptors sensitive to sound

Eyes – receptors sensitive to light

Nose and tongue – receptors sensitive to chemicals for taste and smell

Skin – receptors sensitive to touch, pressure, pain and temperature changes

Figure 2 This cat relies on its sensory receptors to detect changes in the environment

??? Did you know ...?

Some male moths have receptors so sensitive they can detect the scent of a female several kilometres away and follow the scent trail to find her!

How your nervous system works

Once a sensory receptor detects a stimulus, the information (sent as an electrical impulse) passes along special cells called **neurons**. These are usually found in bundles of hundreds or even thousands of neurons known as **nerves**.

The impulse travels along the neuron until it reaches the **central nervous system** or **CNS**. The CNS is made up of the brain and the spinal cord. The cells which carry impulses from your sense organs to your central nervous system are called **sensory neurons**.

> **c** What is the difference between a neuron and a nerve?

Your brain gets huge amounts of information from all the sensory receptors in your body. It coordinates the information and sends impulses out along special cells. These cells carry information from the CNS to the rest of your body. The cells are called **motor neurons**. They carry impulses to make the right bits of your body – the **effector organs** – respond.

Effector organs are muscles or glands. Your muscles respond to the arrival of impulses by contracting. Your glands respond by releasing (**secreting**) chemical substances.

The way your nervous system works can be summed up as:

receptor → sensory neuron → coordinator (CNS) → motor neuron → effector

> **d** What is the difference between a sensory neuron and a motor neuron?

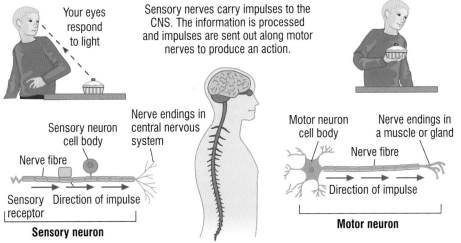

Figure 3 The rapid responses of our nervous system allow us to respond to our surroundings quickly – and in the right way!

Study tip

Make sure you are clear that 'motor' means movement. 'Motor neurons' stimulate the muscles to contract.

Study tip

Be careful to use the terms neuron and nerve correctly.

Talk about **impulses** (*not* messages) travelling along a neuron.

Key points

- The nervous system uses electrical impulses to enable you to react quickly to your surroundings and coordinate what you do.

- Cells called receptors detect stimuli (changes in the environment).

- Like all animal cells, light receptor cells and other receptors have a nucleus, cytoplasm and cell membrane.

- Impulses from receptors pass along sensory neurons to the brain or spinal cord (CNS). Impulses are sent along motor neurons from the brain (CNS) to the effector organs.

Summary questions

1 Copy and complete using the words below:

neurons   receptors   electrical   CNS   environment   nervous

Your .... system carries fast .... impulses. Changes in the .... are picked up by your .... and impulses travel along your .... to your ....

2 Make a table to show the different types of sense receptor. For each one, give an example of the sort of things it responds to, e.g. touch receptors respond to an insect crawling on your skin.

3 Explain what happens in your nervous system when you see a piece of chocolate, pick it up and eat it.

B1 2.2

Reflex actions

Learning objectives

- What is a reflex?
- Why are reflexes so important?

Your nervous system lets you take in information from your surroundings and respond in the right way. However, some of your responses are so fast that they happen without giving you time to think.

When you touch something hot, or sharp, you pull your hand back before you feel the pain. If something comes near your face, you blink. Automatic responses like these are known as **reflexes**.

What are reflexes for?

Reflexes are very important both for human beings and for other animals. They help you to avoid danger or harm because they happen so fast. There are also lots of reflexes that take care of your basic body functions. These functions include breathing and moving food through your gut.

It would make life very difficult if you had to think consciously about those things all the time – and would be fatal if you forgot to breathe!

> **a** Why are reflexes important?

How do reflexes work?

Reflex actions involve just three types of neuron. These are:
- sensory neurons
- motor neurons
- relay neurons – these connect a sensory neuron and a motor neuron. Your relay neurons are in the CNS.

An electrical impulse passes from the sensory receptor along the sensory neuron to the CNS. It then passes along a relay neuron (usually in the spinal cord) and straight back along a motor neuron. From there the impulse arrives at the effector organ. The effector organ will be a muscle or a gland. We call this a **reflex arc**.

The key point in a reflex arc is that the impulse bypasses the conscious areas of your brain. The result is that the time between the stimulus and the reflex action is as short as possible.

> **b** Why is it important that the impulses in a reflex arc do not go to the conscious brain?

How synapses work

Your nerves are not joined up directly to each other. There are junctions between them called **synapses**. The electrical impulses travelling along your neurons have to cross these synapses. They cannot leap the gap. Look at Figure 1 to see what happens next.

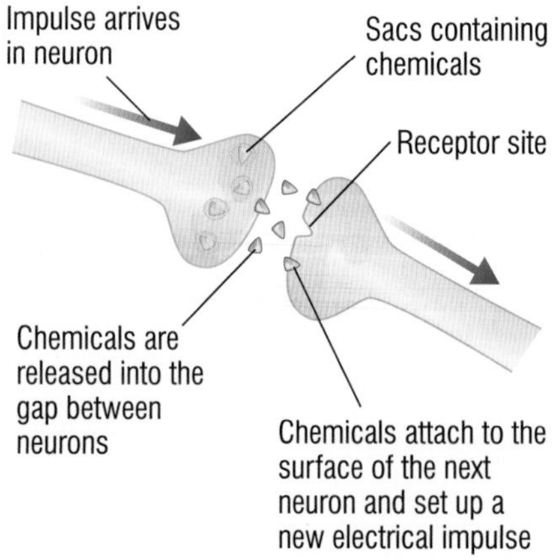

The reflex arc in detail

Look at Figure 2. It shows what would happen if you touched a hot object.

When you touch it, a receptor in your skin is stimulated. An electrical impulse passes along a sensory neuron to the central nervous system – in this case the spinal cord.

Practical

The stick-drop test

You can investigate how quickly nerve impulses travel in your body using metre rules, and either stop clocks or ICT to measure how quickly you catch the ruler OR by standing in a circle holding hands with your eyes closed and measuring how long it takes a hand squeeze to pass around the circle.

Impulse arrives in neuron

Sacs containing chemicals

Receptor site

Chemicals are released into the gap between neurons

Chemicals attach to the surface of the next neuron and set up a new electrical impulse

Figure 1 When an impulse arrives at the junction between two neurons, chemicals are released which cross the synapse and arrive at receptor sites on the next neuron. This starts up an electrical impulse in the next neuron.

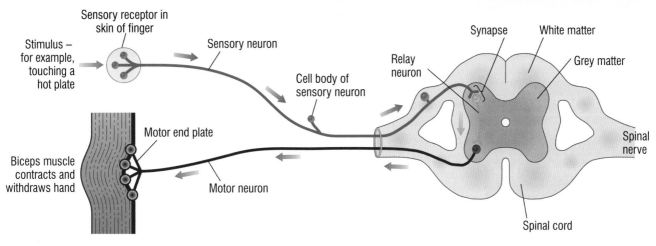

Figure 2 The reflex action which moves your hand away from something hot can save you from being burned. Reflex actions are quick and automatic; you do not think about them.

When an impulse from the sensory neuron arrives in the synapse with a relay neuron, a chemical messenger is released. This chemical crosses the synapse to the relay neuron. There it sets off an electrical impulse that travels along the relay neuron.

When the impulse reaches the synapse between the relay neuron and a motor neuron returning to the arm, another chemical is released.

This chemical crosses the synapse and starts an electrical impulse travelling down the motor neuron. When the impulse reaches the effector organ, it is stimulated to respond.

In this example the impulses arrive in the muscles of the arm, causing them to contract. This action moves the hand rapidly away from the source of pain. If the effector organ is a gland, it will respond by releasing (secreting) chemicals.

Most reflex actions can be shown as follows:

stimulus → receptor → coordinator → effector → response

This is not very different from a normal conscious action. However, in a reflex action the coordinator is a relay neuron either in the spinal cord or in the unconscious areas of the brain. The whole reflex is very fast indeed.

An impulse also travels up the spinal cord to the conscious areas of your brain. You know about the reflex action, but only after it has happened.

Study tip

Make sure you know the correct sequence of links from the receptor to the effector.

Figure 3 Newborn babies have a number of special reflexes which disappear as they grow. This gripping reflex is one of them.

Summary questions

1 Copy and complete using the words below:

conscious   motor   reflex   relay   response   sensory   stimulus

In a .... arc the electrical impulse bypasses the .... areas of your brain. The time between the .... and the .... is as short as possible. Only .... neurons, ... neurons and .... neurons are involved.

2 Explain why some actions, such as breathing and swallowing, are reflex actions, while others such as speaking and eating are under your conscious control.

3 Draw a flow chart to explain what happens when you step on a pin. Make sure you include an explanation of how a synapse works.

Key points

- Some responses to stimuli are automatic and rapid and are called 'reflex actions'.
- Reflex actions run everyday bodily functions and help you to avoid danger.

B1 2.3 Hormones and the menstrual cycle

Hormones are chemical substances that coordinate many processes within your body. Special **glands** make and release (secrete) these hormones into your body. Then the hormones are carried around your body to their target organs in the bloodstream. Hormones regulate the functions of many organs and cells. They can act very quickly, but often their effects are quite slow and long lasting.

A woman's **menstrual cycle** is a good example of control by hormones. Hormones are made in a woman's pituitary gland and her ovaries control her menstrual cycle. The levels of the different hormones rise and fall in a regular pattern. This affects the way her body works.

What is the menstrual cycle?

The average length of the menstrual cycle is about 28 days. Each month the lining of the womb thickens ready to support a developing baby. At the same time an egg starts maturing in the ovary.

About 14 days after the egg starts maturing it is released from the ovary. This is known as **ovulation**. The lining of the womb stays thick for several days after the egg has been released.

If the egg is fertilised by a sperm, then pregnancy may take place. The lining of the womb provides protection and food for the developing embryo. If the egg is not fertilised, the lining of the womb and the dead egg are shed from the body. This is the monthly bleed or **period**.

All of these changes are brought about by hormones. These are made and released by the **pituitary gland** (a pea-sized gland in the brain) and the **ovaries**.

a What controls the menstrual cycle?
b Why does the lining of the womb build up each month?

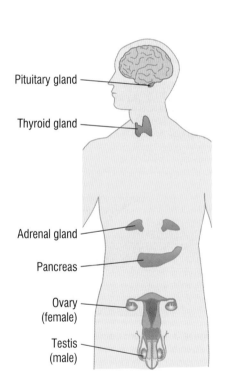

Pituitary gland

Thyroid gland

Adrenal gland

Pancreas

Ovary (female)

Testis (male)

Figure 1 Hormones act as chemical messages. They are made in glands in one part of the body but have an effect somewhere else.

How the menstrual cycle works

Once a month, a surge of hormones from the pituitary gland in the brain starts eggs maturing in the ovaries. The hormones also stimulate the ovaries to produce the female sex hormone **oestrogen**.

- Follicle stimulating hormone (**FSH**) is secreted by the pituitary gland. It makes eggs mature in the ovaries. FSH also stimulates the ovaries to produce oestrogen.

- Oestrogen is made and secreted by the ovaries. It stimulates the lining of the womb to build up ready for pregnancy. It inhibits (slows down) the production of more FSH.

- Other hormones involved in the menstrual cycle are luteinising hormone (LH) and **progesterone**.

The hormones produced by the pituitary gland and the ovary act together to control what happens in the menstrual cycle. As the oestrogen levels rise they inhibit the production of FSH and encourage the production of LH by the pituitary gland. When LH levels reach a peak in the middle of the cycle, they stimulate the release of a mature egg.

Did you know ...?

A baby girl is born with ovaries full of immature eggs, but they do nothing until she has gone through the changes of **puberty**.

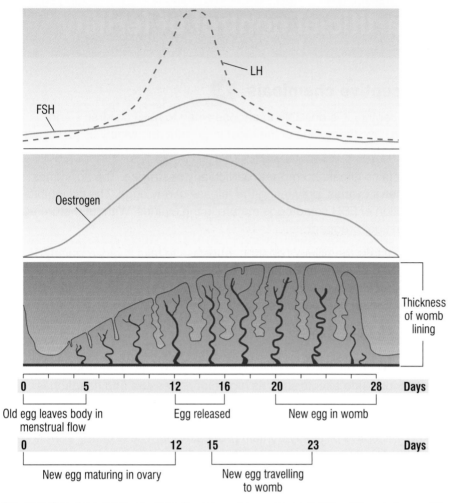

Figure 2 The changing levels of the female sex hormones control the different stages of the menstrual cycle

Study tip

Be clear on the difference between FSH and oestrogen.

FSH

● causes eggs to mature
● stimulates the ovary to produce oestrogen.

Oestrogen

● causes the lining of the uterus to develop
● inhibits FSH production
● stimulates the release of a mature egg.

Study tip

Make sure you know the difference between eggs maturing and eggs being released.

Summary questions

1 Copy and complete using the words below:

   28   hormones   FSH   menstrual   oestrogen   ovary

   During the .... cycle a mature egg is released from the .... about every .... days. The cycle is controlled by several .... including .... and ....

2 Look at Figure 2 above:
   a Explain what happens to FSH.
   b On which days is the female having a menstrual period?
   c Which hormone controls the build-up of the lining of the womb?

3 Produce a poster to explain the events of the menstrual cycle to women who are hoping to start a family. You will need to explain the graphs at the top of this page and show when a woman is most likely to get pregnant. Remember sperm can live for up to three days inside the woman's body.

Key points

● Hormones control the release of an egg from the ovary and the build-up of the lining of the womb in the menstrual cycle.

● Some of the hormones involved are FSH from the pituitary gland and oestrogen from the ovary.

B1 2.4

The artificial control of fertility

Learning objectives

- How can hormones be used to stop pregnancy?

- How can hormones help to solve the problems of infertility?

Contraceptive chemicals

In the 21st century it is possible to choose when to have children – and when not to have them. One of the most important and widely used ways of controlling fertility is to use **oral contraceptives** (the **contraceptive pill**).

The pill contains female hormones, particularly oestrogen. The hormones affect women's ovaries, preventing the release of any eggs. The pill inhibits the production of FSH so no eggs mature in the ovaries. Without mature eggs, women can't get pregnant.

Anyone who uses the pill as a contraceptive has to take it very regularly. If they forget to take it, the artificial hormone levels drop. Then their body's own hormones can take over very quickly. This can lead to the unexpected release of an egg – and an unexpected baby.

a What is a contraceptive?

The first birth control pills contained very large amounts of oestrogen. They caused serious side effects such as high blood pressure and headaches in some women. Modern contraceptive pills contain much lower doses of oestrogen along with some progesterone. They cause fewer side effects. Some contraceptive pills only contain progesterone. These cause even fewer side effects. However, they are not quite so good at preventing pregnancy because they don't stop the eggs from maturing.

b What is the difference between the mixed pill and the progesterone-only pill?

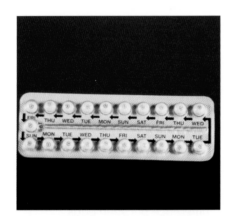

Figure 1 The contraceptive pill contains a mixture of hormones which effectively trick the body into thinking it is already pregnant, so no more eggs are released

Fertility treatments

In the UK as many as one couple in six have problems having a family when they want one. There are many possible reasons for this infertility. It may be linked to a lack of female hormones. Some women want children but do not make enough FSH to stimulate the eggs in their ovaries. Fortunately, artificial FSH can be used as a fertility drug. It stimulates the eggs in the ovary to mature and also triggers oestrogen production.

??? Did you know ...?

In the early days of using fertility drugs there were big problems with the doses used. In 1971 an Italian doctor removed 15 four-month-old fetuses (ten girls and five boys) from the womb of a 35-year-old woman after treatment with fertility drugs. Not one of them survived.

Figure 2 Most people who take fertility drugs end up with one or two babies. But in 1983 the Walton family from Liverpool had six baby girls who all survived.

Fertility drugs are also used in **IVF** (*in vitro* fertilisation). Conception usually takes place in the fallopian tube. This is the tube between the ovary and the uterus that the egg travels along. If the fallopian tubes are damaged, the eggs cannot reach the uterus so women cannot get pregnant naturally.

Fortunately doctors can now help. They collect eggs from the ovary of the mother and fertilise them with sperm from the father outside the body. The fertilised eggs develop into tiny embryos. The embryos are inserted into the uterus (womb) of the mother. In this way they bypass the faulty tubes.

During IVF the woman is given FSH to make sure as many eggs as possible mature in her ovaries. LH is also given at the end of the cycle to make sure all the mature eggs are released. IVF is expensive and not always successful.

1 Fertility drugs are used to make lots of eggs mature at the same time for collection.

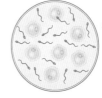

2 The eggs are collected and placed in a special solution in a Petri dish.

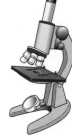

3 A sample of semen is collected and the sperm and eggs are mixed in the Petri dish.

4 The eggs are checked to make sure they have been fertilised and the early embryos are developing properly.

How Science Works

The advantages and disadvantages of fertility treatment

The use of hormones to control fertility has been a major scientific breakthrough. But like most things, there are advantages and disadvantages! Here are some points to think about:

In the developed world, using the pill has helped make families much smaller than they used to be. There is less poverty because with fewer children being born there are fewer mouths to feed and more money to go round.

The pill has also helped to control population growth in countries such as China, where they find it difficult to feed all their people. In many other countries of the developing world the pill is not available because of a lack of money, education and doctors.

The pill can cause health problems so a doctor always oversees its use.

The use of fertility drugs can also have some health risks for the mother and it can be expensive for society and parents. A large multiple birth can be tragic for the parents if some or all of the babies die. It also costs hospitals a lot of money to keep very small premature babies alive.

Controlling fertility artificially also raises many ethical issues for society and individuals. For example, some religious groups think that preventing conception is denying life and ban the use of the pill.

The mature eggs produced by a woman using fertility drugs may be collected and stored, or fertilised and stored, until she wants to get pregnant later. But what happens if the woman dies, or does not want the eggs or embryos any more?

● What, in your opinion, are the main advantages and disadvantages of using artificial hormones to control female fertility?

5 When the fertilised eggs have formed tiny balls of cells, 1 or 2 of the tiny embryos are placed in the uterus of the mother. Then, if all goes well, at least one baby will grow and develop successfully.

Figure 3 New reproductive technology using hormones and IVF has helped thousands of infertile couples to have babies

Key points

● Hormones can be used to control fertility.

● Oral contraceptives contain hormones, which stop FSH production so no eggs mature.

● FSH can be used as a fertility drug for women, to stimulate eggs to mature in their ovaries. These eggs may be used in IVF treatments.

Summary questions

1 Explain the meaning of the following terms: oral contraceptive, fallopian tube, fertility drug, *in vitro* fertilisation.

2 Explain how artificial female hormones can be used to:
   a prevent unwanted pregnancies
   b help people overcome infertility.

B1 2.5

Controlling conditions

Learning objectives

- How are conditions inside your body controlled?

- Why is it so important to control your internal environment?

Figure 1 Everything you do affects your internal environment

Study tip

Sweating causes the body to cool. Energy from the body is used to evaporate the water in sweat.

Figure 2 You can change your behaviour to help control your temperature, for example by adding extra clothing or turning up the heating when it's really cold

The conditions inside your body are known as its **internal environment**. Your organs cannot work properly if this keeps changing. Many of the processes which go on inside your body aim to keep everything as constant as possible. This balancing act is called **homeostasis**.

It involves your nervous system, your hormone systems and many of your body organs.

a Why is homeostasis important?

Controlling water and ions

Water can move in and out of your body cells. How much it moves depends on the concentration of mineral ions (such as salt) and the amount of water in your body. If too much water moves into or out of your cells, they can be damaged or destroyed.

You take water and minerals into your body as you eat and drink. You lose water as you breathe out, and also in your sweat. You lose salt in your sweat as well. You also lose water and salt in your **urine**, which is made in your **kidneys**.

Your kidneys can change the amount of salt and water lost in your urine, depending on your body conditions. They help to control the balance of water and mineral ions in your body. The concentration of the urine produced by your kidneys is controlled by both nerves and hormones.

So, for example, imagine drinking a lot of water all in one go. Your kidneys will remove the extra water from your blood and you will produce lots of very pale urine.

b What do your kidneys control?

Controlling temperature

It is vital that your deep core body **temperature** is kept at 37°C. At this temperature your **enzymes** work best. At only a few degrees above or below normal body temperature the reactions in your cells stop and you will die.

Your body controls your temperature in several ways. For example, you can sweat to cool down and shiver to warm up. Your nervous system is very important in coordinating the way your body responds to changes in temperature.

Once your body temperature drops below 35°C you are at risk of dying from hypothermia. Several hundred old people die from the effects of cold each year. So do a number of young people who get lost on mountains or try to walk home in the snow after a night out.

If your body temperature goes above about 40–42°C your enzymes and cells don't work properly. This means that you may die of heat stroke or heat exhaustion.

c What is the ideal body temperature?

Controlling blood sugar

When you digest a meal, lots of sugar (glucose) passes into your blood. Left alone, your blood glucose levels would keep changing. The levels would be very high straight after a meal, but very low again a few hours later. This would cause chaos in your body.

However, the concentration of glucose in your blood is kept constant by hormones made in your **pancreas**. This means your body cells are provided with the constant supply of energy that they need.

> **d** What would happen to your blood sugar level if you ate a packet of sweets?

Figure 3 Sweets like this are almost all sugar. When you eat them your body has to deal with the effect on your blood.

Summary questions

1 Copy and complete using the words below:

body   constant   homeostasis   hormones   internal   nervous

Conditions in the .... environment of your .... must be kept .... This is called .... The control is given by both your .... and your .... system.

2 Why is it important to control:
   a water levels in the body
   b the body temperature
   c sugar (glucose) levels in the blood?

3 **a** Look at the marathon runners in Figure 1. List the ways in which the running is affecting their:
   i water balance
   ii ion balance
   iii temperature.
   b It is much harder to run a marathon in a costume than in running clothes. Explain why this is.

Key points

● Humans need to maintain a constant internal environment, controlling levels of water, ions and blood sugar, as well as temperature.

● Homeostasis is the result of the coordination of your nervous system, your hormones and your body organs.

Hormones and the control of plant growth

B1 2.6

Learning objectives

- What stimuli do plants respond to?
- How do plants respond to their environment?
- Why do farmers and gardeners use plant hormones?

Figure 1 Seedlings like this radish show you clearly how plant shoots respond to light – they grow towards it

Did you know …?

The first scientists to show the way the shoot of a plant responds to light from one direction were Charles Darwin and his son Francis.

Practical

The effect of light on the growth of seedlings

You can investigate the effect of one-sided light on the growth of seedlings using a simple box with a hole cut in it and cress seedling growing in a Petri dish.

It is easy to see how animals, such as ourselves, take in information about the surroundings and then react to it. But plants also need to be coordinated. They are sensitive to light, water and gravity.

Plants are sensitive

When seeds are spread they may fall any way up in the soil. It is very important that when the seed starts to grow (germinate) the roots grow downwards into the soil. Then they can anchor the seedling and keep it stable. They can also take up the water and minerals needed for healthy growth.

At the same time the shoots need to grow upwards towards the light so they can **photosynthesise** as much as possible.

Plant roots are sensitive to gravity and water. The roots grow towards moisture and in the direction of the force of gravity. **Plant shoots** are sensitive to gravity and light. The shoots grow towards light and against the force of gravity. This means that whichever way up the seed lands, the plant always grows the right way up!

Plant responses

Plant responses happen as a result of plant hormones which coordinate and control growth. These responses are easy to see in young seedlings, but they also happen in adult plants. For example, the stems of a houseplant left on a windowsill will soon bend towards the light. The response of a plant to light is known as **phototropism**. The response of a plant to gravity is called **gravitropism** (also known as geotropism).

The responses of plant roots and shoots to light, gravity and moisture are controlled by a hormone called **auxin**. The response happens because of an uneven distribution of this hormone in the growing shoot or root. This causes an unequal growth rate. As a result the root or shoot bends in the right direction.

Phototropism can clearly be seen when a young shoot responds to light from one side only. The shoot will bend so it is growing towards the light. Auxin moves from the side of the shoot where the light is falling to the unlit side of the shoot. The cells on that side respond to the hormone by growing more – and so the shoot bends towards the light. Once light falls evenly on the shoot, the levels of auxin will be equal on all sides and so the shoot grows straight again.

Gravitropisms can be seen in roots and shoots. Auxin has different effects on root and shoot cells. High levels of auxin make shoot cells grow more but inhibit growth of root cells. This is why roots and shoots respond differently to gravity.

a What is the name of the plant hormone which controls phototropism and gravitropism?

1 A normal young bean plant is laid on its side in the dark. Auxin is equally spread through the tissues.

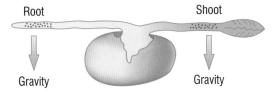

3 The root grows *more* on the side with *least* auxin, making it bend and grow down towards the force of gravity. When it has grown down, the auxin becomes evenly spread again.

The shoot grows *more* on the side with *most* auxin, making it bend and grow up away from the force of gravity. When it has grown up, the auxin becomes evenly spread again.

2 In the root, more auxin gathers on the lower side.

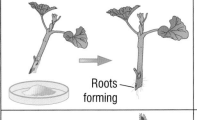

In the shoot, more auxin gathers on the lower side.

Figure 2 Gravitropism (or geotropism) in shoots and roots. The uneven distribution of the hormone auxin causes unequal growth rates so the roots grow down and the shoots grow up.

Using plant hormones 🅺

Plant hormones can be used to manage plants grown in the garden or home. Farmers also use them to grow better crops.

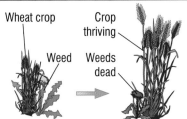 Roots forming	Gardeners and horticulturists rely on taking cuttings to produce lots of identical plants. Plant growth hormones are used as rooting powder. A little placed on the end of a cutting stimulates the growth of new roots and helps the cutting to grow into a new plant.
Wheat crop Crop thriving Weed Weeds dead	You can use high doses of plant hormones as weed killers. Most weeds are broad-leaved plants which absorb a lot of hormone weed killer. This makes them go into rapid, uncontrolled growth which kills them. Narrow-leaved plants such as grasses and cereal are not affected, so the crop or lawn keeps growing well.

Figure 3 Some human uses of plant hormones

Summary questions

1 Copy and complete using the words below:

hormone auxin gravitropism light sensitive moisture

Plants are .... to ...., which is called phototropism. They respond to gravity which is .... Plants are also sensitive to .... These responses are coordinated by the plant .... called ....

2 Why are the responses of shoots and roots so important in the life of plants?

3 Explain carefully, using diagrams, how a plant shoot responds to light shining at it from one side only.

Key points

- Plants are sensitive to light, moisture and gravity.
- Plant responses are brought about by plant hormones (auxin).
- The responses of roots and shoots to stimuli such as light and gravity are the result of the unequal distribution of plant hormones.
- We can use plant growth hormones as weed killers and as rooting hormones on cuttings.

B1 2.7   Using hormones

Learning objectives

- What are some of the issues associated with the use of hormones to control human fertility?

- How much should we use plant hormones to produce our food?

People can control human fertility. Infertile women can have treatment and have babies, while fertile women can choose not to have children. We can make thousands of identical plants and we can kill weeds as they grow. We do these things using hormones. But not everyone agrees with what is being done.

Woman 1

'I married late – I was 40 – and we wanted a family but my periods stopped when I was 41. Now we have a chance again. I haven't got any eggs so doctors will use FSH as a fertility drug to help them take lots of eggs from my donor, a younger woman. We want this child SO much!'

Woman 2

'We've got three lovely children. I decided to donate some of my eggs to help couples who aren't as lucky as we are. I don't mind the age of the woman who gets my eggs as long as she manages to have a baby and loves it!'

Figure 1 Using hormones to control fertility has made it possible for women in their 50s and 60s to become pregnant and have a baby – but not everyone thinks this is a good idea

Man 1

'I think it is disgraceful and unnatural for women to have babies when they are older. We are interfering with nature and with God's will and no good will come of it. The mother might die before the child is an adult.'

Doctor

'All our evidence shows that infertility treatment can be just as successful in older women as it is in younger ones. We have to use artificial hormones to get the womb ready and use donor eggs, but once the women are pregnant their own hormones take over.'

Man 2

'I can't see anything wrong with older women having babies as long as they are fit and well. I know some people object to it, but some women have babies in their fifties naturally – and lots of men father children in their 60s and even their 70s and no one objects to that, do they?'

a Which hormone would doctors use to stimulate the ovaries of an egg donor?

links

For information on plant growth hormones, look back at B1 2.6 Hormones and the control of plant growth.

Activity

Make your mind up!

There is a lot of debate about older mothers. Use what you have learned in this spread to help you write a 2–3-minute report for your school radio. Use the title: 'Older mothers – should science help?'

It will go out in a regular slot called 'Science issues', so students will be expecting some science as well as opinions.

Plant hormones, plant killers

Scientists have discovered that plant hormones such as auxin make effective weed killers. The hormones used in commercial weed killers are not natural hormones extracted from plants. They are made in chemical factories. These weed killers seem effective and safe. People use them on their lawns because the grass is not affected, but the weeds like dandelions and daisies are killed. Golf courses are kept weed-free in the same way.

Farmers around the world use hormone weed killers to kill off the weeds in cereal crops. They are one of the reasons why the yield of cereal crops is now so much bigger all across Europe. That means more food is available at cheaper prices.

But chemicals based on plant hormones (synthetic plant hormones) can cause serious problems. In the Vietnam War, one of these chemicals (Agent Orange) was sprayed on the forests. It works in the same way as natural plant hormones and in high doses it strips all the leaves off the trees. This made it easier for the US soldiers to find enemy fighters. It caused terrible damage to the forests. Not only that, hundreds of thousands of people were badly affected by the powerful chemicals.

b What is Agent Orange?

Practical

The effect of rooting compounds and weed killers on the growth of plants

You can investigate the effect of rooting hormone by taking some cuttings and growing half of them with rooting powder and half without.

Did you know ... ?

Agent Orange has been used to destroy areas of the Amazon rainforest, particularly for the building of the Tucurui dam in Brazil.

Activity

Is it worth it?

Synthetic plant hormones, like many scientific discoveries, can be used both to benefit people and to cause great harm. Some people have suggested that synthetic plant hormones should all be banned. Plan a short speech EITHER supporting this idea OR disagreeing with it.

Use evidence and persuasive writing in your speech.

Key points

- There are benefits and problems associated with the use of hormones to control fertility and these must be evaluated carefully.

- Plant hormones are very useful as weed killers but their use can damage the environment.

Summary questions

1 Make a table to summarise the main points for and against allowing older women to use hormones such as FSH to help them have a baby.

2 How can plant hormones be used to kill plants?

Summary questions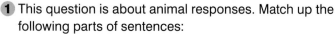

1 This question is about animal responses. Match up the following parts of sentences:

a Many processes in the body …	A … effector organs.
b The nervous system allows you …	B … secreted by glands.
c The cells which are sensitive to light …	C … to react to your surroundings and coordinate your behaviour.
d Hormones are chemical substances …	D … are found in the eyes.
e Muscles and glands are known as …	E … are known as nerves.
f Bundles of neurons …	F … are controlled by hormones.

2 a What is the job of your nervous system?

  b Where in your body would you find nervous receptors which respond to:

    i light

    ii sound

    iii heat

    iv touch?

  c Draw and label a simple diagram of a reflex arc. Explain carefully how a reflex arc works and why it allows you to respond quickly to danger.

3 a What is the menstrual cycle?

  b What is the role of the following hormones in the menstrual cycle:

    i FSH

    ii oestrogen?

4 a Explain carefully the difference between nervous and hormone control of your body.

  b What are synapses and why are they important in your nervous system?

  c How can hormones be used to control the fertility of a woman?

5 The table shows four ways in which water leaves your body, and the amounts lost on a cool day:

Source of water loss	Cool day (water loss in cm³)	Hot day (loss in cm³)
Breath	400	The same
Skin	500	A
Urine	1500	B
Faeces	150	C

  a On a hot day, would the amount of water lost in A, B and C be less, the same or more than the amount of water lost on a cool day?

  b Name the process by which we lose water from the skin.

  c On a cool day the body gained 2550 cm³ of water. 1750 cm³ came directly from drinking. Where did the rest come from?

6 It is very important to keep the conditions inside the body stable. Taking part in school sports on a hot day without a drink for the afternoon would be difficult for your body. Explain how your body would keep the internal environment as stable as possible.

7 a What is gravitropism (geotropism)?

  b Explain carefully how the following respond to gravity, including the part played by plant hormones. Diagrams may help in your explanations.

    i a root

    ii a shoot

8 You are provided with some very young single shoots. Devise an experiment which would demonstrate that shoots grow towards the light.

Practice questions

1 A dog responds to stimuli.

a Link the receptor descriptions to the correct part of the animal by choosing the correct letter (A, B, C or D).
  i Contains receptors to detect chemicals (1)
  ii Contains receptors to detect light (1)
  iii Contains receptors to detect movement of the head and sound. (1)

b The skin of a human contains receptors which are sensitive to touch.
  i Give **one** other stimulus which is detected by human skin. (1)
  ii Suggest why there are many touch receptors in a person's fingertips. (1)

2 a When a person touches a hot surface they move their hand away quickly.

Choose the correct word to complete the sentence.

This is called a .... action. (1)

learned   reflex   thoughtful

b What is the importance of this type of action? (1)

3 This picture shows a Venus flytrap.

a The Venus flytrap catches flies for food. When a fly lands on the leaf the trap closes.

Choose the correct word to complete the sentence.

The shutting of the trap is called a .... (1)

detector   stimulus   response

b Suggest **one** receptor the Venus flytrap has to detect the fly. (1)

4 Hormones are important chemicals which help to control conditions inside living organisms.

a **List A** shows three hormones

List B shows where some hormones are produced.

Match each hormone with where it is produced.

List A	List B
Hormone	**Where produced**
auxin	pituitary gland
oestrogen	kidney
FSH	plant stems and roots
	ovary

(3)

b Choose the correct answer to complete each of the following sentences.
  i The hormone which causes eggs to mature is .... (1)

  auxin   oestrogen   FSH

  ii The hormone which causes growth of the uterus(womb) lining is .... (1)

  auxin   oestrogen   FSH

5 When light is shone in a person's eyes they blink. When a plant is placed near a lamp the stem bends towards the light.

a Choose the correct answer to complete each of the following sentences.

  i The response of the eye to bright light is called a .... action. (1)

  learned   reflex   stimulated

  ii The response of the plant to light is called .... (1)

  gravitropism   hydrotropism   phototropism

b *In this question you will be assessed on using good English, organising information clearly and using specialist terms where appropriate.*

Plants respond to light and gravity. Describe how plant hormones control the growth of roots and shoots. (6)

B1 3.1 Developing new medicines

Learning objectives

● What are the stages in testing and trialling a new drug?

● Why is testing new drugs so important?

Figure 1 The development of a new medicine takes millions of pounds, involves many people and lots of equipment

Study tip

Make sure you are clear that a medical drug is tested to establish:

● its effectiveness

● its toxicity

● the most appropriate dose.

Study tip

Remember, the cells, tissues and animals act as models to predict how the drug may behave in humans.

We are developing new medicines all the time, as scientists and doctors try to find ways of curing more diseases. We test new medicines in the laboratory. Every new medical treatment has to be extensively tested and trialled before it is used. This process makes sure that it works well and is as safe as possible.

A good medicine is:

● **Effective** – it must prevent or cure a disease or at least make you feel better.

● **Safe** – the drug must not be too toxic (poisonous) or have unacceptable side effects for the patient.

● **Stable** – you must be able to use the medicine under normal conditions and store it for some time.

● **Successfully taken into and removed from your body** – it must reach its target and be cleared from your system once it has done its work.

Developing and testing a new drug

When scientists research a new medicine they have to make sure all these conditions are met. It can take up to 12 years to bring a new medicine into your doctor's surgery. It can also cost a lot of money; up to about £350 million!

Researchers target a disease and make lots of possible new drugs. These are tested in the laboratory to find out if they are toxic and if they seem to do their job. They are tested on cells, tissues and even whole organs. Many chemicals fail at this stage.

The small numbers of chemicals which pass the earlier tests are now tested on animals. This is done to find out how they work in a whole living organism. It also gives information about possible doses and side effects. The tissues and animals are used as models to predict how the drugs may behave in humans.

Drugs that pass animal testing will be tested on human volunteers in clinical trials. First very low doses are given to healthy people to check for side effects. Then it is tried on a small number of patients to see if it treats the disease. If it seems to be safe and effective, bigger clinical trials take place to find the optimum dose for the drug.

If the medicine passes all the legal tests it is licensed so your doctor can prescribe it. Its safety will be monitored for as long as it is used.

a What are the important properties of a good new medicine?

Double-blind trials

In human trials, scientists use a **double-blind trial** to see just how effective the new medicine is. Some patients with the target disease agree to take part in the trials. They are either given a **placebo** that does not contain the drug or the new medicine. Neither the doctor nor the patients know who has received the real drug and who has received the placebo until the trial is complete. The patients' health is monitored carefully.

Often the placebo will contain a different drug that is already used to treat the disease. That is so the patient is not deprived of treatment by taking part in the trial.

Why do we test new medicines so thoroughly?

Thalidomide is a medicine which was developed in the 1950s as a sleeping pill. This was before there were agreed standards for testing new medicines. In particular, tests on pregnant animals, which we now know to be essential, were not carried out.

Then it was discovered that thalidomide stopped morning sickness during pregnancy. Because thalidomide seemed very safe for adults, it was assumed to be safe for unborn children. Doctors gave it to pregnant women to relieve their sickness.

Tragically, thalidomide was **not** safe for developing fetuses. It affected the fetuses of many women who took the drug in the early stages of pregnancy. They went on to give birth to babies with severe limb deformities.

The thalidomide tragedy led to a new law being passed. It set standards for the testing of all new medicines. Since the Medicines Act 1968, new medicines must be tested on animals to see if they have an effect on developing fetuses.

There is another twist in the thalidomide story. Doctors discovered it can treat leprosy. They started to use the drug against leprosy in the developing world but again children were born with abnormalities. Its use for leprosy has now been banned by the World Health Organisation (WHO).

However doctors are finding more uses for the drug. It can treat some autoimmune diseases (where the body attacks its own cells) and even some cancers. It is now used very carefully and never given to anyone who is or might become pregnant.

Figure 2 This woman has limb deformities because her mother took thalidomide during her pregnancy. She was just one of thousands of people affected by the thalidomide tragedy, many of whom have gone on to live full and active lives.

b Why was thalidomide prescribed to pregnant women?

Summary questions

1 Copy and complete using the words below:

effective   trialled   safe   medicine   stable   tested

Every new .... has to be extensively .... and .... before you can use it to make sure that it works well. A good medicine can be taken into and removed from your body, and it is ...., ... and ....

2 **a** Testing a new medicine costs a lot of money and can take up to 12 years. Make a flow chart to show the main stages in testing new drugs.

   b Why is an active drug often used as the control in a clinical trial instead of a sugar pill placebo which does nothing?

3 **a** What were the flaws in the original development of thalidomide?

   b Why do you think that the World Health Organisation has stopped the use of thalidomide to treat leprosy but the drug is still being used in the developed world to treat certain rare conditions?

Key points

- When we develop new medicines they have to be tested and trialled extensively before we can use them.

- Drugs are tested to see if they work well. We also make sure they are not too toxic and have no unacceptable side effects.

- Thalidomide was developed as a sleeping pill and was found to prevent morning sickness in early pregnancy. It had not been fully tested and it caused birth defects.

B1 3.2 How effective are medicines?

Learning objectives

- What are statins?
- How good are statins at preventing cardiovascular disease?
- Can drugs you buy over the counter be as good as the drugs that your doctor prescribes?

The statin revolution

As you have seen, high blood cholesterol levels are linked to a higher than average risk of cardiovascular disease. In other words, you are more likely to have a heart attack or a stroke if your blood cholesterol is high.

Doctors now have an amazing weapon against high cholesterol levels and the problems they can cause. They can use a group of drugs called **statins**. Statins are drugs that lower the amount of cholesterol in your blood. They stop your liver producing too much cholesterol. Patients need to keep to a relatively low fat diet as well for the best effects.

a What are statins?

Here are some different opinions about these exciting new drugs:

A GP

'Some people just can't get their cholesterol level right by changing their diet. It doesn't matter how hard they try. I've been very pleased with the results using statins. Almost all my patients have now got healthy cholesterol levels. What's more, we have lost far fewer people to strokes and heart attacks since we started using the drugs.'

A Member of NICE (National Institute for Health and Clinical Excellence)

'We have looked at data from lots of really large, powerful research trials involving over 30 000 patients. The trials all show similar results. Using a statin drug can lower your chances of having a heart attack or stroke by 25 to 40%, and we didn't find too many side effects.'

Patient 1

'I'm so pleased with my new medicine, the pills have brought my cholesterol levels right down and I'm feeling really well.'

Patient 2

'The great thing about these new statins that the doctor has given me is that they control my cholesterol for me. It's back to the cream cakes and chips for me, and I won't have to worry about my heart!'

Patient 3

'I'm very worried about possible side effects with these new tablets – the leaflet said they can cause liver damage. I know my cholesterol levels were very high without the tablets, but I'm not sure about taking these tablets. I don't want my liver to be damaged!'

Study tip

Statins are medical drugs which are used to reduce cholesterol levels in the blood. This reduces the risk of heart disease.

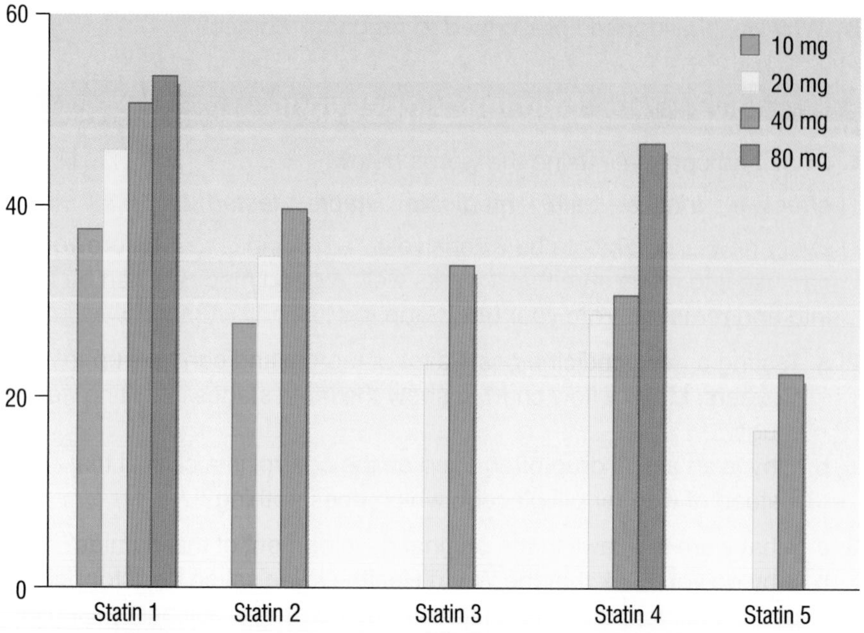

Figure 1 This graph shows the effects of different doses of different statins on the 'bad' cholesterol in the blood

Prescribed drugs *v.* non-prescribed

The medicines your doctor prescribes for you have been thoroughly tested and trialled. However some people choose to use non-prescribed medicines they buy for themselves. Some of these medicines are little more than sugar, flavouring and water. They will not hurt you – but they won't make you better either.

Some non-prescribed medicines can be dangerous. They are made from herbs and 'natural products' but can contain potentially dangerous chemicals. Remember that many of our effective prescribed medicines come from living organisms.

You can only tell if a non-prescribed medicine is as effective as a prescribed medicine if it undergoes double-blind clinical trials. Very few of them are ever evaluated in this way because of the expense.

One example of a non-prescribed medicine which seems to work well is the herb St John's Wort. If you suffer from **depression** your doctor may prescribe Prozac (fluoxetine). Some people prefer to use non-prescribed St John's Wort (*hypericum*). It is a herbal remedy which has been used as an antidepressant for around 2000 years.

There have now been some scientific studies carried out. They compare the effectiveness of St John's Wort to the most commonly used antidepressant medicines. The evidence so far suggests that the herbal treatment is as effective as the most common medicines used to treat depression. It also seems more effective than placebos and it has fewer side effects than the prescribed medicines.

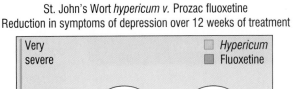

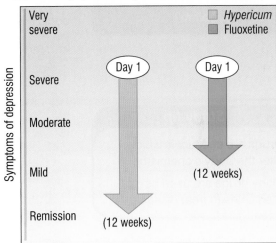

St. John's Wort *hypericum v.* Prozac fluoxetine
Reduction in symptoms of depression over 12 weeks of treatment

Figure 2 Data from one of a number of studies which show that *hypericum* (St John's Wort) is at least as effective as commonly prescribed drugs in treating depression

b Look at Figure 2. How much improvement in depression scores was seen over 12 weeks of using Prozac and St John's Wort?

Activity

Write a short report on statins for the health page of your local paper.

Activity

There are lots of food products which claim to lower blood cholesterol. Investigate one of these claims.

- See how much evidence you can discover.
- Plan an investigation to help show whether this dietary supplement or alternative food can really lower blood cholesterol levels.

links

For more information on cholesterol levels, see B1 1.3 Inheritance, exercise and health.

Summary questions

1 Copy and complete using the words below:

   statins   cholesterol   lower   cardiovascular   risk

   If your blood .... levels are high you have an increased .... of .... disease. Drugs called .... can .... your risk and help to keep you healthy.

2 Explain why is it unwise to think that non-prescribed drugs cannot cause harmful side effects.

Key points

- Statins lower the amount of cholesterol in the blood and can reduce the risk of cardiovascular disease by up to 40%.

- The effectiveness of both prescribed and non-prescribed drugs can only be measured in proper double-blind trials.

B1 3.3 Drugs

Learning objectives

- What is a drug?

- What is addiction?

- Why are drugs such as cannabis, cocaine and heroin such a problem?

Study tip

Drugs are chemicals which alter the body's chemistry.

Many drugs are used as medicines to treat disease.

A **drug** is a substance that alters the way in which your body works. It can affect your mind, your body or both. In every society there are certain drugs which people use for medicine, and other drugs which they use for pleasure.

Many of the drugs that are used both for medicine and for recreation come originally from natural substances, often plants. Many of them have been known to and used by indigenous (long-term inhabitants of an area) peoples for many years. Usually some of the drugs that are used for pleasure are socially acceptable and legal, while others are illegal.

Figure 1 Millions of pounds worth of illegal drugs are brought into the UK every year. It is a constant battle for the police to find and destroy drugs like these.

a What do we mean by 'indigenous peoples'?

Drugs are everywhere in our society. People drink coffee and tea, smoke cigarettes and have a beer, an alcopop or a glass of wine. They think nothing of it. Yet all of these things contain drugs – caffeine, nicotine and alcohol (the chemical ethanol). These drugs are all legal.

Other drugs, such as cocaine, ecstasy and heroin are illegal. Which drugs are legal and which are not varies from country to country. Alcohol is legal in the UK as long as you are over 18, but it is illegal in many Arab states. Heroin is illegal almost everywhere.

b Give an example of one drug which is legal and one which is illegal in the UK.

Because drugs affect the chemistry of your body, they can cause great harm. This is even true of drugs we use as medicines. However, because medical drugs make you better, it is usually worth taking the risk.

But legal recreational drugs, such as alcohol and tobacco, and illegal substances, such as solvents, cannabis and cocaine, can cause terrible damage to your body. Yet they offer no long-term benefits to you at all.

∞ links

For more information on the mental health problems that can be caused by cannabis, see B1 3.5 Does cannabis lead to hard drugs?

What is addiction?

Some drugs change the chemical processes in your body so that you may become addicted to them. You can become dependent on them. If you are addicted to a drug, you cannot manage properly without it. Some drugs, for example heroin and cocaine, are very addictive.

Once addicted, you generally need more and more of the drug to keep you feeling normal. When addicts try to stop using drugs they usually feel very unwell. They often have aches and pains, sweating, shaking, headaches and cravings for their drug. We call these **withdrawal symptoms**.

> c What do we mean by 'addiction'?

The problems of drug abuse

People take drugs for a reason. Drugs can make you feel very good about yourself. They can make you feel happy and they can make you feel as if your problems no longer matter. Unfortunately, because most recreational drugs are addictive, they can soon become a problem themselves.

No drugs are without a risk. Cannabis is often thought of as a relatively 'soft' – and therefore safe – drug. But evidence is growing which shows that cannabis smoke contains chemicals which can cause mental illness to develop in some people.

Hard drugs, such as cocaine and heroin, are extremely addictive. Using them often leads to very severe health problems. Some of these come from the drugs themselves. Others come from the lifestyle that often goes with drugs.

Because these drugs are illegal, they are expensive. Young people often end up turning to crime to pay for their drug habit. They don't eat properly or look after themselves. They can also contract serious illnesses, such as hepatitis, STDs (sexually transmitted diseases) and HIV/Aids especially if drugs are taken intravenously (via a needle).

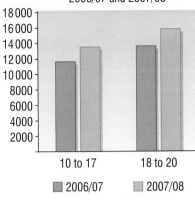

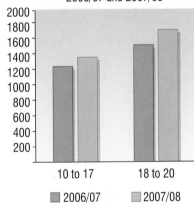

Figure 2 Illegal drugs are often linked with crime. In the UK more and more young people are being arrested for drug offences – using or selling illegal drugs.

Summary questions

1 Copy and complete using the words below:

mind cocaine ecstasy legal alcohol drug body

A .... alters the way in which your body works. It can affect the ...., the .... or both. Some drugs are .... e.g. caffeine and .... Other drugs, such as ...., ... and heroin are illegal.

2 a Why do people often need more and more of a drug?
 b What happens if you stop taking a drug when you are addicted to it?

3 a Why do people take drugs?
 b Explain some of the problems linked with using cannabis, cocaine and heroin.
 c Look at Figure 2. What does this tell you about the difference in drug use between boys and girls?
 d What does Figure 2 tell you about the trend in drug use in young people?
 e Why do you think young people continue to take these drugs when they are well aware of the dangers?

Key points

- Drugs change the chemical processes in your body, so you may become addicted to them.
- Addiction is when you become physically or mentally dependent on a drug.
- Smoking cannabis may cause mental health problems.
- Hard drugs, such as cocaine and heroin, are very addictive and can cause serious health problems.

B1 3.4

Legal and illegal drugs

Learning objectives

- How do drugs like caffeine and heroin affect your nervous system?

- Which has the bigger overall impact on health – legal or illegal drugs?

What is the most widely used drug in the world? It is probably one that most of you will have used at least once today, yet no one really thinks about. The caffeine in your cup of tea, mug of coffee or can of cola is a drug.

Many people find it hard to get going in the morning without a mug of coffee. They are probably addicted to the drug caffeine. It stimulates your brain and increases your heart rate and blood pressure.

Figure 1 NASA scientists have shown that common house spiders spin their webs very differently when given some commonly used legal and illegal drugs. The effect of caffeine on the nervous system of a spider is particularly dramatic!

a What drug is in a can of cola?

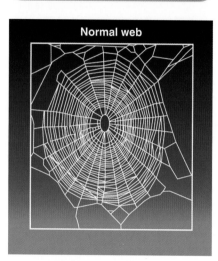

How do drugs affect you?

Many of the drugs used for medical treatments have little or no effect on your nervous system. However, all of the recreational drugs that people use for pleasure affect the way your nervous system works, particularly your brain. It is these changes that people enjoy when they use the drugs. The same changes can cause addiction. Once addicted, your body doesn't work properly without the drug.

Some drugs like caffeine, nicotine and cocaine speed up the activity of your brain. They make you feel more alert and energetic.

Others, like alcohol and cannabis, slow down the responses of your brain. They make you feel calm and better able to cope. Heroin actually stops impulses travelling in your nervous system. Therefore you don't feel any pain or discomfort. Cannabis produces vivid waking dreams. It can make you see or hear things that are not really there.

Why do people use drugs?

People use drugs for a variety of reasons. They feel that caffeine, nicotine and alcohol help them cope with everyday life. Few people who use these legal drugs would think of themselves as addicts. Yet the chemicals they take can have a big physical and psychological impact (see Figure 1).

As for the illegal recreational drugs – people who try them may be looking for excitement or escape. They might want to be part of the crowd or just want to see what happens. Yet many drugs are addictive and your body needs increasingly more to feel the effects.

Study tip

Drugs may be:
- legal or illegal
- addictive or non-addictive.

Learn examples of all of these.

Impact of drugs on health

Some recreational drugs are more harmful than others. Most media reports on the dangers of drugs focus on illegal drugs. But in fact the impact of legal drugs on health is much greater than the impact of illegal drugs. That's because far more people take them. Millions of people in the UK take medicines such as statins, or smoke or drink alcohol. Only a few thousand take heroin.

A recent case history shows you how emotions and politics can be more important than scientific evidence in the way society reacts to drugs. In 2010, several young people died after apparently taking a relatively new legal drug known as 'meow-meow'. The drug was made illegal even though at least one of the 'victims' had not taken meow-meow.

In fact, in the UK, there are around 2000 deaths linked to using illegal drugs each year.

But every year in the UK around 9000 people die as a result of alcohol-abuse. About 90 000 people die from smoking-related diseases. Yet alcohol and nicotine remain completely legal drugs.

Everyone can see the dangers to health of non-prescribed, illegal drugs. However, choosing which drugs to make illegal does not appear to be based on the scientific evidence of health damage alone.

DON'T LET DRUG DEALERS CHANGE THE FACE OF YOUR NEIGHBOURHOOD.
Call Crimestoppers anonymously on 0800 555 111.

Figure 2 Drugs can seem appealing, exciting and fun. Many people use them briefly and then leave them behind. But the risks of addiction are high, and no one can predict whom drugs will affect most.

> **b** Why do legal drugs cause many more health problems than illegal drugs?

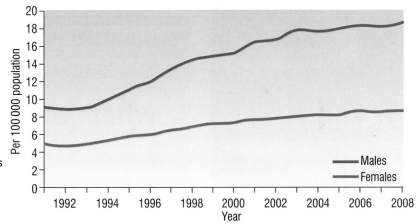

Figure 3 This graph shows how alcohol-related deaths almost doubled between 1990 and 2008 (Source: National Statistics Office)

Summary questions

1 Copy and complete using the words below:

brain   health   illegal   legal   recreation

Drugs which people use for .... all affect the nervous system, particularly the .... Some of these drugs are legal but some of them are .... More people suffer .... problems caused by .... drugs than illegal ones.

2 Use data from Figure 3 to help you answer these:
   a How many men and women died of alcohol-related diseases per 100 000 of the population in 1992?
   b How many men and women died of alcohol-related diseases per 100 000 of the population in 2008?
   c Suggest reasons for this increase in alcohol-related deaths.
   d Why do you think alcohol remains a legal drug when it causes so many deaths?

3 Compare the overall impact of legal and illegal drugs on the nation's health.

Key points

- Many recreational drugs affect the nervous system, particularly the brain. Some are more harmful than others.

- Some recreational drugs are legal and others are illegal.

- The overall impact of legal drugs on health is much greater than illegal drugs because more people use them.

B1 3.5   Does cannabis lead to hard drugs?

Learning objectives

- How do people move from using recreational drugs to hard drugs?
- Is cannabis harmful?

Cannabis – the facts?

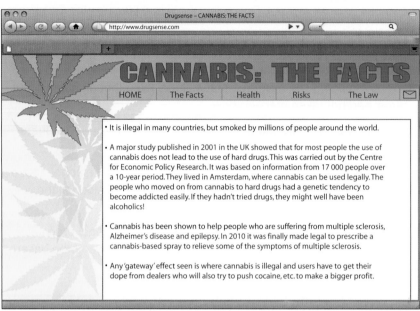

Drugsense – CANNABIS: THE FACTS

http://www.drugsense.com

CANNABIS: THE FACTS

| HOME | The Facts | Health | Risks | The Law |

- It is illegal in many countries, but smoked by millions of people around the world.

- A major study published in 2001 in the UK showed that for most people the use of cannabis does not lead to the use of hard drugs. This was carried out by the Centre for Economic Policy Research. It was based on information from 17 000 people over a 10-year period. They lived in Amsterdam, where cannabis can be used legally. The people who moved on from cannabis to hard drugs had a genetic tendency to become addicted easily. If they hadn't tried drugs, they might well have been alcoholics!

- Cannabis has been shown to help people who are suffering from multiple sclerosis, Alzheimer's disease and epilepsy. In 2010 it was finally made legal to prescribe a cannabis-based spray to relieve some of the symptoms of multiple sclerosis.

- Any 'gateway' effect seen is where cannabis is illegal and users have to get their dope from dealers who will also try to push cocaine, etc. to make a bigger profit.

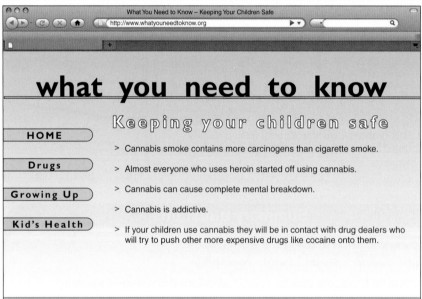

What You Need to Know – Keeping Your Children Safe

http://www.whatyouneedtoknow.org

what you need to know

Keeping your children safe

| HOME |
| Drugs |
| Growing Up |
| Kid's Health |

> Cannabis smoke contains more carcinogens than cigarette smoke.

> Almost everyone who uses heroin started off using cannabis.

> Cannabis can cause complete mental breakdown.

> Cannabis is addictive.

> If your children use cannabis they will be in contact with drug dealers who will try to push other more expensive drugs like cocaine onto them.

Figure 1 How can you find out the truth about cannabis and the effect it might have on you, your friend or – if you are a parent – your child?

a What diseases are helped by the chemicals in cannabis?

Cannabis – where do you stand?

A lot of scientific research has been done into the effects of cannabis on our health. The links between cannabis use and addiction to hard drugs has also been investigated.

Unfortunately many of the studies have been quite small. They have not used large sample sizes, so the evidence is not strong.

Figure 2 In the minds of many people – parents, teachers and politicians – cannabis is a 'gateway' drug. It opens the door to the use of other much harder drugs such as cocaine and heroin. Your health – and indeed your life itself – is at risk. How accurate is this picture?

The UK Government downgraded cannabis to a Class C drug in 2004. Then stronger negative evidence emerged. It found that cannabis use can trigger mental health problems in people who are vulnerable to such problems. In 2009 the decision to downgrade was reversed and cannabis is now a Class B drug again.

What the doctors say

- The evidence is clear that for some people cannabis use can trigger mental illness. This may be serious and permanent. It is particularly the case for people who have a genetic tendency to mental health problems.

- A study has been carried out on 1600 14- to 15-year-old students in Australia. It showed that the youngsters who use cannabis regularly have a significantly higher risk of depression. However it doesn't work the other way round. Children who are already suffering depression are no more likely than anyone else to use cannabis.

- All the evidence suggests that teenagers are particularly vulnerable to mental health problems triggered by cannabis. Consider a teenager who starts smoking cannabis before they are 15. They are four times more likely to develop schizophrenia or another psychotic illness by the time they are 26 than a non-user.

Figure 3 The doctors at the Royal College of Psychiatrists are the people who deal with mental health problems of all kinds. They have some real concerns about cannabis-use.

Untangling the evidence

The evidence shows that almost all heroin users were originally cannabis users. This is not necessarily a case of cannabis use causing heroin addiction. Almost all cannabis users are originally smokers – but we don't claim that smoking cigarettes leads to cannabis use! In fact the vast majority of smokers do not go on to use cannabis. Just as the vast majority of cannabis users do not move on to hard drugs like heroin. Most studies suggest that cannabis can act as a 'gateway' to other drugs. However, that is **not** because it makes people want a stronger drug but because it puts them in touch with illegal drug dealers.

> **b** How much does using cannabis before you are 15 appear to increase your risk of developing serious mental illness?

Summary questions

1 **a** What is meant by a 'gateway' drug?
   b Why is cannabis considered a gateway drug?

2 Cannabis is linked to some mental health problems, but tobacco is known to cause hundreds of thousands of deaths each year through heart disease and lung cancer. Why do you think cannabis is illegal and tobacco is legal?

Activity

You are going to set up a classroom debate. The subject is:

'We believe that cannabis should be made a legal drug.'

You are going to prepare **two** short speeches – one **for** the idea of legalising cannabis and one **against**.

You can use the information on these pages and also look elsewhere for information. Try books and leaflets and on the internet.

In both of your speeches you must base your arguments on scientific evidence as well as considering the social, moral and ethical implications of any change in the law. You have to be prepared to argue your case (both for and against) and answer any questions – so do your research well!

Key points

- People can progress from using recreational drugs such as cannabis to addiction to hard drugs because cannabis is illegal and has to be obtained from a drug dealer.

- Cannabis smoke contains chemicals which may cause mental illness in some people. Teenagers are particularly vulnerable to this effect.

B1 3.6

Drugs in sport

Learning objectives

- Can drugs make you better at sport?

- Is it ethical to use drugs to win?

The world of sport has a big problem with the illegal use of drugs. In theory, the only difference between competitors should be their natural ability and the amount they train. However, there are many performance-enhancing drugs that allow athletes to improve on their natural ability. The people who do this get labelled as cheats if they are caught.

Performance-enhancing drugs

Different sports need different things from the competitors.

Anabolic steroids are drugs that help build up muscle mass. They are used by athletes who need to be very strong, such as weightlifters. Athletes who need lots of muscle to be very fast, such as sprinters, also sometimes use anabolic steroids. Taking anabolic steroids and careful training means you can make much more muscle and build it where you want it.

Strong painkilling drugs can allow an athlete to train and compete with an injury, causing further and perhaps permanent damage. These drugs are illegal for use by people involved in sport.

Different sports need great stamina – marathons and long distance cycling races are two examples. Some cyclists (and other athletes) use a drug to stimulate their body to make more red blood cells. This means they can carry more oxygen to their muscles. The drug is a compound found naturally in the body so drug-testers are looking for abnormally high levels of it.

Fast reactions are vital in many sports, and there are drugs that will make you very alert and on edge. On the other hand, in sports such as darts and shooting, you need very steady hands. Some athletes take drugs to slow down their heart and reduce any shaking in their hands to try and win medals.

a   What are anabolic steroids and why do athletes use them?

Catching the cheats

Athletes found using illegal drugs are banned from competing. The sports authorities keep producing new tests for drugs and run random drugs tests to try and identify the cheats. But some competitors are always looking for new ways to cheat without being found out. So the illegal use of drugs in sport continues. Some medicines contain banned drugs which can enhance performance, so athletes need to be very careful so they don't end up 'cheating' by accident.

Figure 1 Weightlifters need a lot of muscle so it can be tempting to cheat. Eleven Bulgarian weightlifters tested positive for anabolic steroids and were disqualified from the 2008 Olympics.

Figure 2 The Tour de France has had many drug problems. Cyclists have died after using illegal drugs to help them go faster. Floyd Landis, the winner in 2006, was disqualified for using steroids.

The ethics of using drugs in sport

There are lots of ways an athlete can improve their performance. Where does wanting to win end and cheating begin? Is the use of performance-enhancing substances ever acceptable in sport? These are questions scientists cannot answer – society has to decide.

For example, if an athlete lives and trains at high altitude for several weeks, their body makes a hormone which increases their red blood cell count. This is legal. But it is illegal to buy the hormone and inject it to make more red blood cells.

Here are some of the arguments that athletes use to justify the use of substances that are banned and could do them harm:

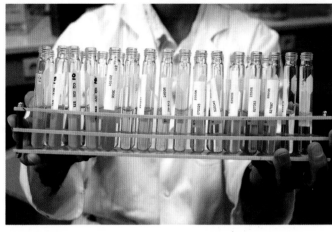

Figure 3 Athletes can be asked to produce a urine sample for a drug test at any time, whether they are competing, training or resting

- They want to win.
- They feel that other athletes are using these substances, and unless they take them they will be left behind.
- They think the health risks are just scare stories.
- Some athletes claim that they did not know they were taking drugs – their coaches supply them hidden in 'supplements'.

There are a number of ethical points that society needs to consider. Top athletes compete for the satisfaction of winning and millions of people enjoy watching them. Most performance-enhancing drugs risk the health of the athlete at the high doses used in training. They can even cause death. Even if the individual is prepared to take the risk, is this ethically acceptable? At the moment most people say 'no'.

Often the substances used by cheats are so expensive, or new, that most competitors can't afford them. This gives the richest competitors an unfair advantage. For example, most athletes could afford anabolic steroids if they wanted to use them, but not the most recent versions that are not detected by the drug-testing process.

There are some people who think that athletes should be able to do what they like with their bodies. At the moment most of society does not agree with this view – what do you think?

> **Study tip**
>
> Make sure you understand why athletes are banned from using some medical drugs.

b Why do athletes use drugs which could cause them harm?

Summary questions

1 Copy and complete using the words below:

   compete   performance-enhancing   muscles   steroids   athletes

   Some .... use .... drugs to help them .... more successfully. Many use anabolic .... which help them to develop bigger ....

2 Suggest the advantages and disadvantages to an athlete of using banned performance-enhancing drugs to help win a competition.

3 It has been suggested that athletes be allowed to use any drugs to improve their performance. Suggest arguments for and against this proposal.

> **Key points**
>
> - Anabolic steroids and other banned performance-enhancing drugs are used by some athletes.
>
> - The use of performance-enhancing drugs is considered unethical by most people.

Summary questions (k)

1 a Why do new medicines need to be tested and trialled before doctors can use them to treat their patients?

b Why is the development of a new medicine so expensive?

c Do you think it would ever be acceptable to use a new medicine before all the trials had been completed?

2 a What is a statin?

b How do statins help reduce the number of people who suffer from cardiovascular disease?

c Which of the statins in Figure 1, B1 3.2 is most effective?

d The most effective drug is not always the one used. Why do you think other statins might be prescribed?

3 Some students decided to test whether drinking coffee could affect heart rate. They asked the class to help them with their investigation. They divided the class into two groups. Both groups had their pulses taken. They gave one group a drink of coffee. They waited for 10 minutes and then took their pulses again. They then followed the same procedure with the second group.

a What do you think the second group were asked to drink?

b State a control variable that should have been used.

c Explain why it would have been a good idea not to tell the two groups exactly what they were drinking.

d Study this table of results that they produced.

Group	Increase in pulse rate (beats per minute)
With caffeine	12, 15, 13, 10, 15, 16, 10, 15, 16, 21, 14, 13, 16
Without caffeine	4, 3, 4, 5, 7, 5, 7, 4, 2, 6, 5, 4, 7

Can you detect any evidence for systematic error in these results? If so, describe this evidence.

e Is there any evidence for a random error in these results? If so, describe this evidence.

f What is the range for the increase in pulse rates without caffeine?

g What is the mean (or average) increase in pulse rate:
  i with caffeine?
  ii without caffeine?

4 Look at Figure 3, B1 3.4. Compare the data in that graph to Figure 2, B1 3.3. Both show impact of drug taking on individuals in society.

a What are the similarities between the two data sets?

b Explain the relative impact of legal and illegal drugs on individuals and on society.

5 a Why do some athletes use illegal drugs, such as anabolic steroids, when they are training or competing?

b What are the arguments for and against the use of these performance-enhancing drugs?

c People sometimes use illegal performance-enhancing drugs on horses. They use pain killers, stimulants and substances which make the skin on their legs very sensitive. Sometimes they are given sedatives so they run slowly. Discuss the ethical aspects of giving performance-changing drugs to animals.

Practice questions (k)

1 People take drugs for many different reasons.

alcohol heroin penicillin statin steroid thalidomide

Choose a word from above to match the following sentences.

a an illegal drug which is highly addictive (1)

b a drug used by athletes to make them perform better (1)

c a medical drug which is used to reduce cholesterol levels (1)

2 A drug company wants to test a new painkiller called PainGo2. The company hope that the new drug will cure headaches quicker than PainGo1.

PainGo2 has to be tested in clinical trials. PainGo2 is twice as strong as PainGo1.

Phase 1 trial – a few healthy people will be given one or two tablets of PainGo2.

Phase 2 trial – a small group (200–300) of patients with headaches will be given PainGo2.

Phase 3 trial – 3 large groups (2000 in each group) of patients with headaches will be given either PainGo2 or PainGo1 or a placebo.

a What is the purpose of the Phase 1 trial? (1)

b Suggest why in Phase 2 the patients were asked to record how they felt after taking the PainGo2.

Suggest why. (1)

c What is a placebo? (1)

d Phase 3 was done as a double-blind trial by doctors who had patients with headaches.

In a double-blind trial who will know who is given the new drug?

Choose your answer from the choices below.

A *the patient only*

B *the doctor only*

C *both the doctor and the patient*

D *neither the doctor nor the patient* (1)

e Why is it important to use the placebo in the Phase 3 trial? (1)

f Why are some patients given PainGo1 in Phase 3? (1)

3 a Give **one** example of:
  i a legal recreational drug (1)
  ii an illegal recreational drug. (1)

b Some recreational drugs are addictive.
  i Give **one** example of a recreational drug that is very addictive. (1)
  ii Explain how the action of a drug makes a person become addicted to it. (1)

c Some doctors think that smoking cannabis causes depression. Doctors investigated the cannabis smoking habits of 1500 young adults.

The table shows the percentage of cannabis smokers in the investigation who became depressed.

How many times the men or women had smoked cannabis in the last 12 months	Percentage of men who became depressed	Percentage of women who became depressed
Less than 5 times	9	16
More than 5 times, but less than once per week	10	17
1–4 times per week	12	31
Every day	15	68

From the data, give **two** conclusions that can be drawn about the relationship between cannabis and depression. (2)

AQA, 2007

4 *In this question you will be assessed on using good English, organising information clearly and using specialist terms where appropriate.*

Read the description of an investigation into the link between smoking cannabis and heroin addiction.

> Six 'teenage' rats were given a small dose of THC – the active chemical in cannabis – every three days between the ages of 28 and 49 days. This is the equivalent of human ages 12 to 18.
>
> The amount of THC given was roughly equivalent to a human smoking one cannabis 'joint' every three days.
>
> A control group of six 'teenage' rats did not receive THC.
>
> After 56 days catheters (narrow tubes) were inserted in all twelve of the now adult rats and they were able to self-administer heroin by pushing a lever.
>
> All the rats began to self-administer heroin frequently. After a while, they stabilised their daily intake at a certain level.
>
> The ones that had been on THC as 'teenagers' stabilised their heroin intake at a much higher level than the others. They appeared to be less sensitive to the effects of heroin. This pattern continued throughout their lives.
>
> Reduced sensitivity to the heroin means that the rats take larger doses. This has been shown to increase the risk of addiction.

Evaluate this investigation with respect to establishing a link between cannabis smoking and heroin addiction in humans. Remember to include a conclusion to your evaluation. (6)

AQA, 2007

B1 4.1 Adapt and survive

Learning objectives

- What do organisms need to live?

- How do organisms survive in many different conditions?

The variety of conditions on the surface of the Earth is huge. It ranges from hot, dry deserts to permanent ice and snow. There are deep, saltwater oceans and tiny freshwater pools. Whatever the conditions, almost everywhere on Earth you will find living organisms able to survive and reproduce.

Survive and reproduce

Living organisms need a supply of materials from their surroundings and from other living organisms so they can survive and reproduce successfully. What they need depends on the type of organism.

- Plants need light, carbon dioxide, water, oxygen and nutrients to produce glucose energy in order to survive.
- Animals need food from other living organisms, water and oxygen.
- Microorganisms need a range of things. Some are like plants, some are like animals and some don't need oxygen or light to survive.

Living organisms have special features known as **adaptations**. These features make it possible for them to survive in their particular habitat, even when the conditions are very extreme.

links

For more information on plant adaptation, see B1 4.3 Adaptation in plants.

Plant adaptations

Plants need to photosynthesise to produce the glucose needed for energy and growth. They also need to have enough water to maintain their cells and tissues. They have adaptations that enable them to live in many different places. For example, most plants get water and nutrients from the soil through their roots. Epiphytes are found in rainforests. They have adaptations which allow them to live high above the ground attached to other plants. They collect water and nutrients from the air and in their specially adapted leaves.

Figure 1 Mangroves are trees that live in soil with very little oxygen, often with their roots covered by salty water. They have special adaptations to get rid of the salt through their leaves, and roots which grow in the air to get oxygen.

Study tip

Practise recognising plant and animal adaptations and try to work out where they might live from the adaptation. This will help in your examination where you may be asked to do the same.

Some plant adaptations are all about reproduction. *Rafflesia arnoldii* produces flowers which are 1 m across, weigh about 11 kg and smell of a rotting corpse. The plants are rare so the dramatic and very smelly flower increases the chances of flies visiting and carrying pollen from one plant to another.

a Why do plants need to photosynthesise?

Animal adaptations

Animals cannot make their own food. They have to eat plants or other animals. Many of the adaptations of animals help them to get the food they need. So you can tell what a mammal eats by looking at its teeth. **Herbivores** have teeth for grinding up plant cells. **Carnivores** have teeth adapted for tearing flesh or crushing bones. Animals also often have adaptations to help them find and attract a mate.

links

For more information on animal adaptation, see B1 4.2 Adaptation in animals.

Adapting to the environment

Some of the adaptations seen in animals and plants help them to survive in a particular environment. Some sea birds get rid of all the extra salt they take in from the sea water by 'crying' very salty tears from a special salt gland. Animals that need to survive extreme winter temperatures often produce a chemical in their cells which acts as antifreeze. It stops the water in the cells from freezing and destroying the cell. Plants such as water lilies have lots of big air spaces in their leaves. This adaptation enables them to float on top of their watery environment and make food by photosynthesis.

Organisms that survive and reproduce in the most difficult conditions are known as **extremophiles**.

Living in extreme environments

Microorganisms are found in more places in the world than any other living thing. These places range from ice packs to hot springs and geysers. Microorganisms have a range of adaptations which make this possible. Many extremophiles are microorganisms.

Some extremophiles live at very high temperatures. Bacteria known as thermophiles can survive at temperatures of over 45 °C and often up to 80 °C or higher. In most organisms the enzymes stop working at around 40 °C. These extremophiles have specially adapted enzymes that do not **denature** and so work at these high temperatures. In fact, many of these organisms cannot survive and reproduce at lower temperatures.

Other bacteria have adaptations so they can grow and reproduce at very low temperatures, down to –15 °C. They are found in ice packs and glaciers around the world.

Most living organisms struggle to survive in a very salty environment because of the problems it causes with water balance. However, there are species of extremophile bacteria that can only live in extremely salty environments such as the Dead Sea and salt flats. They have adaptations to their cytoplasm so that water does not move out of their cells into their salty environment. But in ordinary sea water, they would swell up and burst!

b What is a thermophile?

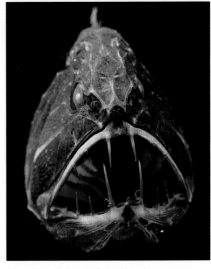

Figure 2 Animals from the deep oceans are adapted to cope with enormous pressure, no light and very cold, salty water. If these extremophiles are brought to the surface too quickly, they explode because of the rapid change in pressure.

Figure 3 Black smoker bacteria live in deep ocean vents, 2500 m down, at temperatures of well over 100 °C, with enormous pressure, no light and an acid pH of about 2.8. They have adaptations to cope with some of the most extreme conditions on Earth.

Summary questions

1 Copy and complete using the words below:

adaptations organisms materials survive extreme

To .... and reproduce, organisms need a supply of .... from their surroundings and the living .... in their habitat. They have .... that enable them to survive in their particular habitat, even when the conditions are very ....

2 Make a list of what plants and animals need from their surroundings to survive and reproduce.

3 **a** What is an extremophile?
   b Give two examples of adaptations found in different extremophiles.

Key points

- Organisms need a supply of materials from their surroundings and from other living organisms to survive and reproduce.

- Organisms have features (adaptations) that enable them to survive in the conditions in which they normally live.

- Extremophiles have adaptations enabling them to live in extreme conditions of salt, temperature or pressure.

B1 4.2

Adaptation in animals

Learning objectives

- How can hair help animals survive in very cold climates?

- What are the advantages – and disadvantages – of lots of body fat?

- How do animals adapt to hot, dry climates?

Animals have adaptations that help them to get the food and mates they need to survive and reproduce. They also have adaptations for survival in the conditions where they normally live.

Animals in cold climates

To survive in a cold environment you must be able to keep yourself warm. Animals which live in very cold places, such as the Arctic, are adapted to reduce the energy they lose from their bodies. You lose body heat through your body surface (mainly your skin). The amount of energy you lose is closely linked to your surface area : volume (SA : V) ratio.

Study tip

Remember, the *larger* the animal, the *smaller* the surface area : volume (SA : V) ratio.

Animals often have *increased* surface areas in *hot* climates, and *decreased* surface areas in *cold* climates.

Maths skills

Surface area : volume ratio

The surface area : volume ratio is very important when you look at the adaptations of animals that live in cold climates. It explains why so many Arctic mammals, such as seals, walruses, whales and polar bears, are relatively large.

The ratio of surface area to volume falls as objects get bigger. You can see this clearly in the diagram. The larger the surface area : volume ratio, the larger the rate of energy loss. So mammals in a cold climate grow to a large size. This keeps their surface area : volume ratio as small as possible and so helps them hold on to their body heat.

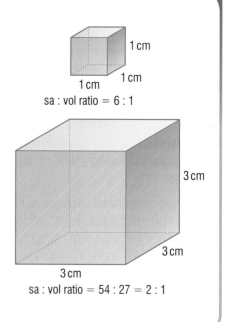

sa : vol ratio = 6 : 1

sa : vol ratio = 54 : 27 = 2 : 1

a Why are so many Arctic animals large?

Animals in very cold climates often have other adaptations too. The surface area of the thinly skinned areas of their bodies, like their ears, is usually very small. This reduces their energy loss.

Many Arctic mammals have plenty of insulation, both inside and out. Inside they have blubber (a thick layer of fat that builds up under the skin). On the outside a thick fur coat will insulate an animal very effectively. These adaptations really reduce the amount of energy lost through their skin.

The fat layer also provides a food supply. Animals often build up their fat in the summer. Then they can live off their body fat through the winter when there is almost no food.

Figure 1 The Arctic is a cold and bleak environment. However, the animals that live there are well adapted for survival. Notice the large size, small ears, thick coat and white camouflage of this polar bear.

b List three ways in which Arctic animals keep warm in winter.

Camouflage

Camouflage is important both to predators (so their prey doesn't see them coming) and to prey (so they can't be seen). The colours that would camouflage an Arctic animal in summer against plants would stand out against the snow in winter. Many Arctic animals, including the Arctic fox, the Arctic hare and the stoat, have grey or brown summer coats that change to pure white in the winter. Polar bears don't change colour. They have no natural predators on the land. They hunt seals all year round in the sea, where their white colour makes them less visible among the ice.

The colour of the coat of a lioness is another example of effective camouflage. The sandy brown colour matches perfectly with the dried grasses of the African savannah. Her colour hides the lioness from the grazing animals which are her prey.

Surviving in dry climates

Dry climates are often also hot climates – like deserts. Deserts are very difficult places for animals to live. There is scorching heat during the day, followed by bitter cold at night. Water is also in short supply.

The biggest challenges if you live in a desert are:

- coping with the lack of water
- stopping body temperature from getting too high.

Many desert animals are adapted to need little or nothing to drink. They get the water they need from the food they eat.

Mammals keep their body temperature the same all the time. So as the environment gets hotter, they have to find ways of keeping cool. Sweating means they lose water, which is not easy to replace in the desert.

> **c** Why do mammals try to cool down without sweating in hot, dry conditions?

Animals that live in hot conditions adapt their behaviour to keep cool. They are often most active in the early morning and late evening, when it is not so hot. During the cold nights and the heat of the day they rest in burrows where the temperature doesn't change much.

Many desert animals are quite small, so their surface area is large compared to their volume. This helps them to lose heat through their skin. They often have large, thin ears to increase their surface area for losing energy.

Another adaptation of many desert animals is to have thin fur. Any fur they do have is fine and silky. They also have relatively little body fat stored under the skin. These features make it easier for them to lose energy through the surface of the skin.

Figure 2 Jerboas are very small and elephants are very big. They both show clear adaptations that help them survive in the hot, dry places where they live.

Key points

- All living things have adaptations that help them to survive in the conditions where they live.
- Animals that are adapted for cold environments are often large, with a small surface area : volume (SA : V) ratio. They have thick insulating layers of fat and fur.
- Changing coat colour in the different seasons gives animals year-round camouflage.
- Adaptations for hot, dry environments include a large SA : V ratio, thin fur, little body fat and behaviour patterns that avoid the heat of the day.

Summary questions

1 **a** List the main problems that face animals living in cold conditions like the Arctic.
  b List the main problems that face animals living in the desert.

2 Animals that live in the Arctic are adapted to keep warm through the winter. Describe three of these adaptations and explain how they work.

3 **a** Using Figure 2, describe the visible adaptations of a jerboa and an elephant to keeping cool in hot conditions.
  b Suggest other ways in which animals might be adapted to survive in hot, dry conditions.

B1 4.3

Adaptation in plants

Learning objectives

- How do plants lose water?

- How are plants adapted to live in dry conditions?

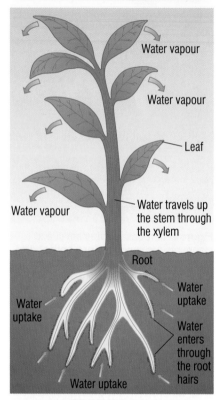

Figure 1 Plants lose water vapour from the surface of their leaves. When the conditions are hot and dry, they may lose water very quickly.

∞ links

For information on surface area : volume ratio, look back at B1 4.2 Adaptation in animals.

Figure 2 Marram grass grows on sand dunes. It has tightly curled leaves to reduce the surface area for water loss so it can survive the dry conditions.

Plants need light, water, space and nutrients to survive. There are some places where plants cannot grow. In deep oceans no light penetrates and so plants cannot photosynthesise. In the icy wastes of the Antarctic it is simply too cold for plants to grow.

Almost everywhere else, including the hot, dry areas of the world, you find plants growing. Without them there would be no food for the animals. But plants need water for photosynthesis and to keep their tissues supported. If a plant does not get the water it needs, it wilts and eventually dies.

a Why do plants need water?

Plants take in water from the soil through their roots. It moves up through the plant and into the leaves. There are small openings called **stomata** in the leaves of a plant. These open to allow gases in and out for photosynthesis and **respiration**. At the same time water vapour is lost through the stomata by evaporation.

The rate at which a plant loses water is linked to the conditions it is growing in. When it is hot and dry, photosynthesis and respiration take place quickly. As a result, plants also lose water vapour very quickly. Plants that live in very hot, dry conditions need special adaptations to survive. Most of them either reduce their surface area so they lose less water or store water in their tissues. Some do both!

b How do plants lose water from their leaves?

Changing surface area

When it comes to stopping water loss through the leaves, the surface area : volume ratio is very important to plants. A few desert plants have broad leaves with a large surface area. These leaves collect the dew that forms in the cold evenings. They then funnel the water towards their shallow roots.

Some plants in dry environments have curled leaves. This reduces the surface area of the leaf. It also traps a layer of moist air around the leaf. This reduces the amount of water the plant loses by evaporation.

Most plants that live in dry conditions have leaves with a very small surface area. This adaptation cuts down the area from which water can be lost. Some desert plants have small fleshy leaves with a thick cuticle to keep water loss down. The cuticle is a waxy covering on the leaf that stops water evaporating.

The best-known desert plants are the cacti. Their leaves have been reduced to spines with a very small surface area indeed. This means the cactus only loses a tiny amount of water. Not only that, its sharp spines also put animals off eating the cactus.

c Why do plants often reduce the surface area of their leaves?

Collecting water

Many plants that live in very dry conditions have specially adapted and very big root systems. They may have extensive root systems that spread over a very wide area, roots that go down a very long way, or both. These adaptations allow the plant to take up as much water as possible from the soil. The mesquite tree has roots that grow as far as 50 m down into the soil.

Storing water

Some plants cope with dry conditions by storing water in their tissues. When there is plenty of water after a period of rain, the plant stores it. Some plants use their fleshy leaves to store water. Others use their stems or roots.

For example, cacti don't just rely on their spiny leaves to help them survive in dry conditions. The fat green body of a cactus is its stem, which is full of water-storing tissue. These adaptations make cacti the most successful plants in a hot, dry climate.

Figure 3 A large saguaro cactus in the desert loses less than one glass of water a day. A UK apple tree can lose a whole bath of water in the same amount of time!

> **d** In which parts can a plant store its water?

Study tip

Remember that plants need their stomata open for photosynthesis and respiration. This is why they lose water by evaporation from their leaves.

Summary questions

1 Copy and complete using the words below:

   adaptations desert plants spines stems water

   Cacti are .... that live in the .... They have two main .... to help them survive. Their leaves have become .... and they store .... in their ....

2 **a** Explain why plants lose water through their leaves all the time.
   b Why does this make living in a dry place such a problem?

3 Plants living in dry conditions have adaptations to reduce water loss from their leaves. Give three of these and explain how they work.

Key points

- Plants lose water vapour from the surface of their leaves.

- Plant adaptations for surviving in dry conditions include reducing the surface area of the leaves, having water-storage tissues and having extensive root systems.

B1 4.4

Competition in animals

Learning objectives

- What is competition?
- What makes an animal a good competitor?

Figure 1 Some herbivores only feed on one particular plant. Pandas only eat bamboo, so they are open to competition from other animals or to diseases that damage bamboo.

Figure 2 The coral snake (top) is poisonous but the milk snake (bottom) is not. The milk snake is a mimic – it looks like the coral snake. As long as the two species live in the same area the milk snake is protected. Other animals and people leave it alone thinking it is a poisonous coral snake!

Animals and plants grow alongside lots of other living things. Some will be from the same species and others will be completely different. In any area there is only a limited amount of food, water and space, and a limited number of mates. As a result, living organisms have to compete for the things they need.

The best adapted organisms are most likely to win the **competition** for resources. They will be most likely to survive and produce healthy offspring.

a Why do living organisms compete?

What do animals compete for?

Animals compete for many things, including:

- food
- **territory**
- mates.

Competition for food

Competition for food is very common. Herbivores sometimes feed on many types of plant, and sometimes on only one or two different sorts. Many different species of herbivores will all eat the same plants. Just think how many types of animals eat grass!

The animals that eat a wide range of plants are most likely to be successful. If you are a picky eater, you risk dying out if anything happens to your only food source. An animal with wider tastes will just eat something else for a while!

Competition is common among carnivores. They compete for prey. Small mammals like mice are eaten by animals like foxes, owls, hawks and domestic cats. The different types of animals all hunt the same mice. So the animals which are best adapted to the area will be most successful.

Carnivores have to compete with their own species for their prey as well as with different species. Some successful predators are adapted to have long legs for running fast and sharp eyes to spot prey. These features will be passed on to their offspring.

Animals often avoid direct competition with members of other species when they can. It is the competition between members of the same species which is most intense.

Prey animals compete with each other too – to be the one that *isn't* caught! Their adaptations help prevent them becoming a meal for a predator. Some animals contain poisons which make anything that eats them sick or even kills them. Very often these animals also have bright warning colours so that predators quickly learn which animals to avoid. Poison arrow frogs are a good example.

b Give one useful adaptation for a herbivore and one for a carnivore.

Competition for territory

For many animals, setting up and defending a territory is vital. A territory may simply be a place to build a nest. It could be all the space needed for an animal to find food and reproduce. Most animals cannot reproduce successfully if they have no territory. So they will compete for the best spaces. This helps to make sure they will be able to find enough food for themselves and for their young.

Competition for a mate

Competition for mates can be fierce. In many species the male animals put a lot of effort into impressing the females. The males compete in different ways to win the privilege of mating with a female.

In some species – like deer and lions – the males fight between themselves. Then the winner gets the females.

Many male animals display to the females to get their attention. Some birds have spectacular adaptations to help them stand out. Male peacocks have the most amazing tail feathers. They use them for displaying to other males (to warn them off) and to females (to attract them).

What makes a successful competitor?

A successful competitor is an animal that is adapted to be better at finding food or a mate than the other members of its own species. It also needs to be better at finding food than the members of other local species. It must be able to breed successfully.

Many animals are successful because they avoid competition with other species as much as possible. They feed in a way that no other local animals do, or they eat a type of food that other animals avoid. For example, one plant can feed many animals without direct competition. While caterpillars eat the leaves, greenfly drink the sap, butterflies suck nectar from the flowers and beetles feed on pollen.

Figure 3 The territory of a gannet pair may be small but without a space they cannot build a nest and reproduce

Figure 4 The spectacular display of a male peacock attracts females. Unlike deer and lions he doesn't need to fight and risk injury.

> ### Study tip
>
> Learn to look at an animal and spot the adaptations that make it a successful competitor.

Summary questions

1 **a** Give an example of animals competing with members of the same species for food.
  b Give an example of animals competing with members of other species for food.
  c Animals that rely on a single type of food can easily become extinct. Explain why.

2 **a** Give two ways in which animals compete for mates.
  b Suggest the advantages and disadvantages of the methods chosen in part **a**.

3 Explain the adaptations you would expect to find in:
  a an animal that hunts mice
  b an animal that eats grass
  c an animal that hunts and eats other animals
  d an animal that feeds on the tender leaves at the top of trees.

> ### Key points
>
> - Animals often compete with each other for food, territories and mates.
> - Animals have adaptations that make them good competitors.

B1 4.5 Competition in plants

Practical

Investigating competition in plants

Carry out an investigation to look at the effect of competition on plants. Set up two trays of seeds – one crowded and one spread out. Then monitor the plants' height and wet mass (mass after watering). Keep all of the conditions – light level, the amount of water and nutrients available and the temperature – exactly the same for both sets of plants. The differences in their growth will be the result of overcrowding and competition for resources in one of the groups. The data show growth of tree seedlings. You can get results in days rather than months by using cress seeds.

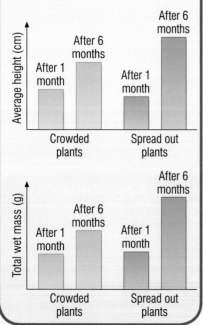

Plants compete fiercely with each other. They compete for:

- light for photosynthesis, to make food using energy from sunlight
- water for photosynthesis and to keep their tissues rigid and supported
- nutrients (minerals) so they can make all the chemicals they need in their cells
- space to grow, allowing their roots to take in water and nutrients, and their leaves to capture light.

a What do plants compete with each other for?

Why do plants compete?

Just like animals, plants are in competition both with other species of plants and with their own species. Big, tall plants such as trees take up a lot of water and nutrients from the soil. They also prevent light from reaching the plants beneath them. So the plants around them need adaptations to help them to survive.

When a plant sheds its seeds they might land nearby. Then the parent plant will be in direct competition with its own seedlings. Because the parent plant is large and settled, it will take most of the water, nutrients and light. So the plant will deprive its own offspring of everything they need to grow successfully. The roots of some desert plants even produce a chemical that stops seeds from germinating, killing the competition even before it begins to grow!

Sometimes the seeds from a plant will all land close together, a long way from their parent. They will then compete with each other as they grow.

b Why is it important that seeds are spread as far as possible from the parent plant?

Coping with competition

Plants that grow close to other species often have adaptations which help them to avoid competition.

Small plants found in woodlands often grow and flower very early in the year. This is when plenty of light gets through the bare branches of the trees. The dormant trees take very little water out of the soil. The leaves shed the previous autumn have rotted down to provide nutrients in the soil. Plants like snowdrops, anemones and bluebells are all adapted to take advantage of these things. They flower, set seeds and die back again before the trees are in full leaf.

Another way plants compete successfully is by having different types of roots. Some plants have shallow roots taking water and nutrients from near the surface of the soil. Others have long, deep roots, which go far underground. Both compete successfully for what they need without affecting the other.

If one plant is growing in the shade of another, it may grow taller to reach the light. It may also grow leaves with a bigger surface area to take advantage of all the light it does get.

Some plants are adapted to prevent animals from eating them. They may have thorns, like the African acacia or the blackberry. They may make poisons that mean they taste very bitter or make the animals that eat them ill. Either way they compete successfully because they are less likely to be eaten than other plants.

c How can short roots help a plant to compete successfully?

Spreading the seeds 🅚

To reproduce successfully, a plant has to avoid competition with its own seedlings. Many plants use the wind to help them spread their seeds as far as possible. They produce fruits or seeds with special adaptations for flight to carry their seeds away. Examples of this are the parachutes of the dandelion 'clock' and the winged seeds of the sycamore tree.

d How do the fluffy parachutes of dandelion seeds help the seeds to spread out?

Some plants use mini-explosions to spread their seeds. The pods dry out, twist and pop, flinging the seeds out and away.

Juicy berries, fruits and nuts are adaptations to tempt animals to eat them. The fruit is digested and the tough seeds are deposited well away from the parent plant in their own little pile of fertiliser!

Fruits that are sticky or covered in hooks get caught up in the fur or feathers of a passing animal. They are carried around until they fall off hours or even days later.

Sometimes the seeds of several different plants land on the soil and start to grow together. The plants that grow fastest will compete successfully against the slower-growing plants. For example:

● The plants that get their roots into the soil first will get most of the available water and nutrients.

● The plants that open their leaves fastest will be able to photosynthesise and grow faster still, depriving the competition of light.

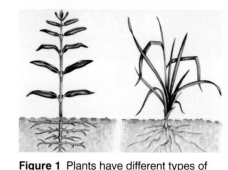
Figure 1 Plants have different types of roots to compete for water and nutrients in the soil

Figure 2 The winged seeds of the sycamore tree

Figure 3 Coconuts will float for weeks or even months on ocean currents, which can carry them hundreds of miles from their parents – and any other coconuts!

Summary questions

1 a How can plants overcome the problems of growing in the shade of another plant?
b How do bluebell plants grow and flower successfully in spite of living under large trees in a wood?

2 a Why is it so important that plants spread their seeds successfully?
b Give three examples of successful adaptations for spreading seeds.

3 The dandelion is a successful weed. Carry out some research and evaluate the adaptations that make it a better competitor than other plants on a school field.

Key points

● Plants often compete with each other for light, for water and for nutrients (minerals) from the soil.

● Plants have many adaptations that make them good competitors.

B1 4.6

How do you survive?

Learning objectives

- How do organisms survive in very unusual conditions?

- What factors are organisms competing for in a habitat?

Figure 1 A fig tree

So far in this chapter we have looked at lots of different ways in which living organisms are adapted. This helps them to survive and reproduce wherever they live. We have looked at why they need to compete successfully against their own species and others. Now we are going to consider three case studies of adaption in living organisms.

Figs and fig wasps

There are about 700 different species of fig trees. Each one has its own species of pollinating wasps, without which the trees will die. The fig flowers of the trees are specially adapted so that they attract the right species of wasp.

Female fig wasps have specially shaped heads for getting into fig flowers. They also have **ovipositors** that allow them to place their eggs deep in the flowers of the fig tree.

Male fig wasps vary. Some species can fly but others are adapted to live in a fig fruit all their life. If they are lucky, a female wasp will arrive in the flower and the male will fertilise her. After this he digs an escape tunnel for the female through the fruit and dies himself! The male wasp has special adaptations (such as the loss of his wings and very small eyes) which help him move around inside the fig fruit to find a female.

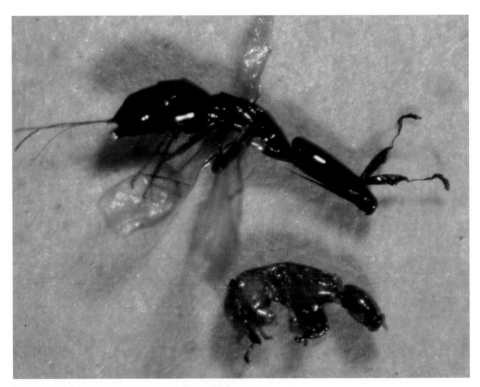

Figure 2 A female (top) and male (bottom) fig wasp

If a fig tree cannot attract the right species of wasp, it will never be able to reproduce. In fact in some areas the trees are in danger of extinction because the wasp populations are being wiped out.

The fastest predator in the world?

It takes you about 650 milliseconds to react to a crisis. But the star-nosed mole takes only 230 milliseconds from the moment it first touches its prey to gulping it down. That's faster than the human eye can see!

What makes this even more amazing is that star-nosed moles live underground and are almost totally blind. Their main sense organ is a crown of fleshy tendrils around the nose – incredibly sensitive to touch and smell but very odd to look at. The ultra-sensitive tendrils can try out 13 possible targets every second.

It seems likely that they have adapted to react so quickly because they can't see what is going on. They need to grab their prey as soon as possible after they touch it. If they don't it might move away or try to avoid them, and they wouldn't know where it had gone.

Figure 3 The star-nosed mole

A carnivorous plant

Venus flytraps are plants that grow on bogs. Bogs are wet and their peaty soil has very few nutrients in it. This makes it a difficult place for plants to live.

The Venus flytrap has special 'traps' that contain sweet smelling nectar. They sit wide open showing their red insides. Insects are attracted to the colour and the smell. Inside the trap are many small, sensitive hairs. As the insect moves about to find the nectar, it will brush against these hairs. Once the hairs have been touched, the trap is triggered. It snaps shut and traps the insect inside.

Special enzymes then digest the insect inside the trap. The Venus flytrap uses the nutrients from the digested bodies of its victims. This is in place of the nutrients that it cannot get from the poor bog soil. After the insect has been digested, the trap reopens ready to try again.

Figure 4 The Venus flytrap – an insect-eating plant

Activity

Case studies

- For each of these three case studies, list how the organisms are adapted for their habitat and how these adaptations help them to compete successfully against both their own species and other species.

- Choose three organisms that you know something about – or find out about three organisms which interest you. Make your own fact file on their adaptations and how these adaptations help them to compete successfully. Include at least one plant.

Summary questions

1 Explain how both of the animals featured compete successfully for food.

2 Why could any species of fig tree or fig wasp easily die out? Give a reason for each.

3 Carry out research to explain the adaptations of a giraffe and why they help it to compete successfully with other animals living in the same area.

Key points

- Organisms have adaptations which enable them to survive in the conditions in which they normally live.

- Plants often compete with each other for light, water and nutrients from the soil.

- Animals often compete with each other for food, mates and territory.

B1 4.7 Measuring environmental change

Learning objectives

- What affects the distribution of living things?

- What causes environmental changes?

- How can we measure environmental changes?

Have you noticed different types of animals and plants when you travel to different places? The distribution of living organisms depends on the environmental conditions and varies around the world.

Factors affecting the distribution of organisms

Non-living factors have a big effect on where organisms live. The average temperature or average rainfall will have a big impact on what can survive. You don't find polar bears in countries where the average temperature is over 20 °C, for example! The amount of rainfall affects the distribution of both plants and animals. Light, pH and the local climate all influence where living organisms are found.

The distribution of different species of animals in water is closely linked to the oxygen levels. Salmon can only live in water with lots of dissolved oxygen, but bloodworms can survive in very low oxygen levels.

Living organisms also affect the distribution of other living organisms. So, for example, koala bears are only found where eucalyptus trees grow. Parasites only live where they can find a host.

One species of ant eats nectar produced by the flowers of the swollen-thorn acacia tree. The ants hollow out the vicious thorns and live in them. So any animal biting the tree not only gets the sharp thorns, they get a mouth full of angry ants as well. The distribution of the ants depends on the trees.

Figure 1 The distribution of bullhorn acacia ants depends on where the swollen-thorn acacia trees grow

> **a** Which non-living environmental factor affects the distribution of polar bears?

Environmental changes

When the environment changes, it can cause a change in the distribution of living organisms in the area. Non-living factors often cause these changes in an environment.

The average temperature may rise or fall. The oxygen concentration in water may change. A change in the amount of sunlight, the strength of the wind or the average rainfall may affect an environment. Any of these factors can affect the distribution of living organisms.

Living factors can also cause a change in the environment where an organism lives, affecting distribution. A new type of predator may move into an area. A new disease-causing pathogen may appear and wipe out a species of animal or plant. Different plants may appear and provide food or a home for a whole range of different species.

> **b** Give an example of a living and a non-living factor that can change an environment.

Measuring environmental change

When an environment changes, the living organisms in it are affected. If the change is big enough, the distribution of animals or plants in an area may change.

You can measure environmental change using non-living indicators. You can measure factors such as average rainfall, temperature, oxygen levels, pH and pollutant levels in water or the air, and much more. All sorts of different instruments are available to do these measurements. These range from simple rain gauges and thermometers to oxygen meters and dataloggers used in schools.

You can also use the changing distribution of living organisms as an **indicator** of environmental change. Living organisms are particularly good as indicators of pollution.

Lichens grow on places like rocks, roofs and the bark of trees. They are very sensitive to air pollution, particularly levels of sulfur dioxide in the atmosphere. When the air is clean, many different types of lichen grow. The more polluted the air, the fewer lichen species there will be. So a field survey on the numbers and types of lichen can be used to give an indication of air pollution. The data can be used to study local sites or to compare different areas of the country.

In the same way you can use invertebrate animals as water pollution indicators. The cleaner the water, the more species you will find. Some species of invertebrates are only found in the cleanest waters. Others can be found even in very polluted waters. Counting the different types of species gives a good indication of pollution levels, and can be used to monitor any changes.

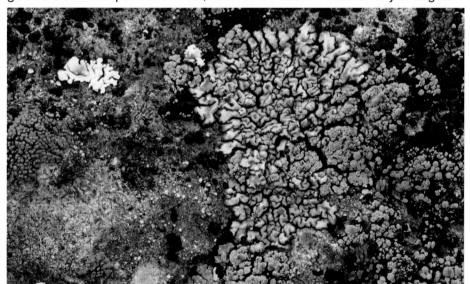

Practical

Indicators of pollution levels

Investigate both the variety of lichens in your local area and the number of invertebrate species in your local pond or stream. This will give you an idea of pollution levels in your area if you compare them to national figures.

Figure 2 Lichens grow well where the air is clean. In an area polluted with sulfur dioxide there would be fewer species. Lichens are good indicators of pollution.

Key points

- Animals and plants may be adapted to cope with specific features of their environment, e.g. thorns, poisons and warning colours.

- Environmental changes may be caused by living or non-living factors.

- Environmental changes can be measured using non-living indicators.

- Living organisms can be used as indicators of pollution.

Summary questions

1 Copy and complete these sentences using the words below:

indicators   distribution   pollution   organisms

Changes in the environment affect the .... of living .... This means living organisms can be used as .... of ....

2 Give three different methods you could use to collect environmental data. For each method, comment on its reliability and usefulness as a source of evidence of environmental change.

B1 4.8   The impact of change

Learning objectives

- How do changes in the environment affect the distribution of living organisms?

- How reproducible are the data about the effect of environmental change on living organisms?

Figure 1 The Dartford warbler

Changing birds of Britain

Temperatures in the UK seem to be rising. Many people like the idea – summer barbeques and low heating bills. But rising temperatures will have a big impact on many living organisms. We could see changes in the distribution of many species. Food plants and animals might become more common, or die out, in different conditions.

The Dartford warbler is small brown bird that breeds mainly in southern Europe. A small population lived in Dorset and Hampshire. By 1963, two very cold winters left just 11 breeding pairs in the UK. But temperatures have increased steadily since. Dartford warblers are now found in Wales, the Midlands and East Anglia. If climate change continues, Dartford warblers could spread through most of England and Ireland. However, in Spain the numbers are dropping rapidly – 25% in the last 10 years – as it becomes too warm. Scientists can simulate the distribution of birds as the climate changes. They predict that by the end of the century Spain could lose most of its millions of Dartford warblers.

Scientists predict that by the end of this century, if climate change continues at its present rate, the range of the average bird species will move nearly 550 km north-east. About 75% of all the birds that nest in Europe are likely to have smaller ranges as a result and many species will be lost for good.

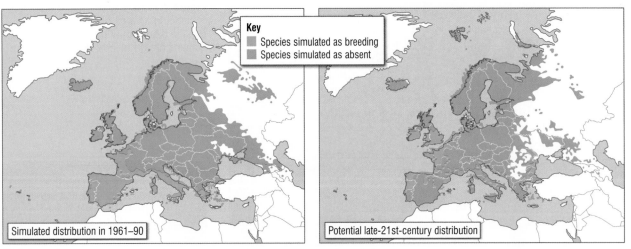

Key
■ Species simulated as breeding
■ Species simulated as absent

Simulated distribution in 1961–90

Potential late-21st-century distribution

Figure 2 The maps show how scientists think the distribution of these birds might change in the future

Table 1 Numbers of breeding pairs of Dartford warblers in the UK

Year	Number of breeding pairs
1961	450
1963	11
1974	560
1984	420
1994	1890
2010	3208

Activity

- Plot a bar graph to show the change in population of the Dartford warbler from 1961 to the present day. Draw an extra bar to show what you would expect the population to be in 2030 if climate change continues in the same way.

- Investigate the effect of climate change on the way birds migrate from one country to another and write a report for a wildlife programme or magazine.

Where are all the bees?

All around the world honey bees are disappearing. In the UK alone, around one in five bee hives has been lost in the last few years. In the United States, around 2 million colonies of bees were lost in 3 years. The bees had been struck down by a new, mystery disease called Colony Collapse Disorder or CCD. The bees either die, or simply fail to return to the hive. Without the mass of worker bees, those bees left in the hive quickly die.

Members of the British Beekeepers Association are alarmed. They say that if hives continue to be lost at the same rate there will be no honey bees left in Britain by 2018. You might think that having fewer bees doesn't really matter. It also means honey is more expensive to buy.

In fact, bees are vitally important in plant ecology. Honey bees pollinate flowers as they collect the nectar. Without them, flowers are not pollinated and fruit does not form. Without bees as pollinators we would have no apples, raspberries, cucumbers, strawberries, peaches … the list goes on and on. There would be cereal crops, because they are pollinated by the wind, but not much else.

No one yet fully understands what is happening to the bees and what is changing their distribution. Scientists think that viral diseases, possibly spread by a parasitic mite, are a major cause. So living factors – the agents of disease – are causing a major change in the environment of the honey bee. This in turn is affecting their distribution.

Other living and non-living factors affecting the environment have also been suggested. Flowering patterns are changing as temperatures vary with climate change. This may affect the food supply of the bees. Farmers spray chemicals that may build up in the bees. Some people have even suggested that mobile phones affect the navigation system of the bees.

Research is continuing all over the world. Disease-resistant strains of bees are being bred. Collecting the evidence to show exactly what environmental change is affecting the honey bee population is proving to be difficult. But until we can find out, the decline of the honey bee looks as if it will continue. There is a little good news – UK numbers have recovered slightly as more people have started keeping bees, probably as a result of all the publicity.

Figure 3 Honey bees are vital pollinators. Bee-pollinated fruits are worth about £50 billion of trade every year.

Activity

- List the main suggested causes for the decline of the honey bee. Use secondary sources to investigate the current state of the research findings for each cause.
- Produce a slide show to justify the investment of research funds into the loss of honey bees. Show what is happening to the bees, the main theories about what is causing the problem and how the problem is being tackled.

Summary questions

1 Using the information on this spread, what aspect of climate change seems to be linked to a change in the distribution of British birds?

2 a Why is the loss of honey bees so important?
  b Why is it important to find out whether the environmental cause of the problem is a living or non-living factor?

Key points

- Both living and non-living factors can cause changes in the environment that affect the distribution of living organisms.
- Reproducible data on the effect of environmental change are not always easy to collect or interpret.

Summary questions ⓚ

1 Match the following words to their definitions:

a	competition	A	an animal that eats plants
b	carnivore	B	an area where an animal lives and feeds
c	herbivore	C	an animal that eats meat
d	territory	D	the way animals compete with each other for food, water, space and mates

2 Cold-blooded animals like reptiles and snakes absorb heat from their surroundings and cannot move until they are warm.

a Why do you think that there are no reptiles and snakes in the Arctic?

b What problems do you think reptiles face in desert conditions and what adaptations could they have to cope with them?

c Most desert animals are quite small. How does this help them survive in the heat?

3 a What are the main problems for plants living in a hot, dry climate?

b Why does reducing the surface area of their leaves help plants to reduce water loss?

c Describe **two** ways in which the surface area of the leaves of some desert plants is reduced.

d How else are some plants adapted to cope with hot, dry conditions?

e Why are cacti such perfect desert plants?

4 a How does marking out and defending a territory help an animal to compete successfully?

b Bamboo plants all tend to flower and die at the same time. Why is this such bad news for pandas, but doesn't affect most other animals?

5 Why is competition between animals of the same species so much more intense than the competition between different species?

6 Use the bar charts from the practical activity on B1 4.5 to answer these questions.

a Describe what happens to the height of both sets of seedlings over the first six months and explain why the changes take place.

b The total wet mass of the seedlings after one month was the same whether or not they were crowded. After six months there was a big difference.

i Why do you think both types of seedling had the same mass after one month?

ii Explain why the seedlings that were more spread out each had more wet mass after six months.

c When scientists carry out experiments such as the one described, they try to use large sample sizes. Why?

d i Name a control variable mentioned in the practical.

ii Why were the other variables kept constant?

7 a Give **three** living factors that can change the environment and affect the distribution of living organisms.

b Give **three** non-living factors that can change the environment and affect the distribution of living organisms.

8 Maize is a very important crop plant. It has many uses – it is made into cornflakes and it is also grown for animal feed. The most important part of the plant is the cob, which fetches the most money. In an experiment to find the best growing conditions, three plots of land were used. The young maize plants were grown in different densities in the three plots.

The results were as follows:

	Planting density (plants/m²)		
	10	**15**	**20**
Dry mass of shoots (kg/m²)	9.7	11.6	13.5
Dry mass of cobs (kg/m²)	6.1	4.4	2.8

a What was the independent variable in this investigation?

b Draw a graph to show the effect of the planting density on the mass of the cobs grown.

c What is the pattern shown in your graph?

d This was a fieldwork investigation. What would the experimenter have taken into account when choosing the location of the three plots?

e Did the experimenter choose enough plots? Explain your answer.

f What is the relationship between the mass of cobs and the mass of shoots at different planting densities?

g The experimenter concluded that the best density for planting the maize is 10 plants per m². Do you agree with this as a conclusion? Explain your answer.

Practice questions ⓚ

1 The picture shows a solenodon.

Solenodons have lived on earth since the Age of the Dinosaurs. They are only found in forests in Haiti and are the only mammals which have a poisonous bite. They are rarely seen because they feed at night. They mainly eat insects and spiders.

a The solenodon has adaptations which help it to survive.
Match the adaptation to the correct letter (A,B, C, D or E) for the following:
i This helps the solenodon to dig its burrow. (1)
ii This helps the solenodon to detect its food. (1)

b The solenodon is at risk of dying out since new animals have been taken to the islands.
Use the information and the picture to help answer the following questions.
i The solenodon is not adapted to flee from predators. Suggest why. (1)
ii If the solenodon is caught by a predator it can defend itself. Suggest how. (1)

2 Trees that live in the rainforests are very tall and often have broad leaves. This is a problem for young trees, which do not get much light.

a Choose the correct answer to complete the sentence.
light   nutrients   space
Rainforest trees have broad leaves so they can compete for .... (1)

b Choose the correct answer to complete the sentence.
larger trees   large seeds with stored food
Trees in the rainforest have adapted to lack of light near the ground by having .... (1)

3 The gemsbok is a large herbivore living in dry desert regions of South Africa. It feeds on grasses that are adapted to the dry conditions by obtaining moisture from the air as it cools at night. The table below shows the water content of these grasses and the feeding activity of the gemsbok over a 24-hour period.

Time of day	% water content of grasses	% of gemsboks feeding
03.00	18	40
06.00	23	60
09.00	25	20
12.00	08	17
15.00	06	16
18.00	05	19
21.00	07	30
24.00	14	50

a i Name the independent variable investigated. (1)
ii Is this a categoric, ordered, discrete or continuous variable? (1)

b How does the water content of the grasses change throughout the 24-hour period? (1)

c Between which recorded times are more than 30% of the gemsboks feeding? (1)

d Suggest **three** reasons why the gemsboks benefit from feeding at this time. (3)
AQA, 2008

B1 5.1 Pyramids of biomass

Learning objectives

- Where does biomass come from?
- What is a pyramid of biomass?

??? Did you know ... ?

Only about 1% of all the light energy falling on the Earth is used by plants and algae for photosynthesis.

Figure 1 Plants can produce a huge mass of biological material in just one growing season

Radiation from the Sun (**solar** or **light energy**) is the source of energy for all groups of living organisms on Earth.

Light (solar) energy pours out continually on to the surface of the Earth. Green plants and algae absorb some of this light energy using chlorophyll for photosynthesis. During photosynthesis some of the light energy is transferred to chemical energy. This energy is stored in the substances that make up the cells of the plants and algae. This new material adds to the **biomass**.

Biomass is the mass of material in living organisms. Ultimately all biomass is built up using energy from the Sun. Biomass is often measured as the dry mass of biological material in grams.

> **a** What is the source of all the energy in the living things on Earth?

The biomass made by plants is passed on through food chains or food webs. It goes into the animals that eat the plants. It then passes into the animals that eat other animals. No matter how long the food chain or complex the food web, the original source of all the biomass involved is the Sun.

In a food chain, there are usually more producers (plants) than primary consumers (herbivores). There are also more primary consumers than secondary consumers (carnivores). If you count the number of organisms at each level you can compare them. However, the number of organisms often does not accurately reflect what is happening to the biomass.

Pyramids of biomass

The amount of biomass at each stage of a food chain is less than it was at the previous stage. We can draw the total amount of biomass in the living organisms at each stage of the food chain. When this biomass is drawn to scale, we can show it as a **pyramid of biomass**.

> **b** What is a pyramid of biomass?

Organism	Number	Biomass – dry mass in g
Oak tree	1	500 000
Aphids	10 000	1000
Ladybirds	200	50

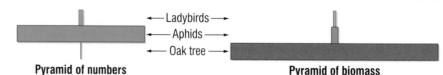

Figure 2 Using a pyramid of biomass shows us the amount of biological material involved at each level of this food chain much more effectively than a pyramid of numbers

Interpreting pyramids of biomass

The amount of material and energy contained in the biomass of organisms at each stage of a food chain is less than it was at the previous stage.

This is because:

● not all organisms at one stage are eaten by the stage above

● some material and energy taken in is passed out as waste by the organism

● when a herbivore eats a plant, lots of the plant biomass is used in respiration by the animal cells to release energy. Only a relatively small proportion of the plant material is used to build new herbivore biomass by making new cells, building muscle tissue etc. This means that very little of the plant biomass eaten by the herbivore in its lifetime is available to be passed on to any carnivore that eats it.

So, at each stage of a food chain the amount of energy in the biomass that is passed on gets less. A large amount of plant biomass supports a smaller amount of herbivore biomass. This in turn supports an even smaller amount of carnivore biomass.

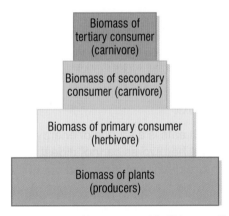

Figure 3 Any food chain can be turned into a pyramid of biomass like this

Study tip

Make sure you can draw pyramids of biomass when you are given the data.

Summary questions

1 **a** What is biomass?

b Why is a pyramid of biomass more useful for showing what is happening in a food chain than a pyramid of numbers?

2

Organism	Biomass, dry mass (g)
grass	100 000
sheep	5000
sheep ticks	30

Draw a pyramid of biomass for this grassland ecosystem.

3 Using the data in Figure 2, calculate the percentage biomass passed on from:

a the producers to the primary consumers

b the primary consumers to the secondary consumers.

Key points

● Radiation from the Sun (solar or light energy) is the main source of energy for all living things. The Sun's light energy is captured and used by green plants and algae during photosynthesis, to make new biomass.

● Biomass is the dry mass of living material in an animal or plant.

● The mass of living material at each stage of a food chain is less than at the previous stage. The biomass at each stage can be drawn to scale and shown as a pyramid of biomass.

B1 5.2

Energy transfers

Learning objectives

- What happens to the material and energy in the biomass of organisms at each stage of a food chain?

- How is some energy transferred to the environment?

Study tip

Make sure you can explain the different ways in which energy is lost between the stages of a food chain.

Figure 2 Animals such as horses produce very large quantities of dung made up of all the biomass they can't digest

Figure 3 These sea anemones don't move much so they don't need to eat much

The amounts of biomass and energy contained in living things get less as you progress up a food chain. Only a small amount of the biomass taken in gets turned into new animal material. What happens to the rest?

Figure 1 The amount of biomass in a lion is a lot less than the amount of biomass in the grass that feeds the zebra it preys on. But where does all the biomass go?

Energy loss in waste

The biomass that an animal eats is a source of energy, but not all of the energy can be used. Firstly, herbivores cannot digest all of the plant material they eat. The material they can't digest is passed out of the body in faeces.

The meat that carnivores eat is easier to digest than plants. This means that carnivores need to eat less often and produce less waste. But like herbivores, most carnivores cannot digest all of their prey, such as hooves, claws, bones and teeth. Therefore some of the biomass that they eat is lost in their faeces.

When an animal eats more protein than it needs, the excess is broken down. It gets passed out as **urea** in the urine. This is another way biomass – and energy – are transferred from the body to the surroundings.

> **a** Why is biomass lost in faeces?

Energy loss due to movement

Part of the biomass eaten by an animal is used for respiration in its cells. This supplies all the energy needs for the living processes taking place within the body, including movement.

Movement uses a great deal of energy. The muscles use energy to contract and also get hot. So the more an animal moves about the more energy (and biomass) it uses from its food.

> **b** Why do animals that move around a lot use up more of the biomass they eat than animals that don't move much?

Keeping a constant body temperature

Much of the energy animals release from their food in cellular respiration is eventually transferred heating their surroundings. Some of this heat is produced by the muscles as the animals move.

Energy transfers to the surroundings are particularly large in mammals and birds. That is because they use energy to keep their bodies at a constant temperature. They use energy all the time, to keep warm when it's cold or to cool down when it's hot. So mammals and birds need to eat far more food than animals such as fish and amphibians to get the same increase in biomass.

Practical

Investigating the energy released by respiration

Even plants transfer energy by heating their surroundings in cellular respiration. You can investigate this using germinating peas in a vacuum flask.

- What would be the best way to monitor the temperature continuously?
- Plan the investigation.

Figure 4 Only between 2% and 10% of the biomass eaten by an animal such as this horse will get turned into new horse. The rest of the stored energy will be used for movement or transferred, heating the surroundings, or lost in waste materials.

Summary questions

1 Copy and complete using the words below:

biomass temperature energy chain growth movement producers respiration waste

The amounts of .... and .... contained in living things always get less at each stage of a food .... from .... onwards. Biomass is lost as .... products and used to release energy in .... This is used for .... and to control body .... Only a small amount is used for ....

2 Explain why so much of the energy from the Sun that lands on the surface of the Earth is not turned into biomass in animals.

Key points

- The amounts of biomass and energy get less at each successive stage in a food chain.

- This is because some material and energy are always lost in waste materials, and some are used for respiration to supply energy for living processes, including movement. Much of the energy is eventually transferred by heating to the surroundings.

B1 5.3

Decay processes

Learning objectives

- Why do things decay?
- Why are decay processes so important?
- How are materials cycled in a stable community?

Study tip

You need to know the type of organisms that cause decay, the conditions needed for decay and the importance of decay in recycling nutrients.

Figure 1 This tomato is slowly being broken down by the action of decomposers. You can see the fungi clearly but the bacteria are too small to be seen.

Did you know ...?

The 'Body Farm' is a US research site where scientists have buried human bodies in many different conditions. They are studying every stage of human decay to help police forces all over the world work out when someone died and if they were murdered.

Plants take nutrients from the soil all the time. These nutrients are passed on into animals through food chains and food webs. If this was a one-way process the resources of the Earth would have been exhausted long ago.

Fortunately all these materials are recycled. Many trees shed their leaves each year, and most animals produce droppings at least once a day. Animals and plants eventually die as well. A group of organisms known as the **decomposers** then break down the waste and the dead animals and plants. In this process decomposers return the nutrients and other materials to the environment. The same material is recycled over and over again. This often leads to very stable communities of organisms.

a Which group of organisms take materials out of the soil?

The decay process

Decomposers are a group of microorganisms that include bacteria and fungi. They feed on waste droppings and dead organisms.

Detritus feeders, such as maggots and some types of worms, often start the process of decay. They eat dead animals and produce waste material. The bacteria and fungi then digest everything – dead animals, plants and detritus feeders plus their waste. They use some of the nutrients to grow and reproduce. They also release waste products.

The waste products of decomposers are carbon dioxide, water, and nutrients that plants can use. When we say that things decay, they are actually being broken down and digested by microorganisms.

The recycling of materials through the process of decay makes sure that the soil contains the mineral ions that plants need to grow. The decomposers also 'clean up' the environment, removing the bodies of all the dead organisms.

b What type of organisms are decomposers?

Conditions for decay

The speed at which things decay depends partly on the temperature. Chemical reactions in microorganisms, like those in most living things, work faster in **warm conditions**. They slow down and might even stop if conditions are too cold. Decay also stops if it gets too hot. The enzymes in the decomposers change shape and stop working.

Most microorganisms also grow better in **moist conditions**. The moisture makes it easier for them to dissolve their food and also prevents them from drying out. So the decay of dead plants and animals – as well as leaves and dung – takes place far more rapidly in warm, moist conditions than it does in cold, dry ones.

Although some microbes survive without oxygen, most decomposers respire like any other organism. This means they need oxygen to release energy, grow and reproduce. This is why decay takes place more rapidly when there is **plenty of oxygen** available.

c Why are water, warmth and oxygen needed for the process of decay?

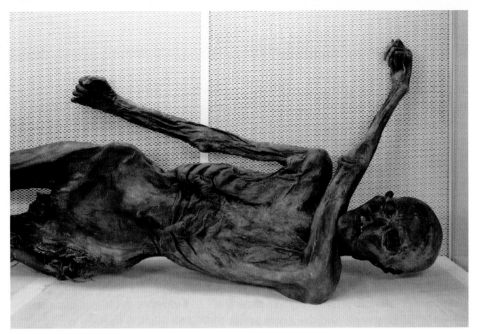

Figure 2 The decomposers cannot function at low temperatures so if an organism – such as this 4000-year-old man – is frozen as it dies, it will be preserved with very little decay

Practical

Investigating decay

Plan an investigation into the effect of temperature on how quickly things decay.

- Write a question that can be used as the title of this investigation.
- Identify the independent variable in the investigation.

The importance of decay in recycling

Decomposers are vital for recycling resources in the natural world. What's more, we can take advantage of the process of decay to help us recycle our waste.

In **sewage treatment plants** we use microorganisms to break down the bodily waste we produce. This makes it safe to release into rivers or the sea. These sewage works have been designed to provide the bacteria and other microorganisms with the conditions they need. That includes a good supply of oxygen.

Another place where the decomposers are useful is in the garden. Many gardeners have a **compost heap**. Grass cuttings, vegetable peelings and weeds are put onto the compost heap. It is then left to allow decomposing microorganisms break all the plant material down. It forms a brown, crumbly substance known as compost which can be used as a fertiliser.

Key points

- Living things remove materials from the environment as they grow. They return them when they die through the action of the decomposers.

- Materials decay because they are broken down (digested) by microorganisms. Microorganisms digest materials faster in warm, moist conditions. Many of them also need oxygen.

- The decay process releases substances that plants need to grow.

- In a stable community the processes that remove materials (particularly plant growth) are balanced by the processes that return materials.

Summary questions

1 Copy and complete using the words below:

bacteria carbon dead decomposers digest microorganisms nutrients waste water

.... are a group of .... that includes fungi and .... They feed on .... droppings and .... organisms. They .... them and use some of the .... They also release waste products which include .... dioxide and ...., which plants can use.

2 Explain why the processes of decay are so important in keeping the soil fertile.

B1 5.4 The carbon cycle

Learning objectives

- What is the carbon cycle in nature?

- Which processes remove carbon dioxide from the atmosphere – and which processes return it?

Figure 1 Within the natural cycle of life and death in the living world, mineral nutrients are cycled between living organisms and the physical environment

Did you know ...?

Every year about 166 gigatonnes of carbon are cycled through the living world. That's 166 000 000 000 tonnes – an awful lot of carbon!

Imagine a stable community of plants and animals. The processes that remove materials from the environment are balanced by processes that return materials. Materials are constantly cycled through the environment. One of the most important of these is carbon.

All of the main **molecules** that make up our bodies (carbohydrates, proteins, fats and DNA) are based on carbon atoms combined with other **elements**.

The amount of carbon on the Earth is fixed. Some of the carbon is 'locked up' in **fossil fuels** like coal, oil and gas. It is only released when we burn them.

Huge amounts of carbon are combined with other elements in carbonate rocks like limestone and chalk. There is a pool of carbon in the form of carbon dioxide in the air. It is also found dissolved in the water of rivers, lakes and oceans. All the time a relatively small amount of available carbon is cycled between living things and the environment. We call this the **carbon cycle**.

> **a** What are the main sources of carbon on Earth?

Photosynthesis

Green plants and algae remove carbon dioxide from the atmosphere for photosynthesis. They use the carbon from carbon dioxide to make carbohydrates, proteins and fats. These make up biomass of the plants and algae. The carbon is passed on to animals that eat the plants. The carbon goes on to become part of the carbohydrates, proteins and fats in these animal bodies.

This is how carbon is taken out of the environment. But how is it returned?

> **b** What effect does photosynthesis have on the distribution of carbon levels in the environment?

Respiration

Living organisms respire all the time. They use oxygen to break down glucose, providing energy for their cells. Carbon dioxide is produced as a waste product. This is how carbon is returned to the atmosphere.

When plants, algae and animals die their bodies are broken down by decomposers. These are animals and microorganisms such as blowflies, moulds and bacteria that feed on the dead bodies. The animals which feed on dead bodies and waste are called *detritus feeders*. They include animals such as worms, centipedes and many insects.

Carbon is released into the atmosphere as carbon dioxide when these organisms respire. All of the carbon (in the form of carbon dioxide) released by the various living organisms is then available again. It is ready to be taken up by plants and algae in photosynthesis.

Combustion

Fossil fuels contain carbon, which was locked away by photosynthesising organisms millions of years ago. When we burn fossil fuels, carbon dioxide is produced, so we release some of that carbon back into the atmosphere:

Photosynthesis: carbon dioxide + water (+ light energy) → glucose+ oxygen

Respiration: glucose + oxygen → carbon dioxide + water (+ energy)

Combustion: fossil fuel or wood + oxygen → carbon dioxide + water
(+ energy)

The constant cycling of carbon is summarised in Figure 2.

Study tip

Make sure you can label the processes in a diagram of the carbon cycle.

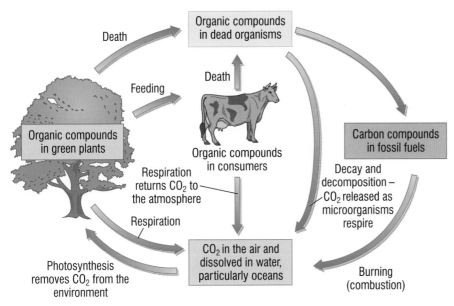

Figure 2 The carbon cycle in nature

Energy transfers

When plants and algae photosynthesise, they transfer light energy into chemical energy can be the food that they make. This chemical energy is transferred from one organism to another through the carbon cycle. Some of the energy can be used for movement or transferred as energy to the organisms and its surroundings at each stage. The decomposers break down all the waste and dead organisms and cycle the materials as plant nutrients. By this time all of the energy originally absorbed by green plants and algae during photosynthesis has been transferred elsewhere.

For millions of years the carbon cycle has regulated itself. However, as we burn more fossil fuels we are pouring increasing amounts of carbon dioxide into the atmosphere. Scientists fear that the carbon cycle may not cope. If the levels of carbon dioxide in our atmosphere increase it may lead to global warming.

Figure 3 Fossil fuels such as coal contain large amounts of carbon

Summary questions

1 a What is the carbon cycle?
  b What are the main processes involved in the carbon cycle?
  c Why is the carbon cycle so important for life on Earth?

2 a Where does the carbon come from that is used in photosynthesis?
  b Explain carefully how carbon is transferred through an ecosystem.

Key points

- The constant cycling of carbon in nature is known as the carbon cycle.

- Carbon dioxide is removed from the atmosphere by photosynthesis. It is returned to the atmosphere through respiration and combustion.

B1 5.5   Recycling organic waste ⓚ

Learning objectives

- Why should we recycle organic kitchen and garden waste?
- How can we investigate the most effective way to recycle this organic waste?

Figure 1 Some landfill sites now collect the methane that is produced as organic material decays and use it to generate electricity. But if everyone recycled their own organic waste, we would need far fewer landfill sites and there would be no problem.

 Did you know ... ?

One tonne of organic kitchen and garden waste produces 200 to 400 m³ of gas. Around 27% of the methane produced in the UK each year comes from landfill sites.

Activity

Plan an assembly to be used with students in Years 7–9 suggesting that the school introduces a scheme to recycle all the organic waste from the kitchens and the school grounds to make compost. The compost could then be sold to the local community for charity. Remember, you need to explain why and how this should be done as well as recruit volunteers to help run the compost bins.

The problem of waste

People produce lots of waste – and getting rid of it is a big problem. Whenever we prepare food we produce **organic waste** to throw away, such as vegetable peelings. Gardening produces lots of organic waste too, including the grass cuttings when we mow the lawn. We put about 100 million tonnes of waste a year into landfill sites and about two thirds of that is organic matter. By recycling our organic waste we can reduce this mountain of waste material.

The kitchen and garden waste we put into landfill sites doesn't rot easily in the conditions there. It forms a smelly liquid which soaks into the ground and can pollute local rivers and streams. In these conditions the microorganisms that break down the plant and animal material produce mainly methane gas. This is a **greenhouse gas** that adds to the problem of global warming.

a Give two examples of the organic waste you might put into a compost bin.

The simplest way to recycle kitchen and garden waste is to make compost. Natural decomposing organisms break down all the plant material to make a brown, crumbly substance. This compost is full of the nutrients that have been released by the decomposers. The process takes from a few months to over a year. The compost forms a really good, natural fertiliser. It also greatly reduces the amount of rubbish you need to send to the landfill site.

b Which greenhouse gas, other than carbon dioxide, is given off as organic material decays in landfill sites?

Making compost

Composting can be done on a small scale or on a large scale. There are several different factors which are important in making successful compost:

- Compost can be made with or without oxygen – mixing your compost regularly helps air get in. If the microorganisms have oxygen they generate energy, which kills off weed seeds and speeds up the process. Without oxygen the process releases little energy and is slower.
- The warmer the compost mixture, the faster the compost will be made (up to about 70°C, at which point the microorganisms stop working properly).
- The decay process is faster in moist conditions than in dry ones. (In fact, decay does not take place at all in perfectly dry conditions.)

Practical

Investigating the decay of organic matter

We have seen that the presence of oxygen and moisture, as well as the temperature, affect the rate of decay. Choose one of these factors to investigate. Carry out any tests on the sort of materials that might go into a garden compost bin.

- Plan to find out what effect your chosen factor has on the rate at which the material decays.
- Pool the conclusions of each group to decide on the ideal conditions for composting organic waste.
- Comment on the limitations of the conclusions you can draw.

A Compost heap: The simplest and cheapest method. Kitchen and garden waste is put in a pile, with new material added to the top, and left to rot down.

C Council composting: Local councils may collect garden or kitchen waste and use large-scale bins to recycle the material to make compost. They may shred the material before adding it to the bins to increase the surface area. You can buy the compost from the schemes to put on your garden.

B Compost bin: Bins are often made of plastic and may be sold cheaply by local councils to encourage people to recycle their organic waste. Instructions include watering the bin in dry weather and mixing the contents from time to time.

D Black bag composting: A black plastic bag is filled with kitchen and garden waste and sealed. The microorganisms work slowly as they have little or no oxygen, but in about a year the contents will have decomposed and formed compost.

Figure 2 Different methods of composting

Summary questions

1. Why is it important to recycle organic kitchen and garden waste?

2. Evaluate each of the four methods of making compost shown in Figure 2, giving advantages and disadvantages of each.

3. How do mixing the compost regularly, adding a variety of different types of organic waste and watering in dry weather improve the composting process?

Key points

- Recycling organic kitchen and garden waste is necessary to reduce landfill, reduce the production of methane and to recycle the minerals and nutrients in the organic material.

- Composting organic waste can be done in a variety of different ways.

Summary questions (k)

1

Biomass measured in g dry biomass/m²

Top carnivore	25
Secondary consumer	200
Primary consumer	2500
Producer	25 000

a From this diagram, calculate the percentage biomass passed on:

  i from producers to primary consumers

  ii from primary to secondary consumers

  iii from secondary consumers to top carnivores.

b In any food chain or food web the biomass of the producers is much larger than that of any other level of the pyramid. Why is this?

c In any food chain or food web there are only small numbers of top carnivores. Use your calculations to help you explain why.

d All of the animals in the pyramid of biomass shown here are cold blooded. What difference would it have made to the average percentage of biomass passed on between the levels if mammals and birds had been involved? Explain the difference.

2 The world population is increasing and there are food shortages in many parts of the world. Explain, using pyramids of biomass to help you, why it would make a better use of resources if people everywhere ate much less meat and more plant material.

3 Chickens for us to eat are often farmed intensively to provide meat as cheaply as possible. The birds arrive in the broiler house as 1-day-old chicks. They are slaughtered at 42 days of age when they weigh about 2 kg. The temperature, amount of food and water and light levels are carefully controlled. About 20 000 chickens are reared together in one house. The table below shows their weight gain.

Age (days)	1	7	14	21	28	35	42
Mass (g)	36	141	404	795	1180	1657	1998

a Plot a graph to show the growth rate of one of these chickens.

b Explain why the temperature is so carefully controlled in the broiler house.

c Explain why so many birds are reared together in a relatively small area.

d Why are birds for eating reared like this?

4 Microorganisms decompose organic waste and dead bodies. We preserve food to stop this decomposition taking place. Use your knowledge of decomposition to explain how each method stops the food going bad:

a Food may be frozen.

b Food may be cooked – cooked food keeps longer than fresh food.

c Food may be stored in a vacuum pack – with all the air sucked out.

d Food may be tinned – it is heated and sealed in an airtight container.

5

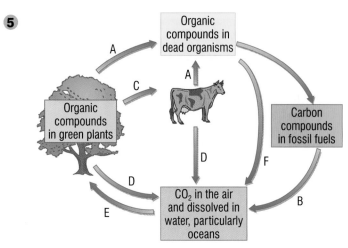

a How is carbon dioxide removed from the atmosphere in the carbon cycle?

b How does carbon dioxide get into the atmosphere?

c Where is most of the carbon stored?

d Why is the carbon cycle so important and what could happen if the balance of the reactions was disturbed?

6 a The temperature in the middle of a compost heap will be quite warm. Heat is produced as microbes respire. How does this help the compost to be broken down more quickly?

b In sewage works oxygen is bubbled through the tanks containing sewage and microorganisms. How does this help make sure the human waste is broken down completely?

Practice questions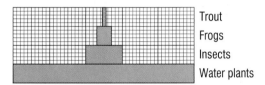

1 Rabbits eat very large amounts of grass. A single hawk eats a few rabbits.

  a Draw a pyramid of biomass for the rabbits, grass and the hawk. (2)

  b Much of the energy from the grass is not transferred to the hawk.
Suggest **two** reasons why. (2)

2 Choose words from below to complete each sentence.

*carbon dioxide cool dry insects microorganisms
moist nitrogen oxygen rats warm*

  a Plant waste in a compost heap is decayed by .... (1)

  b The plant waste decays faster in conditions which are .... and .... (2)

  c The plant waste will also decay faster when the air contains plenty of .... (1)

3 The diagram shows what happens to the energy in the food a calf eats.

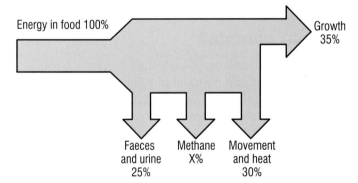

In the calculations show clearly how you work out your answer.

  a Calculate the percentage of energy lost in methane (X). (2)

  b The energy in the food the calf eats in one day is 10 megajoules.
Calculate the amount of this energy that would be lost in faeces and urine. (2)

  c Name the process which transfers the energy from the food into movement. (1)

  d The farmer decides to move his calf indoors so that it will grow quicker.
Suggest **two** reasons why. (2)

4 *In this question you will be assessed on using good English, organising information clearly and using specialist terms where appropriate.*

The constant cycling of carbon in nature is called 'The carbon cycle'.

Each autumn, trees lose their leaves.

Describe how the carbon in the leaves is recycled so that the trees can use it again. (6)

5 The diagram shows a pyramid of biomass drawn to scale.

Trout
Frogs
Insects
Water plants

  a What is the source of energy for the water plants? (1)

  b The ratio of the biomass of water plants to the biomass of insects is 5 : 1.
Calculate the ratio of the biomass of insects to the biomass of frogs.
Show clearly how you work out your answer. (2)

  c Give **two** reasons why the biomass of the frog population is smaller than the biomass of the insect population. (2)

  d Some insects die.
Describe how the carbon in the dead insect bodies may be recycled. (4)

AQA, 2006

B1 6.1

Inheritance

Learning objectives

- How do parents pass on genetic information to their offspring?

- In which part of a cell is the genetic information found?

Young animals and plants resemble their parents. Horses have foals and people have babies. Chestnut trees produce conkers that grow into little chestnut trees. Many of the smallest organisms that live in the world around us are actually identical to their parents. So what makes us the way we are?

Figure 1 This mother cat and her kittens are not identical, but they are obviously related

Why do we resemble our parents?

Most families have characteristics that we can see clearly from generation to generation. People like to comment when one member of a family looks very much like another. Characteristics like nose shape, eye colour and dimples are inherited. They are passed on to you from your parents.

Your resemblance to your parents is the result of information carried by **genes**. These are passed on to you in the sex cells (**gametes**) from which you developed. This genetic information determines what you will be like.

a Why do you look like your parents?

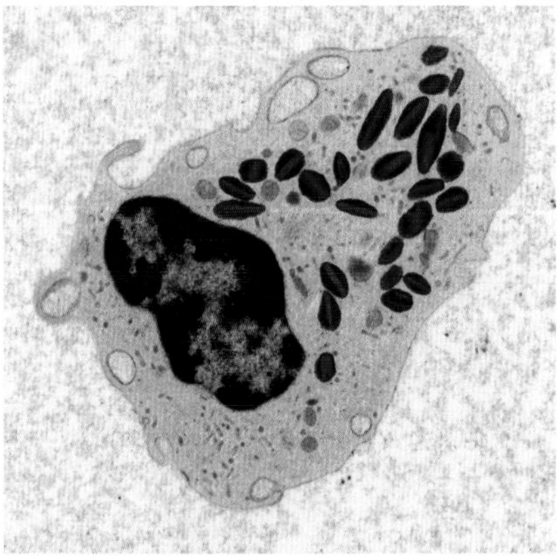

Chromosomes and genes

The genetic information is carried in the nucleus of your cells. It is passed from generation to generation during reproduction. The nucleus contains all the plans for making and organising a new cell. What's more, the nucleus contains the plans for a whole new you!

b In which part of a cell is the genetic information found?

Inside the nucleus of all your cells there are thread-like structures called **chromosomes**. The chromosomes are made up of a special chemical called **DNA** (deoxyribonucleic acid). This is where the genetic information is actually stored.

DNA is a long molecule made up of two strands that are twisted together to make a spiral. This is known as a double helix – imagine a ladder that has been twisted round.

Study tip

Make sure you know the difference between chromosomes, genes and DNA.

Figure 2 The nucleus of each of your cells contains your chromosomes. The chromosomes carry the genes, which control the characteristics of your whole body.

Each different type of organism has a different number of chromosomes in their body cells. Humans have 46 chromosomes while turkeys have 82. You inherit half your chromosomes from your mother and half from your father, so chromosomes come in pairs. You have 23 pairs of chromosomes in all your normal body cells.

Each of your chromosomes contains thousands of genes joined together. These are the units of inheritance.

Each gene is a small section of the long DNA molecule. Genes control what an organism is like. They determine its size, its shape and its colour. Genes work at the level of the molecules in your body to control the development of all the different characteristics you can see. They do this by controlling all the different enzymes and other proteins made in your body.

Your chromosomes are organised so that both of the chromosomes in a pair carry genes controlling the same things. This means your genes also come in pairs – one from your father and one from your mother.

c Where would you find your genes?

Some of your characteristics are decided by a single pair of genes. For example, there is one pair of genes which decides whether or not you will have dimples when you smile. However, most of your characteristics are the result of several different genes working together. For example, your hair and eye colour are both the result of several different genes.

Did you know that scientists are still not sure exactly how many genes we have? At the moment they think it is between 20 000 to 25 000.

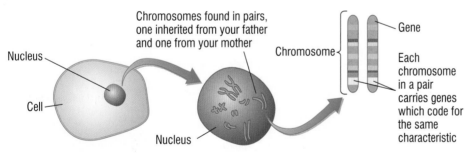

Figure 3 The nucleus of your cell contains the chromosomes that carry the genes which control the characteristics of your whole body

Summary questions

1 Copy and complete using the words below:

chromosomes   genes   genetic   gametes   nucleus

Offspring look like their parents because of .... information passed on to them in the .... (sex cells) from which they developed. The information is contained in the ...., which are found in the .... of the cell. The information is carried by the ....

2 **a** What is the basic unit of inheritance?
  b Offspring inherit information from their parents, but do not look exactly like them. Why not?

3 **a** Why do chromosomes come in pairs?
  b Why do genes come in pairs?
  c How many genes do scientists think humans have?

Key points

- Parents pass on genetic information to their offspring in the sex cells (gametes).

- The genetic information is found in the nucleus of your cells. The nucleus contains chromosomes, and chromosomes carry the genes that control the characteristics of your body.

- Different genes control the development of different characteristics.

B1 6.2

Types of reproduction

Learning objectives

- What is a clone?
- Why does asexual reproduction result in offspring that are identical to their parents?
- How does sexual reproduction produce variety?

Reproduction is very important to living things. It is during reproduction that genetic information is passed on from parents to their offspring. There are two very different ways of reproducing – **asexual reproduction** and **sexual reproduction**.

Asexual reproduction

Asexual reproduction only involves one parent. There is no joining of special sex cells and there is no variety in the offspring.

Asexual reproduction gives rise to identical offspring known as **clones**. Their genetic material is identical both to the parent and to each other.

> **a** Why is there no variety in offspring from asexual reproduction?

Asexual reproduction is very common in the smallest animals and plants and in bacteria. However, many bigger plants like daffodils, strawberries and brambles do it too. The cells of your body reproduce asexually all the time. They divide into two identical cells for growth and to replace worn-out tissues.

Figure 1 A mass of daffodils like this can contain hundreds of identical flowers. This is because they come from bulbs that reproduce asexually. They also reproduce sexually using their flowers.

Sexual reproduction

Sexual reproduction involves a male sex cell and a female sex cell from two parents. These two special sex cells (gametes) join together to form a new individual.

The offspring that result from sexual reproduction inherit genetic information from both parents. This means you will have some characteristics from both of your parents, but won't be exactly like either of them. This introduces variety. The offspring of sexual reproduction show much more variation than the offspring from asexual reproduction. In plants the gametes involved in sexual reproduction are found within ovules and pollen. In animals they are called ova (eggs) and sperm.

Sexual reproduction is risky because it relies on the sex cells from two individuals meeting but it also introduces variety. That's why we find sexual reproduction in organisms ranging from bacteria to people.

> **b** How does sexual reproduction cause variety in the offspring?

Study tip

- asexual reproduction – one parent → clones
- sexual reproduction – two parents → variety

Variation

Why is sexual reproduction so important? The variety it produces is a great advantage in making sure a species survives. Variety makes it more likely that at least a few of the offspring will have the ability to survive difficult conditions.

If you take a closer look at how sexual reproduction works, you can see how variation appears in the offspring.

Different genes control the development of different characteristics about you. Most things about you, such as your hair and eye colour, are controlled by several different pairs of genes. A few of your characteristics are controlled by one single pair of genes. For example, there are genes that decide whether:

- your earlobes are attached closely to the side of your head or hang freely
- your thumb is straight or curved
- you have dimples when you smile
- you have hair on the second segment of your ring finger.

We can use these genes to help us understand how inheritance works.

Figure 2 Although these young people have some family likenesses, the variety caused by the mixing of their parents' genetic information is clear

c Why is variety important?

Curved thumb

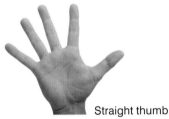
Straight thumb

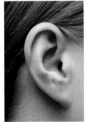
Attached ear lobe

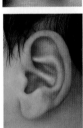
Unattached ear lobe

Dimples

No dimples

Figure 3 These are all human characteristics that are controlled by a single pair of genes. They can help us to understand how sexual reproduction introduces variety and how inheritance works.

You will get a random mixture of genetic information from your parents, which is why you don't look exactly like either of them!

Summary questions

1 Define the following words:
   a asexual reproduction
   b sexual reproduction
   c gamete
   d variation.

2 Compare the advantages and disadvantages of sexual reproduction with asexual reproduction.

3 A daffodil reproduces asexually using bulbs and sexually using flowers.
   a How does this help to make them very successful plants?
   b Explain the genetic differences between a daffodils's sexually and asexually produced offspring.

Key points

- In asexual reproduction there is no joining of gametes and only one parent. There is no genetic variety in the offspring.

- The genetically identical offspring of asexual reproduction are known as clones.

- In sexual reproduction male and female gametes join. The mixture of genetic information from two parents leads to genetic variety in the offspring.

Genetic and environmental differences

Figure 1 However much this Falabella eats, it will never be as tall as the Shire. It just isn't in the genes.

Have a look at the ends of your fingers and notice the pattern of your fingerprints. No one else in the world will have exactly the same fingerprints as you. Even identical twins have different fingerprints. What factors make you so different from other people?

Nature – genetic variety

The genes you inherit determine a lot about you. An apple tree seed will never grow into an oak tree. Environmental factors, such as the weather or soil conditions do not matter. The basic characteristics of every species are determined by the genes they inherit.

Certain human characteristics are clearly inherited. Features such as eye colour, the shape of your nose and earlobes, your sex and dimples are the result of genetic information inherited from your parents. But your genes are only part of the story.

a Where do the genes you inherit come from?

Nurture – environmental variety

Some differences between you and other people are completely due to the environment you live in. For example, if a woman drinks heavily when she is pregnant, her baby may be very small when it is born and have learning difficulties. These characteristics are a direct result of the alcohol the fetus has to deal with as it develops. You may have a scar as a result of an accident or an operation. These characteristics are all environmental, not genetic.

Genes certainly play a major part in deciding how an organism will look. However, the conditions in which it develops are important too. Genetically identical plants can be grown under different conditions of light or soil nutrients. The resulting plants do not look identical. Plants deprived of light, carbon dioxide or nutrients do not make as much food as plants with plenty of everything. The deprived plants will be smaller and weaker. They have not been able to fulfil their 'genetic potential'.

b Why are genetically identical plants so useful for showing the effect of the environment on appearance?

Combined causes of variety

Many of the differences between individuals of the same species are the result of both their genes and the environment. For example, you inherit your hair colour and skin colour from your parents. However, whatever your inherited skin colour, it will be darker if you live in a sunny environment. If your hair is brown or blonde, it will be lighter if you live in a sunny country.

Your height and weight are also affected by both your genes and the conditions in which you grow up. You may have a genetic tendency to be overweight. However, if you never have enough to eat you will be underweight.

Figure 2 The differences in these cows are partly genetic and partly down to their environment, from the milk they drank as calves to the quality of the grass they eat each day

Investigating variety

It is quite easy to produce genetically identical plants to investigate variety. You can then put them in different situations to see how the environment affects their appearance. Scientists also use groups of animals that are genetically very similar to investigate variety. You cannot easily do this in a school laboratory.

The only genetically identical humans are identical twins who come from the same fertilised egg. Scientists are very interested in identical twins, to find out how similar they are as adults.

It would be unethical to take identical twins away from their parents and have them brought up differently just to investigate environmental effect. But there are cases of identical twins who have been adopted by different families. Some scientists have researched these separated identical twins.

Often identical twins look and act in a remarkably similar way. Scientists measure features such as height, weight and IQ (a measure of intelligence). The evidence shows that human beings are just like other organisms. Some of the differences between us are mainly due to genetics and some are largely due to our environment.

In one study, scientists compared four groups of adults:
- separated identical twins
- identical twins brought up together
- non-identical, same sex twins brought up together
- same sex, non-twin siblings brought up together.

The differences between the pairs were measured. A small difference means the individuals in a pair are very alike. If there was a big difference between the identical twins the scientists could see that their environment had more effect than their genes.

links

For more information on producing genetically identical plants, see B1 6.4 Cloning.

Figure 3 Whether identical twins are brought up together or apart, they are often very similar as adults

Table 1 Differences in pairs of adults

Measured difference in:	Identical twins brought up together	Identical twins brought up apart	Non-identical twins	Non-twin siblings
height (cm)	1.7	1.8	4.4	4.5
mass (kg)	1.9	4.5	4.6	4.7
IQ	5.9	8.2	9.9	9.8

Summary questions

1 Copy and complete using the words below.

   combination   identical   developed   genes

   Everybody is different, even .... twins. Some of the differences are caused by our .... Some differences are caused by the conditions in which we have .... Many differences are caused by a .... of both.

2 **a** Using the data from Table 1, explain which human characteristic appears to be mostly controlled by genes and which appears to be most affected by the environment.

   b Why do you think non-twin siblings reared together were included in the study as well as twins reared together and apart?

3 You are given 20 pots containing identical cloned seedlings, all the same height and colour. Explain how you would investigate the effect of temperature on the growth of these seedlings compared to the impact of their genes.

Study tip

- Genes control the development of characteristics.
- Characteristics may be changed by the environment.

Key points

- The different characteristics between individuals of a family or species may be due to genetic causes, environmental causes or a combination of both.

B1 6.4

Cloning

Learning objectives

- How do we clone plants?
- How do we clone animals?
- Why do we want to create clones?

A clone is an individual that has been produced asexually and is genetically identical to the parent. Many plants reproduce naturally by cloning and this has been used by farmers and gardeners for many years.

Cloning plants

Gardeners can produce new plants by taking cuttings from older plants. How do you take a cutting? First you remove a small piece of a plant. This is often part of the stem or sometimes just part of the leaf. If you keep it in the right conditions, new roots and shoots will form. It will grow to give you a small, complete new plant.

Using this method you can produce new plants quickly and cheaply from old plants. The cuttings will be genetically identical to the parent plants.

∞ links

For information on taking plant cuttings, look back at B1 2.6 Hormones and the control of plant growth.

Many growers now use hormone rooting powders to encourage cuttings to grow. Cuttings are most likely to develop successfully if you keep them in a moist atmosphere until their roots develop. We produce plants such as orchids and many conifer trees commercially by cloning in this way.

a Why does a cutting look the same as its parent plant?

Cloning tissue

Taking cuttings is a form of artificial asexual reproduction. It has been carried out for hundreds of years. In recent years scientists have come up with a more modern way of cloning plants called **tissue culture**. It is more expensive but it allows you to make thousands of new plants from one tiny piece of plant tissue.

The first step is to use a mixture of plant hormones to make a small group of cells from the plant you want to clone produce a big mass of identical plant cells.

Figure 1 Simple cloning by taking cuttings is a technique used by gardeners and nurserymen all around the world.

Then, using a different mixture of hormones and conditions, you can stimulate each of these cells to form a tiny new plant. This type of cloning guarantees that you can produce thousands of offspring with the characteristics you want from one individual plant.

b What is the advantage of tissue culture over taking cuttings?

Cloning animals

In recent years cloning animals has become quite common in farming, particularly transplanting cloned cattle embryos. Cows normally produce only one or two calves at a time. If you use embryo cloning, your best cows can produce many more top-quality calves each year.

Figure 2 Tissue culture makes it possible to produce thousands of identical plants quickly and easily

How does embryo cloning work? You give a top-quality cow fertility hormones so that it produces a lot of eggs. You fertilise these eggs using sperm from a really good bull. Often this is done inside the cow and the embryos that are produced are then gently washed out of her womb. Sometimes the eggs are collected and you add sperm in a laboratory to produce the embryos.

At this very early stage of development every cell of the embryo can still form all of the cells needed for a new cow. They have not become specialised.

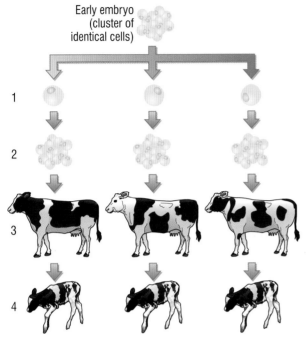

Figure 3 Cloning cattle embryos

1 Divide each embryo into several individual cells.

2 Each cell grows into an identical embryo in the lab.

3 Transfer embryos into their host mothers, which have been given hormones to get them ready for pregnancy.

4 Identical cloned calves are born. They are not biologically related to their mothers.

Cloning cattle embryos and transferring them to host cattle is skilled and expensive work. It is worth it because using normal reproduction, a top cow might produce 8–10 calves during her working life. Using embryo cloning she can produce more calves than that in a single year.

Cloning embryos means we can transport high-quality embryos all around the world. They can be carried to places where cattle with a high milk yield or lots of meat are badly needed for breeding with poor local stock. Embryo cloning is also used to make lots of identical copies of embryos that have been **genetically modified** to produce medically useful compounds.

Study tip

- Remember clones have identical genetic information.
- Make sure you are clear about the difference between a tissue and an embryo.

∞ links

For more information on cloning embryos, see B1 6.5 Adult cell cloning.

Summary questions

1 Define the following words:
   a cuttings
   b tissue cloning
   c asexual reproduction
   d embryo cloning.

2 Make a table to compare the similarities and differences between tissue cloning and taking cuttings.

3 **a** Cloning cattle embryos is very useful. Why?
   b Draw a flow chart to show the stages in the embryo cloning of cattle.
   c Suggest some of the economic and ethical issues raised by embryo cloning in cattle.

Key points

- New plant clones can be produced quickly and cheaply by taking cuttings from mature plants. The new plants are genetically identical to the older ones.

- A modern technique for cloning plants is tissue culture using cells from a small part of the original plant.

- Transplanting cloned embryos is one way in which animals are cloned.

B1 6.5   Adult cell cloning

Learning objectives

- How did scientists clone a sheep?
- What are the steps in the techniques of adult cell cloning?

True cloning of animals, without sexual reproduction involved at all, has been a major scientific breakthrough. It is the most complicated form of asexual reproduction you can find.

Adult cell cloning

To clone a cell from an adult animal is easy. The cells of your body reproduce asexually all the time to produce millions of identical cells. However, to take a cell from an adult animal and make an embryo or even a complete identical animal is a very different thing.

When a new whole animal is produced from the cell of another adult animal, it is known as **adult cell cloning**. This is still relatively rare. You place the nucleus of one cell into the empty egg cell of another animal of the same species. Then you place the resulting embryo into the uterus of another adult female where it develops until it is born.

Here are the steps involved:

- The nucleus is removed from an unfertilised egg cell.
- At the same time the nucleus is taken from an adult body cell, e.g. a skin cell of another animal of the same species.
- The nucleus from the adult cell is inserted (placed) in the empty egg cell.
- The new cell is given a tiny electric shock that makes it start dividing to form embryo cells. These contain the same genetic information as the original adult cell and the original adult animal.
- When the embryo has developed into a ball of cells it is inserted into the womb of an adult female to continue its development.

Adult cell cloning has been used to produce a number of whole animal clones. The first large mammal ever to be cloned from the cell of another adult animal was Dolly the sheep, born in 1997.

Figure 1 Dolly the sheep was the first large mammal to be cloned from another adult mammal. She went on to have lambs of her own in the normal way.

> **a** What is the name of the technique that produced Dolly the sheep?

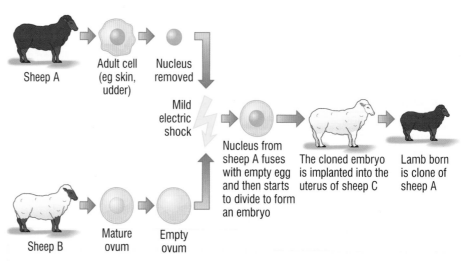

Sheep A — Adult cell (eg skin, udder) — Nucleus removed

Mild electric shock

Sheep B — Mature ovum — Empty ovum

Nucleus from sheep A fuses with empty egg and then starts to divide to form an embryo

The cloned embryo is implanted into the uterus of sheep C

Lamb born is clone of sheep A

Figure 2 Adult cell cloning is still a very difficult technique – but scientists hope it may bring benefits in the future

Study tip

Animals can be cloned by using embryo transplants or by adult cell cloning.

When Dolly was produced she was the only success from hundreds of attempts. The technique is still difficult and unreliable, but scientists hope that it will become easier in future.

links

For more information on adult cell cloning, see B1 6.7 Making choices about technology.

How Science Works

The benefits and disadvantages of adult cell cloning

One big hope for adult cell cloning is that animals that have been genetically engineered to produce useful proteins in their milk can be cloned. This would give us a good way of producing large numbers of cloned, medically useful animals.

This technique could also be used to help save animals from extinction, or even bring back species of animals that died out years ago. The technique could be used to clone pets or prized animals so that they continue even after the original has died. However, some people are not happy about this idea.

There are some disadvantages to this exciting science as well. Many people fear that the technique could lead to the cloning of human babies. This could be used to help infertile couples, but it could also be abused. At the moment this is not possible, but who knows what might be possible in the future?

Another problem is that modern cloning techniques produce lots of plants or animals with identical genes. In other words, cloning reduces variety in a population. This means the population is less able to survive any changes in the environment that might happen in the future. That's because if one of them does not contain a useful characteristic, none of them will.

In a more natural population, at least one or two individuals can usually survive change. They go on to reproduce and restock. This could be a problem in the future for cloned crop plants or for cloned farm animals.

b How might adult cell cloning be used to help people?

Did you know …?

The only human clones alive at the moment are natural ones known as identical twins! But the ability to clone mammals such as Dolly the sheep has led to fears that some people may want to have a clone of themselves produced – whatever the cost.

Key points

- Scientists cloned Dolly the sheep using adult cell cloning.

- In adult cell cloning the nucleus of a cell from an adult animal is transferred to an empty egg cell from another animal. A small electric shock causes the egg cell to begin to divide and starts embryo development. The embryo is placed in the womb of a third animal to develop. The animal that is born is genetically identical to the animal that donated the original adult cell.

Summary questions

1 Copy and complete using the words below:

mammal adult technique genetic Dolly

In .... cell cloning an animal is produced that is an exact .... copy of another adult animal. .... the sheep was the first large .... to be produced using this modern cloning ....

2 Produce a flow chart to show how adult cell cloning works.

3 What are the main advantages and disadvantages of the development of adult cell cloning techniques?

B1 6.6 Genetic engineering

- What is genetic engineering?
- How are genes transferred from one organism to another?
- What are the issues involved in genetic engineering?

What is genetic engineering?

Genetic engineering involves changing the genetic material of an organism. You take a gene from one organism and transfer it to the genetic material of a completely different organism. So, for example, genes from the chromosomes of a human cell can be 'cut out' using enzymes and transferred to the cell of a bacterium. The gene carries on making a human protein, even though it is now in a bacterium.

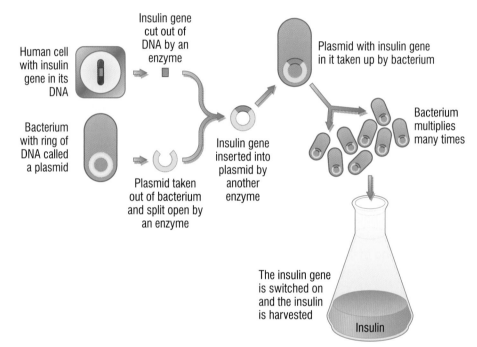

Figure 1 The principles of genetic engineering. A bacterial cell receives a gene from a human being so it makes the human hormone insulin.

a How is a gene taken out of one organism to be put into another?

If genetically engineered bacteria are cultured on a large scale they will make huge quantities of protein from other organisms. We now use them to make a number of drugs and hormones used as medicines.

Transferring genes to animal and plant cells

There is a limit to the types of proteins that bacteria are capable of making. As a result, genetic engineering has moved on. Scientists have found that genes from one organism can be transferred to the cells of another type of animal or plant at an early stage of their development. As the animal or plant grows it develops with the new desired characteristics from the other organism. For example, glowing genes from jellyfish have been used to produce crop plants which give off a blue light when they are attacked by insects. Then the farmer knows when they need spraying.

b Why are genes inserted into animals and plants as well as into bacteria?

The benefits of genetic engineering

Genetically engineered bacteria can make exactly the proteins we need, in exactly the amounts needed and in a very pure form. For example, people with diabetes need supplies of the hormone insulin. In the past people used animal insulin extracted from the pancreases of pigs and cattle. Now they can use pure human insulin produced by genetically engineered bacteria (see Figure 1).

We can use engineered genes to improve the growth rates of plants and animals. They can be used to improve the food value of crops as genetically modified (GM) crops usually have much bigger yields than ordinary crops. They can also be designed to grow well in dry, hot or cold parts of the world so could help to solve the problems of world hunger. Crops can be engineered to produce plants which make their own pesticide or are resistant to herbicides used to control weeds.

Human engineering

If there is a mistake in your genetic material, you may have a genetic disease. These can be very serious. Many people hope that genetic engineering can solve the problem.

It might become possible to put 'healthy' genes into the affected cells by genetic engineering, so they work properly. Perhaps the cells of an early embryo can be engineered so that the individual develops into a healthy person. If these treatments become possible, many people would have new hope of a normal life for themselves or their children.

c What do we mean by a 'genetic disease'?

The disadvantages of genetic engineering

Genetic engineering is still a very new science. No one knows what all of the long-term effects might be. For example, insects may become pesticide-resistant if they eat a constant diet of pesticide-forming plants.

Some people are concerned about the effect of eating GM food on human health. Genes from genetically modified plants and animals might spread into the wildlife of the countryside. GM crops are often made infertile, which means farmers in poor countries have to buy new seed each year.

People might want to manipulate the genes of their future children. This may be to make sure they are born healthy, but there are concerns that people might want to use it to have 'designer' children with particular characteristics such as high intelligence. Genetic engineering raises issues for us all to think about.

Figure 2 You can't tell that food is genetically modified just by looking at it! In the UK, few GM foods are sold and they have to be clearly labelled. Many other countries, including the USA, are less worried and use GM food widely.

Key points

- Genes can be transferred to the cells of animals and plants at an early stage of their development so they develop desired characteristics. This is genetic engineering.

- In genetic engineering, genes from the chromosomes of humans and other organisms can be 'cut out' using enzymes and transferred to the cells of bacteria and other organisms.

- There are advantages and disadvantages associated with genetic engineering.

Summary questions

1 Copy and complete using the words below:

cell engineering enzymes gene genetic transfer

Genetic .... involves changing the .... material of an organism. You cut a .... from one organism using .... and .... it to the .... of a completely different organism.

2 a Make a flow chart that explains the stages of genetic engineering.

b Make two lists, one to show the possible advantages of genetic engineering and the other to show the possible disadvantages.

B1 6.7 Making choices about technology

- What sort of economic, social and ethical issues are there about new techniques such as cloning and genetic engineering?

Cloning pets

Cc, or Copycat, was the first cloned cat to be produced. Most of the research into cloning had been focused on farm and research animals – but cats are thought of first and foremost as pets.

Much of the funding for cat cloning in the US comes from companies who are hoping to be able to clone people's dying or dead pets for them. It has already been shown that a successful clone can be produced from a dead animal. Cells from beef from a slaughter house were used to create a live cloned calf.

It took one hundred and eighty-eight attempts to make Cc, producing 87 cloned embryos, only one of which resulted in a kitten. Cloning your pet won't be easy or cheap. The issue is, should people be cloning their dead cats, or would it be better to give a home to one of the thousands of unwanted cats already in existence? Even if a favourite pet cat is cloned, it may look nothing like the original because the coat colour of many cats is the result of genes switching on and off at random in the skin cells. The clone will develop and grow in a different environment to the original cat as well. This means other characteristics that are affected by the environment will probably be different too.

Figure 1 The cat on the left is Rainbow. The cat on the right is Cc, Rainbow's clone. Rainbow and Cc share the same DNA – but they don't look the same.

To some people these are exciting events. To others they are a waste of time, money and the lives of all the embryos that don't make it. What do you think?

??? Did you know … ?

Dogs have also been cloned. In 2009, an American couple paid more than £100 000 to have a clone of their much-loved pet Labrador. The new dog is called Lancelot encore (encore means 'again').

Figure 2 Lancelot encore, a clone of a much-loved pet, and a portrait of the original dog

Activity

In B1 6.4 and B1 6.5 there is information about cloning animals and plants for farming. Here you have two different stories about cloning animals for money (Cc and Lancelot encore).

There is talk of a local company setting up a laboratory to clone cats, dogs and horses for anyone in the country who wants to do this.

Write a letter or post a blog either *for* the application or *against* it. Make sure you use clear, sensible arguments and put the science of the situation across clearly.

The debate about GM foods

Ever since genetically modified foods were first introduced there has been controversy and discussion about them. For example, varieties of GM rice known as 'golden rice' and 'golden rice 2' have been developed. These varieties of rice produce large amounts of vitamin A. Up to 500 000 children go blind each year as a result of lack of vitamin A in their diets. In theory golden rice offers a solution to this problem. In fact, many people objected to the way trials of the rice were run and the cost of the product. No golden rice is yet being grown in countries affected by vitamin A blindness.

There is a lot of discussion about genetically modified crops. Here are some commonly expressed opinions.

Figure 3 The amount of beta carotene in golden rice and golden rice 2 is reflected in the depth of colour of the rice

John, 49, plumber, UK

'I'm very concerned about GM foods. Who knows what we're all eating nowadays. I don't want strange genes inside me, thank you very much. We've got plenty of fruit and vegetables as it is – why do we need more?'

Ali, 26, shop assistant, UK

'I think GM food is such a good idea. If the scientists can modify crops so they don't go off so quickly, food should get cheaper, and there will be more to go around. And what about these plants that produce pesticides? That'll stop a lot of crop spraying, so that should make our food cleaner and cheaper. It's typical of us in the UK that we moan and panic about it all.'

Tilahun, 35, farmer, Ethiopia

'I have some real worries about the GM crops that don't form fertile seeds. In the past, farmers in poorer countries just kept seeds from the previous year's crops, so it was cheap and easy. With the GM crops we have to buy new seeds every year – although I hear that won't be the case with golden rice. On the other hand, these GM crops don't need spraying very much. They grow well in our dry conditions, they give a much bigger crop yield and keep well too – so there are some advantages.'

Activity

You are going to produce a 5-minute slot for a daytime television show on '**Genetic engineering – a good thing or not?**' Using the information here and on B1 6.6 Genetic engineering (and extra research if you have time), plan out a script for your time on air, remembering that you have to inform the public about genetic engineering, entertain them and make them think about the issues involved.

Key points

- There are a number of economic, social and ethical issues concerning cloning and genetic engineering which need to be considered when making judgements about the use of this science.

Summary questions

1 People get very concerned about cloning. Do you think these fears are justified? Explain your answer.

2 Summarise the main advantages and disadvantages of genetic engineering expressed here.

Summary questions k

1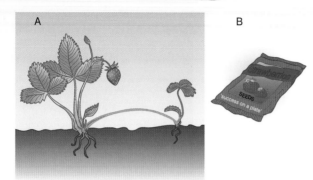

A   B

a How has the small plant shown in diagram A been produced?

b What sort of reproduction is this?

c How were the seeds in B produced?

d How are the new plants that you would grow from the packet of seeds shown in B different from the new plants shown in A?

2 a What is a gene?

b Where do you find genes?

c What is a gamete?

3 Tissue culture techniques mean that 50 000 new raspberry plants can be grown from one old one instead of two or three by taking cuttings. Cloning embryos from the best bred cows means that they can be genetically responsible for 30 or more calves each year instead of two or three.

a How does tissue culture differ from taking cuttings?

b How can one cow produce 30 or more calves in a year?

c What are the similarities between cloning plants and cloning animals in this way?

d What are the differences in the techniques for cloning animals and plants?

e Why do you think there is so much interest in finding different ways to make the breeding of farm animals and plants increasingly efficient?

4 a Describe the process of adult cell cloning.

b There has been a great deal of media interest and concern about cloning animals but very little about cloning plants. Why do you think there is such a difference in the way people react to these two different technologies?

5 Human growth is usually controlled by growth hormones produced by the pituitary gland in your brain. If you don't make enough hormones, you don't grow properly and remain very small. This condition affects 1 in every 5000 children. Until recently the only way to get growth hormone was from the pituitary glands of dead bodies. Genetically engineered bacteria can now make plenty of pure growth hormone.

a Draw and label a diagram to explain how a healthy human gene for making growth hormone can be taken from a human chromosome and put into a working bacterial cell.

b What are the advantages of producing substances like growth hormone using genetic engineering?

6 In 2003 two mules called Idaho Gem and Idaho Star were born in America. They were clones of a famous racing mule. They both seem very healthy. They were separated and sent to different stables to be reared and trained for racing. So far Idaho Gem has been more successful than his cloned brother, winning several races against ordinary racing mules. There is a third clone, Utah Pioneer, which has not been raced.

a The mules are genetically identical. How do you explain the fact that Idaho Gem has beaten Idaho Star in several races?

b Why do you think one of the clones is not being raced?

c Their progress is being carefully monitored by scientists. What type of data do you think will be available from these animals?

7 One concern people have about GM crops is that they might cross pollinate with wild plants. Scientists need to find out how far pollen from a GM crop can travel to be able to answer these concerns.

Describe how a trial to investigate this might be set up.

Practice questions

1 Strawberries are able to reproduce many plants from one parent plant.

Choose the correct answer to complete each sentence.
a Producing new plants with one parent is called .... (1)

*asexual reproduction   genetic engineering
sexual reproduction*

b The advantage of this is that all the strawberry fruits will .... (1)

be bigger   all taste better   all taste the same

c A disadvantage of this to the strawberry plants is that .... (1)

*there is more variation   they are genetically identical
they cannot mate*

2 Read the passage. Use the information and your own knowledge to answer the questions.

> At one time, the boll weevil destroyed cotton crops. Farmers sprayed the crops with a pesticide.
>
> The weevil died out but another insect, the bollworm moth, became resistant to this pesticide.
>
> In the 1990s large crops of the cotton plant were destroyed by the bollworm moth. The pesticides then used to kill the moth were expensive and very poisonous, resulting in deaths to humans.
>
> Scientists investigated alternative ways to control the bollworm moth. They found out that a type of bacterium produced a poison which killed bollworm larvae (grubs).
>
> A GM cotton crop plant was developed which produced the poison to kill bollworms. This proved to be very effective and farmers were able to stop using pesticide sprays.
>
> Now farmers have another problem. Large numbers of other insects have multiplied because they were not killed when the farmers stopped using pesticides. Some of these insects have started to destroy the GM cotton and farmers are beginning to use pesticides again!

a **i** Give **one** advantage of spraying crops with pesticides. (1)
  ii Give **two** disadvantages of spraying crops with pesticides. (2)
  iii Give **one** economic advantage of using GM cotton. (1)

  iv Some people object to using GM crops. Suggest **one** reason why. (1)

b *In this question you will be assessed on using good English, organising information clearly and using specialist terms where appropriate.*

The GM cotton was genetically engineered to produce the same poison as the bacterium.
Describe fully how this is done. (6)

3 The use of cloned animals in food production is controversial.

> It is now possible to clone 'champion' cows.
>
> Champion cows produce large quantities of milk.

a Describe how adult cell cloning could be used to produce a clone of a 'champion' cow. (4)

b Read the passage about cloning cattle.

> The Government has been accused of 'inexcusable behaviour' because a calf of a cloned American 'champion' cow has been born on a British farm. Campaigners say it will undermine trust in British food because the cloned cow's milk could enter the human food chain.
>
> But supporters of cloning say that milk from clones and their offspring is as safe as the milk we drink every day.
>
> Those in favour of cloning say that an animal clone is a genetic copy. It is not the same as a genetically engineered animal. Opponents of cloning say that consumers will be uneasy about drinking milk from cloned animals.

Use the information in the passage and your own knowledge and understanding to evaluate whether the government should allow the production of milk from cloned 'champion' cows.

Remember to give a conclusion to your evaluation. (5)

AQA, 2006

B1 7.1

Theories of evolution

Learning objectives

- What is the theory of evolution?
- What is the evidence that evolution has taken place?

We are surrounded by an amazing variety of life on planet Earth. Questions such as 'Where has it all come from?' and 'When did life on Earth begin?' have puzzled people for many generations.

The theory of **evolution** tells us that all the species of living things alive today have evolved from the first simple life forms. Scientists think these early forms of life developed on Earth more than 3 billion years ago. Most of us take these ideas for granted – but they are really quite new.

Up to the 18th century most people in Europe believed that the world had been created by God. They thought it was made, as described in the Christian Bible, a few thousand years ago. However, by the beginning of the 19th century scientists were beginning to come up with new ideas.

Lamarck's theory of evolution

Jean-Baptiste Lamarck was a French biologist. He thought that all organisms were linked by what he called a 'fountain of life'. He made the great step forward of suggesting that individual animals adapted and evolved to suit their environment. His idea was that every type of animal evolved from primitive worms. The change from worms to other organisms was caused by the **inheritance of acquired characteristics**.

Lamarck's theory was that the way organisms behaved affected the features of their body – a case of 'use it or lose it'. If animals used something a lot over a lifetime he thought it would grow and develop. Any useful changes that took place in an organism during its lifetime would be passed from a parent to its offspring. The neck of the giraffe is a good example. If a feature wasn't used, Lamarck thought it would shrink and be lost.

Lamarck's theory influenced the way **Charles Darwin** thought. However, there were several problems with Lamarck's ideas. There was no evidence for his 'fountain of life' and people didn't like the idea of being descended from worms. People could also see quite clearly that changes in their bodies – such as big muscles, for example – were not passed on to their children.

We now know that in the great majority of cases Lamarck's idea of inheritance cannot happen. However, scientists have discovered that in a few cases the way an animal behaves actually changes its genes. This results in the next generation behaving in the same way.

Figure 1 In Lamarck's model of evolution, giraffes have long necks because each generation stretched up to reach the highest leaves. So each new generation had a slightly longer neck.

> **a** What do you think is meant by the phrase 'inheritance of acquired characteristics'?

Charles Darwin and the origin of species

Our modern ideas about evolution began with the work of one of the most famous scientists of all time – Charles Darwin. Darwin set out in 1831 as the ship's naturalist on *HMS Beagle*. He was only 22 years old at the start of the voyage to South America and the South Sea Islands.

Darwin planned to study geology on the trip. But as the voyage went on he became as excited by his collection of animals and plants as by his rock samples.

> **b** What was the name of the ship that Darwin sailed on?

Study tip

Remember the basic key stages in natural selection:
survive → breed →
pass on genes

In South America, Darwin discovered a new form of the common rhea, an ostrich-like bird. Two different types of the same bird living in slightly different areas set Darwin thinking.

On the Galapagos Islands he was amazed by the variety of species. He noticed that they differed from island to island. Darwin found strong similarities between types of finches, iguanas and tortoises on the different islands. Yet each was different and adapted to make the most of local conditions.

Darwin collected huge numbers of specimens of animals and plants during the voyage. He also made detailed drawings and kept written observations. The long journey home gave him plenty

Figure 2 Darwin was very impressed by the giant tortoises he found on the Galapagos Islands. The tortoises on each island had different-shaped shells and a slightly different way of life. Darwin made detailed drawings of them all.

of time to think about what he had seen. Charles Darwin returned home after five years with some new and different ideas forming in his mind.

> **c** What is the name of the famous islands where Darwin found so many interesting species?

After returning to England, Darwin spent the next 20 years working on his ideas. Darwin's theory of evolution by natural selection is that all living organisms have evolved from simpler life forms. This evolution has come about by a process of natural selection.

Reproduction always gives more offspring than the environment can support. Only those that have inherited features most suited to their environment – the 'fittest' – will survive. When they breed, they pass on the genes for those useful inherited characteristics to their offspring. This is natural selection.

When Darwin suggested how evolution took place, no one knew about genes. He simply observed that useful inherited characteristics were passed on. Today, we know it is useful genes which are passed from parents to their offspring in natural selection.

Figure 3 Darwin worked here in his study for around 20 years, carrying out experiments and organising his ideas on evolution by natural selection

Study tip

Avoid confusion between:
- the *theory of evolution* and
- the *process of natural selection*.

Summary questions

1 Explain what is meant by the following terms:
   a evolution
   b natural selection

2 What was the importance of the following in the development of Darwin's ideas?
   a South American rheas
   b Galapagos tortoises, iguanas and finches
   c the long voyage of *HMS Beagle*
   d the 20 years from his return to the publication of his book *The Origin of Species*.

Key points

- The theory of evolution states that all the species which are alive today – and many more which are now extinct – evolved from simple life forms that first developed more than 3 billion years ago.

- Darwin's theory is that evolution takes place through natural selection.

Accepting Darwin's ideas

Learning Objectives

● Why was Darwin's theory of evolution only gradually accepted?

Charles Darwin came back from his trip on *HMS Beagle* with new ideas about the variety of life on Earth. He read many books and thought about the ideas of many other people such as Lamarck, Lovell and Malthus. He gradually built up his theory of evolution by natural selection.

He knew his ideas would be controversial. He expected a lot of opposition both from fellow scientists and from religious leaders.

Building up the evidence

Darwin realised he would need lots of evidence to support his theories. This is one of the reasons why it took him so long to publish his ideas. He spent years trying to put his evidence together in order to convince other scientists.

He used the amazing animals and plants he had seen on his journeys as part of that evidence. They showed that organisms on different islands had adapted to their environments by natural selection. So they had evolved to be different from each other.

Darwin carried out breeding experiments with pigeons at his home. He wanted to show how features could be artificially selected. Darwin also studied different types of barnacles (small invertebrates found on seashore rocks) and where they lived. This gave him more evidence of organisms adapting and forming different species.

Darwin built up a network of friends, fellow scientists and pigeon breeders. He didn't travel far from home (he was often unwell) but he spent a lot of time discussing his ideas with this group of friends. They helped him get together the evidence he needed and he trusted them as he talked about his ideas.

Figure 1 The finches found on the different Galapagos islands look very different but all evolved from the same original type of finch by natural selection

Why did people object?

In 1859, Darwin published his famous book *On the Origin of Species by means of Natural Selection* (often known as *The Origin of Species*). The book caused a sensation. Many people were very excited by his ideas and defended them enthusiastically. Others were deeply offended, or simply did not accept them.

There were many different reasons why it took some scientists a long time to accept Darwin's theory of natural selection. They include:

- The theory of evolution by natural selection challenged the belief that God made all of the animals and plants that live on Earth. This religious view was the generally accepted belief among most people in early Victorian England.

- In spite of all Darwin's efforts, many scientists felt there was not enough evidence to convince them of his theory.

- There was no way to explain how variety and inheritance happened. The mechanism of how inheritance happens – by genes and genetics – was not known until 50 years *after* Darwin published his ideas. Because there was no mechanism to explain how characteristics could be inherited, it was much harder for people to accept and understand.

The arguments raged and it took some time before the majority of scientists accepted Darwin's ideas. However, by the time of his death in 1882 he was widely regarded as one of the world's great scientists. He is buried in Westminster Abbey along with other great people like Sir Isaac Newton.

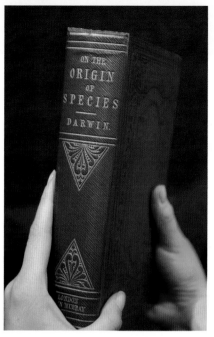

Figure 2 Darwin's famous book – it sold out on the first day of publication!

?? Did you know … ?

Darwin let his children use the back of his original manuscript of *The Origin of Species* as drawing paper. Not many of these original pages exist. Darwin kept the ones that remain because of his children's drawings rather than his own writing!

Figure 3 It wasn't just scientists who were interested in Darwin's ideas. Cartoonists loved the idea of evolution too.

Key points

- Darwin's theory of evolution by natural selection was only gradually accepted for a number of reasons. These include:
  - a conflict with the widely held belief that God made all the animals and plants on the Earth
  - insufficient evidence
  - no mechanism for explaining variety and inheritance – genetics were not understood for another 50 years.

Summary questions

1 **a** Darwin set out in *HMS Beagle* in 1831. How many years later did he publish *The Origin of Species*?
   b What was Darwin's big idea?

2 What type of evidence did Darwin put together to convince other scientists his ideas were right?

3 Why did it take some time before most people accepted Darwin's ideas?

B1 7.3 Natural selection

Learning objectives

- How does natural selection work?
- What is mutation?

Scientists explain the variety of life today as the result of a process called natural selection. The idea was first suggested about 150 years ago by Charles Darwin.

Animals and plants are always in competition with each other. Sometimes an animal or plant gains an advantage in the competition. This might be against other species or against other members of its own species. That individual is more likely to survive and breed. This is known as natural selection.

> **a** Who first suggested the idea of natural selection?

Survival of the fittest

Charles Darwin was the first person to describe natural selection as the 'survival of the fittest'. Reproduction is a very wasteful process. Animals and plants always produce more offspring than the environment can support.

The individual organisms in any species show lots of variation. This is because of differences in the genes they inherit. Only the offspring with the genes best suited to their habitat manage to stay alive and breed successfully. This is natural selection at work.

Think about rabbits. The rabbits with the best all-round eyesight, the sharpest hearing and the longest legs will be the ones that are most likely to escape being eaten by a fox. They will be the ones most likely to live long enough to breed. What's more, they will pass those useful genes on to their babies. The slower, less alert rabbits will get eaten and their genes are less likely to be passed on.

> **b** Why would a rabbit with good hearing be more likely to survive than one with less keen hearing?

The part played by mutation

New forms of genes result from changes in existing genes. These changes are known as mutations. They are tiny changes in the long strands of DNA.

Mutations occur quite naturally through mistakes made in copying DNA when the cells divide. Mutations introduce more variety into the genes of a species. In terms of survival, this is very important.

> **c** What is a mutation?

Many mutations have no effect on the characteristics of an organism, and some mutations are harmful. However, just occasionally a mutation has a good effect. It produces an adaptation that makes an organism better suited to its environment. This makes it more likely to survive and breed.

Whatever the adaptation, if it helps an organism survive and reproduce it will get passed on to the next generation. The mutant gene will gradually become more common in the population. It will cause the species to evolve.

When new forms of a gene arise from mutation, there may be a relatively more rapid change in a species. This is particularly true if the environment changes. If the mutation gives the organism an advantage in the changed environment, it will soon become common.

links

For more information on the competition between plants and animals in the natural world, look back at B1 4.4 Competition in animals and B1 4.5 Competition in plants.

Figure 1 The natural world is often brutal. Only the best adapted predators capture prey – and only the best adapted prey animals escape.

Did you know ...?

Fruit flies can produce 200 offspring every two weeks. The yellow star thistle, an American weed, produces around 150 000 seeds per plant per year. If all those offspring survived we'd be overrun with fruit flies and yellow star thistles!

links

For information on genes, see B1 6.1 Inheritance.

Natural selection in action

Malpeque Bay in Canada has some very large oyster beds. In 1915, the oyster fishermen noticed a few small, flabby oysters with pus-filled blisters among their healthy catch.

By 1922 the oyster beds were almost empty. The oysters had been wiped out by a destructive new disease (soon known as Malpeque disease).

Fortunately a few of the oysters had a mutation which made them resistant to the disease. These were the only ones to survive and breed. The oyster beds filled up again and by 1940 they were producing more oysters than ever.

A new population of oysters had evolved. As a result of natural selection, almost every oyster in Malpeque Bay now carries a gene that makes them resistant to Malpeque disease. So the disease is no longer a problem.

Figure 2 The tiny number of dandelion seeds that survive and grow into plants have a combination of genes that gives them an edge over all the others

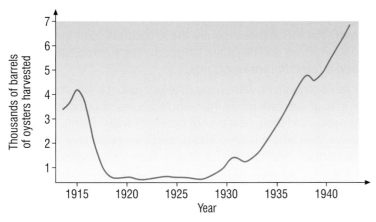

Figure 3 Oyster yields from Malpeque Bay 1915–40. As you can see, disease devastated the oyster beds. However, thanks to the process of natural selection, a healthy population of oysters managed to survive and reproduce again.

d What is Malpeque disease?

Summary questions

1 Copy and complete using the words below:

*adaptation   breed   environment   generation
mutation   selection   organism   survive*

When a .... has a good effect it produces an .... that makes an .... better suited to it's .... This makes it more likely to .... and .... The mutation then gets passed on to the next .... This is natural ....

2 Many features that help animals and plants survive are the result of natural selection. Give three examples, e.g. all-round eyesight in rabbits.

3 Explain how the following characteristics of animals and plants have come about in terms of natural selection.
   a Male red deer have large sets of antlers.
   b Cacti have spines instead of leaves.
   c Camels can tolerate their body temperature rising far higher than most other mammals.

Key points

- Natural selection works by selecting the organisms best adapted to a particular habitat.

- Different organisms in a species show a wide range of variation because of differences in their genes.

- The individuals with the characteristics most suited to their environment are most likely to survive and breed successfully.

- The genes that have produced these successful characteristics are then passed on to the next generation.

- Mutation is a change in the genetic material (DNA) which results in a new form of a gene.

B1 7.4

Classification and evolution

Learning objectives

- What is classification?
- How does classification help us understand evolution?

??? Did you know ...?

The most widely accepted kingdoms of microorganisms are Monera, Protista and Fungi. However, there is still a lot of argument between scientists as to exactly which organisms fit into each kingdom.

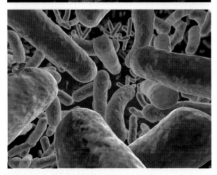

Figure 1 Animals, plants and microorganisms are identified by the differences between them rather than the similarities

How are organisms classified?

Classification is the organisation of living things into groups according to their similarities.

There are millions of different types of living organisms. Biologists classify living things to make it easier to study them. Classification allows us to make sense of the living world. It also helps us to understand how life began and how the different groups of living things are related to each other.

Living things are classified by studying their similarities and differences. By looking at similarities and differences between organisms we can decide which should be grouped together.

The system we use for classifying living things is known as the **natural classification system**. The biggest groups are the **kingdoms**, and the best known are the animal kingdom and the plant kingdom. The microorganisms are then split between three different kingdoms.

Kingdoms contain lots of organisms with many differences but a few important similarities. For example, all animals move their whole bodies about during at least part of their life cycle, and their cells do not have cell walls. Plants on the other hand do not move their whole bodies about, and their cells have cell walls. Also some plant cells contain chloroplasts full of chlorophyll for photosynthesis.

The smallest group is a **species**. Members of the same species are very similar. Any differences are small variations of the same feature. A species is a group of organisms that can breed together and produce fertile offspring. Orang-utans, dandelions and brown trout are all examples of species of living organisms.

a What is classification?

Classification and evolutionary relationships

In the past, we relied on careful observation of organisms to decide which group they belonged to. Out in the field, this is still the main way we identify an organism. However, scientists develop models to suggest relationships between living organisms.

Since Darwin's time, scientists have used classification to show the evolutionary links between different organisms. These models are called **evolutionary trees**. They are built up by looking at the similarities and differences between different groups of organisms. One of the most famous evolutionary trees was produced by Darwin himself. It was found in one of the notebooks that he used to plan his book *The Origin of Species*. It starts off with the words 'I think'. Then it shows how Darwin was beginning to see relationships between different groups of living organisms (see Figure 2).

However, observation may not tell you the whole story. Some organisms look very different but are closely related. Others look very similar but come from very different groups. Now scientists are increasingly using DNA evidence to decide what species an animal belongs to. They look for differences as well as similarities in the DNA. This allows them to work out the **evolutionary relationships** between organisms. It also means they can see how long ago different organisms had a common ancestor.

b What is an evolutionary tree?

Evolutionary and ecological relationships

Classifying organisms helps us to understand how they evolved. It can also help us understand how species have evolved together in an environment. We call this their ecological relationships and it is another way of modelling relationships between organisms.

For example, pandas have a thumb which they use to grip bamboo. However it is not like a human thumb – it has evolved from specialised wrist bones. The only other animals to have a similar 'wrist thumb' are the red pandas. Both red pandas and giant pandas eat bamboo. Based on their modern ecological feeding relationships, it looks as if they are closely related in evolution. However, based on their anatomy and DNA, giant pandas are closely related to other species of bears. Red pandas are much more closely related to racoons.

Recently scientists found a fossil ancestor of red pandas which also had a 'wrist thumb'. There is also evidence from the ecological relationships of this fossil animal. This suggests the thumb evolved as an adaptation for a quick escape into trees carrying prey stolen from sabre-toothed tigers. This is rather different from the giant panda evolving to feed on bamboo.

Now the ecological models and the evolutionary models match – the two species had a common ancestor a very long time ago, but the special 'wrist thumb' evolved separately as adaptations to solve two different ecological problems.

Figure 2 This evolutionary tree was found in one of the notebooks that Darwin used to plan his book *The Origin of Species*

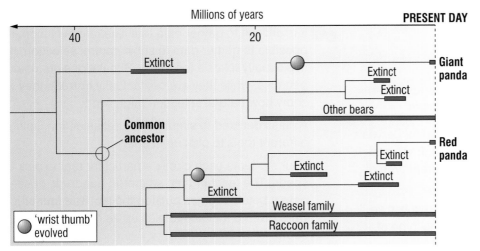

Figure 3 Evolutionary trees like this show us the best model of the evolutionary relationships between organisms

Figure 4 Both the giant panda and the red panda use the 'wrist thumb' to eat bamboo

Summary questions

1 Copy and complete using the words below:

kingdoms animals organisms species classify microorganisms similarities

Scientists .... living .... by studying .... and differences between them. The big groups are called .... and the smallest are called .... All living organisms are either ...., plants or ....

2 What observations can be made to compare living organisms?

3 How are evolutionary trees useful to us?

Key points

- Studying the similarities and differences between organisms allows us to classify them into animals, plants and microorganisms.

- Classification also helps us to understand evolutionary and ecological relationships.

Summary questions ⓚ

1 What was Jean-Baptiste Lamarck's theory of evolution?

2 a What started Charles Darwin thinking about the variety of life and how it has come about?

b Explain Darwin's theory of evolution.

3 a Summarise the similarities and differences between Darwin's and Lamarck's theories of evolution.

b Why do you think Lamarck's theory was so important to the way Darwin's theory was subsequently received?

4

1. *Geospiza magnirostris.*   2. *Geospiza fortis.*
3. *Geospiza parvula.*   4. *Certhidea olivaзea.*

Figure 1 Darwin's finches – more evidence for evolution

Look at the birds in Figure 1. They are known as Darwin's finches. They live on the Galapagos Islands. Each one has a slightly different beak and eats a different type of food.

Explain carefully how natural selection can result in so many different beak shapes from one original type of founder finch.

5 Alfred Russel Wallace came from a poor family but he was a gifted naturalist. He went on a collecting expedition to Borneo, an island in South East Asia that has a rich variety of unique animal and plant life. While he was there, Wallace became ill with a fever and while he was unwell he developed his theory. He had the idea that if species exist in various forms, the organisms that are not well adapted to change are likely to die out. This would leave only the better-adapted forms to survive and breed. Wallace put his ideas down in a paper and sent it to Charles Darwin for advice. Darwin and Wallace both published papers together on their ideas in London at the same time. It was Wallace's work that shocked Darwin into finally writing *The Origin of Species*.

Wallace's ideas were not as well thought out as Darwin's and he did not have the evidence to back them up, which is why it is largely Darwin who is remembered for the theory of evolution by natural selection.

a What was Wallace's theory?

b What are the similarities between Borneo and the Galapagos and how would this have helped Wallace develop his theory?

c Why do you think the arrival of Wallace's letter and paper was such a shock to Darwin?

d Wallace's theories were not strongly supported by evidence. What sort of evidence did Darwin bring forward to support his ideas in *The Origin of Species*?

6 a What is classification?

b Explain two alternative ways of deciding how to classify an organism.

c What are the differences and similarities between an evolutionary relationship and an ecological relationship between organisms?

7 It is difficult to gather data that illustrate evolution. It is possible to gather data to show natural selection, but this usually takes a long time. Simulations are useful because, while they are not factually correct, they do show how natural selection might work.

A class decided to simulate natural selection, using different tools to pick up seeds.

Four students each chose a particular tool to pick up seeds. The teacher then scattered hundreds of seeds onto a patch of grass outside the lab. The four students were given 5 minutes to pick up as many seeds as they could.

James, who was using a spoon, picked up 23 seeds, whilst Farzana, using a fork, could only pick up two. Claire managed seven seeds with the spatula, but Jenny struggled to pick up her two seeds with a pair of scissors.

a Put the essential data into a table.

b How would the data be best presented? Explain your choice.

c Was this a fair test? Explain your answer.

d What conclusion can you draw from this simulation?

e How does this simulation model the situation with the finches on the Galapagos Islands, which evolved into many different species?

Practice questions (k)

1 a This diagram shows a timeline for the evolution of some dinosaurs. The mass of each dinosaur is shown in the brackets by its name.
Choose the correct answer to complete each sentence.

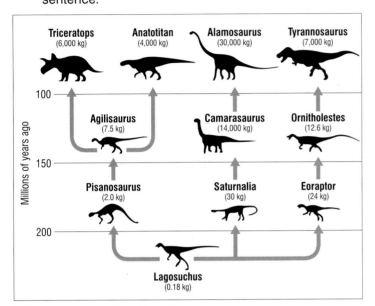

Dinosaur evolution timeline

i A dinosaur which lived between 100 and 150 million years ago is .... (1)

Agilisaurus   Saturnalia   Tyrannosaurus

ii Camarasaurus evolved from .... (1)

Agilisaurus   Alamosaurus   Saturnalia

iii The difference in mass between Agilisaurus and the smallest dinosaur is .... kg. (1)

1.82   5.5   7.32

b i The earliest life forms developed on Earth more than 3 .... years ago. (1)

billion   million   thousand

ii The earliest life forms can be described as .... (1)

bony   complex   simple

c Darwin suggested a theory of evolution. His theory is described as evolution by .... (1)

acquired characteristics   a god   natural selection

2 Giraffes have developed their long necks over millions of years.

Two scientists tried to explain why the giraffes have long necks. They are called Darwin and Lamarck.

Match the name in List A to the correct statement in List B.

List A	List B
Name	**Statement**
Darwin only	Noticed that the neck of the giraffe changed over time
Lamark only	Had enough evidence to prove why the giraffe's neck got longer
Both Lamark and Darwin	Thought that natural selection worked on variations in neck length present at birth
	Thought the giraffe stretched its neck while eating leaves in trees. Then its young inherited the longer neck

(3)

3 The photograph shows a snake eating a toad.

Cane toads were first introduced into Australia in 1935. The toads contain toxins and most species of Australian snake die after eating the toad. The cane toad toxin does not affect all snakes the same way. Longer snakes are less affected by toad toxin. Scientists investigated how red-bellied black snakes had changed in the 70 years since cane toads were introduced into their area. They found that red-bellied black snakes had become longer by around 3–5%.

Suggest an explanation for the change in the body length of the red-bellied black snakes since the introduction of the cane toads. (4)

AQA, 2005

1 The diagrams show some biological processes.

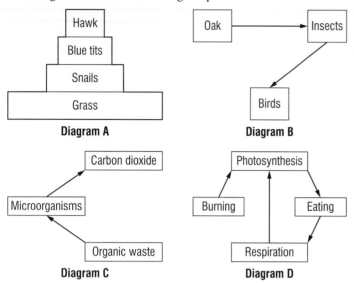

a Write **one** letter, A, B, C or D to complete each sentence
  i   The diagram which shows the process of decay is   (1)
  ii The diagram which shows a pyramid of biomass is   (1)
  iii The diagram which shows the carbon cycle is   (1)
b The grass and the oak tree can photosynthesise.
   What is the source of energy for photosynthesis?   (1)
c In Diagram A, the bar for grass is bigger than the bar for the hawk.
   Suggest **two** reasons why.   (2)

2 Read the passage.

> Some microorganisms live in very unusual environments.
>
> Scientists have discovered bacteria that can survive in temperatures between 80 °C and 105 °C and at pressures 1000 times higher than atmospheric pressure.
>
> Other bacteria, called *halophiles* live in super salty conditions.
>
> Some bacteria need large amounts of iron in quantities that would kill most other organisms.
>
> These unusual environments may be similar to early times on Earth, when the first life forms lived in very extreme conditions, such as very hot water containing high concentrations of salt, iron and other minerals.

a What is the name given to all the bacteria which live in these unusual environments?
                                                                                      (1)
b Which chemical in human cells would not work at temperatures between 80 °C and 105 °C?   (1)
c How long ago did early life forms appear on Earth?   (1)
d Halophiles breed in very salty conditions. The offspring of halophiles can also live in very salty conditions.
   Choose the correct answer to complete the sentence.
   The offspring can live in salty conditions because the parents pass on their ....
   genes   iron   salt   (1)

3 Influenza is a disease caused by a virus.

a Suggest **two** reasons why it is difficult to treat diseases caused by viruses. (2)

b In some years there are influenza epidemics.

The graph shows the death rate in Liverpool during three influenza epidemics.

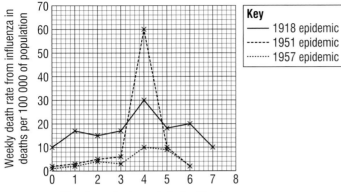

Key
— 1918 epidemic
----- 1951 epidemic
········ 1957 epidemic

Weekly death rate from influenza in deaths per 100 000 of population

Time in weeks from start of epidemic

i The population of Liverpool in 1951 was approximately 700 000.

Calculate the approximate number of deaths from influenza in week 4 of the 1951 epidemic.

Show clearly how you work out your answer. (2)

ii In most years, the number of deaths from influenza in Liverpool is very low.

Suggest, in terms of the influenza virus and the body's immune system, **three** reasons why there were large numbers of deaths in years such as 1918 and 1951. (3)

4 *In this question you will be assessed on using good English, organising information clearly and using specialist terms where appropriate.*

Hormones are used in contraceptive pills.

Read the information about the trialling of the first contraceptive pill.

> The Pill was developed by a team of scientists led by Gregory Pincus. The team needed to carry out large-scale trials on humans.
>
> In the summer of 1955, Pincus visited the island of Puerto Rico. Puerto Rico is one of the most densely populated areas in the world. Officials supported birth control as a form of population control. Pincus knew that if he could demonstrate that the poor, uneducated women of Puerto Rico could use the Pill correctly then so could women anywhere in the world.
>
> The scientists selected a pill with a high dose of hormones to ensure that no pregnancies would occur while test subjects were taking the drug. The Pill was found to be 100% effective when taken properly. But 17% of the women in the study complained of side effects. Pincus ignored these side effects.
>
> The women in the trial had been told only that they were taking a drug that prevented pregnancy. They had not been told that the Pill was experimental or that there was a chance of dangerous side effects.

Evaluate the methods used by Pincus in trialling the contraceptive pill. (6)

Study tip

Before answering a graph question be sure to read all the information before the graph and then read the axis labels. When a graph has several lines, as in 3b, check the key before you take any readings.

Next, read the question instructions. Why has the examiner told you the population is 700 000? You need this information because the numbers of deaths is given per 100 000.

Before doing the calculation, check again that you are reading the point on the correct graph line.

Study tip

When you see the following instruction

In this question you will be assessed on using good English, organising information clearly and using specialist terms where appropriate.

Be aware that you will lose marks if you do not:

● write in a logical order

● use the correct scientific terms and spell them correctly

● PLAN before you WRITE.

C1 1.1 Atoms, elements and compounds

Learning objectives

- What are elements made of?
- How do we represent atoms and elements?
- What is the basic structure of an atom?

Figure 1 An element contains only **one** type of atom – in this case bromine

Look at the things around you and the substances that they are made from. You will find wood, metal, plastic, glass ... the list is almost endless. Look further and the number of different substances is mind-boggling.

All substances are made of **atoms**. There are about 100 different types of atom found naturally on Earth. These can combine in a huge variety of ways. This gives us all those different substances.

Some substances are made up of only one type of atom. We call these substances elements. As there are only about 100 different types of atom, there are only about 100 different elements.

> **a** How many different types of atom are there?
> **b** Why can you make millions of different substances from these different types of atom?

Elements can have very different properties. Elements such as silver, copper and gold are shiny **solids**. Other elements such as oxygen, nitrogen and chlorine are **gases**.

Atoms have their own symbols

The name we use for an element depends on the language being spoken. For example, sulfur is called *Schwefel* in German and *azufre* in Spanish! However, a lot of scientific work is international. So it is important that we have symbols for elements that everyone can understand. You can see these symbols in the **periodic table**.

Group numbers

1	2											3	4	5	6	7	0
						H 1 Hydrogen											**He** 2 Helium
Li 3 Lithium	**Be** 4 Beryllium											**B** 5 Boron	**C** 6 Carbon	**N** 7 Nitrogen	**O** 8 Oxygen	**F** 9 Fluorine	**Ne** 10 Neon
Na 11 Sodium	**Mg** 12 Magnesium											**Al** 13 Aluminium	**Si** 14 Silicon	**P** 15 Phosphorus	**S** 16 Sulfur	**Cl** 17 Chlorine	**Ar** 18 Argon
K 19 Potassium	**Ca** 20 Calcium	**Sc** 21 Scandium	**Ti** 22 Titanium	**V** 23 Vanadium	**Cr** 24 Chromium	**Mn** 25 Manganese	**Fe** 26 Iron	**Co** 27 Cobalt	**Ni** 28 Nickel	**Cu** 29 Copper	**Zn** 30 Zinc	**Ga** 31 Gallium	**Ge** 32 Germanium	**As** 33 Arsenic	**Se** 34 Selenium	**Br** 35 Bromine	**Kr** 36 Krypton
Rb 37 Rubidium	**Sr** 38 Strontium	**Y** 39 Yttrium	**Zr** 40 Zirconium	**Nb** 41 Niobium	**Mo** 42 Molybdenum	**Tc** 43 Technetium	**Ru** 44 Ruthenium	**Rh** 45 Rhodium	**Pd** 46 Palladium	**Ag** 47 Silver	**Cd** 48 Cadmium	**In** 49 Indium	**Sn** 50 Tin	**Sb** 51 Antimony	**Te** 52 Tellurium	**I** 53 Iodine	**Xe** 54 Xenon
Cs 55 Caesium	**Ba** 56 Barium	Lanthanum see below	**Hf** 72 Hafnium	**Ta** 73 Tantalum	**W** 74 Tungsten	**Re** 75 Rhenium	**Os** 76 Osmium	**Ir** 77 Iridium	**Pt** 78 Platinum	**Au** 79 Gold	**Hg** 80 Mercury	**Tl** 81 Thallium	**Pb** 82 Lead	**Bi** 83 Bismuth	**Po** 84 Polonium	**At** 85 Astatine	**Rn** 86 Radon
Fr 87 Francium	**Ra** 88 Radium	Actinium see below															

The transition metals

The halogens   The noble gases

The alkali metals   The alkaline earth metals

La 57 Lanthanum	**Ce** 58 Cerium	**Pr** 59 Praseodymium	**Nd** 60 Neodymium	**Pm** 61 Promethium	**Sm** 62 Samarium	**Eu** 63 Europium	**Gd** 64 Gadolinium	**Tb** 65 Terbium	**Dy** 66 Dysprosium	**Ho** 67 Holmium	**Er** 68 Erbium	**Tm** 69 Thulium	**Yb** 70 Ytterbium	**Lu** 71 Lutetium
Ac 89 Actinium	**Th** 90 Thorium	**Pa** 91 Protactinium	**U** 92 Uranium	**Np** 93 Neptunium	**Pu** 94 Plutonium	**Am** 95 Americium	**Cm** 96 Curium	**Bk** 97 Berkelium	**Cf** 98 Californium	**Es** 99 Einsteinium	**Fm** 100 Fermium	**Md** 101 Mendelevium	**No** 102 Nobelium	**Lr** 103 Lawrencium

Lanthanides

Actinides

Figure 2 The periodic table shows the symbols for the elements

The symbols in the periodic table represent atoms. For example, O represents an atom of oxygen; Na represents an atom of sodium. The elements in the table are arranged in columns, called **groups**. Each group contains elements with similar chemical properties. The 'staircase' drawn in bold is the dividing line between metals and non-metals. The elements to the left of the line are metals. Those on the right of the line are non-metals.

??? Did you know ...?

Only 92 elements occur naturally on Earth. The other heavier elements in the periodic table have to be made artificially and might only exist for fractions of a second before they decay into other, lighter elements.

c Why is it useful to have symbols for atoms of different elements?
d Sort these elements into metals and non-metals: phosphorus (P), barium (Ba), vanadium (V), mercury (Hg) and krypton (Kr).

Atoms, elements and compounds

Most of the substances we come across are not pure elements. They are made up of different types of atom joined together. These are called **compounds**. Chemical bonds hold the atoms tightly together in compounds. Some compounds are made from just two types of atom (e.g. water, made from hydrogen and oxygen). Other compounds consist of more different types of atom.

An atom is made up of a tiny central **nucleus** with **electrons** around it.

links

For more information on what is inside an atom, see C1 1.2 Atomic structure and 1.3 The arrangement of electrons in atoms.

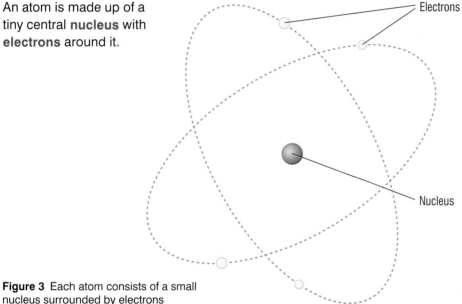

Figure 3 Each atom consists of a small nucleus surrounded by electrons

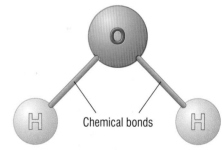

Figure 4 A grouping of two or more atoms bonded together is called a **molecule**. Chemical bonds hold the hydrogen and oxygen atoms together in the water molecule. Water is an example of a compound.

Summary questions

1 Copy and complete using the words below:

atoms   bonds   molecule   compounds

All elements are made up of .... When two or more atoms join together a .... is formed. The atoms in elements and .... are held tightly to each other by chemical ....

2 Explain why when we mix two elements together we can often separate them again quite easily. However, when two elements are chemically combined in a compound, they can be very difficult to separate.

3 Draw diagrams to explain the difference between an element and a compound. Use a hydrogen molecule (H_2) and a hydrogen chloride molecule (HCl) to help explain.

4 Draw a labelled diagram to show the basic structure of an atom.

Key points

● All substances are made up of atoms.

● Elements contain only one type of atom.

● Compounds contain more than one type of atom.

● An atom has a tiny nucleus in its centre, surrounded by electrons.

C1 1.2 Atomic structure

Learning objectives

- What is the charge on a proton, a neutron and an electron?

- What can we say about the number of protons in an atom compared with its number of electrons?

- What is the 'atomic number' and 'mass number' of an atom?

- How are atoms arranged in the periodic table?

?? Did you know ... ?

In 1808, a chemist called John Dalton published a theory of atoms. It explained how atoms joined together to form new substances (compounds). Not everyone liked his theory though – one person wrote 'Atoms are round bits of wood invented by Mr Dalton!'

In the middle of an atom there is a very small nucleus. This contains two types of particles, which we call **protons** and **neutrons**. A third type of particle orbits the nucleus. We call these really tiny particles electrons.

Any atom has the same number of electrons orbiting its nucleus as it has protons in its nucleus.

Protons have a positive charge. Neutrons have no charge – they are neutral. So the nucleus itself has an overall positive charge.

The electrons orbiting the nucleus are negatively charged. The relative charge on a proton is +1 and the relative charge on an electron is −1.

Because any atom contains equal numbers of protons and electrons, the positive and negative charges cancel out. So there is no overall charge on any atom. Its charge is zero. For example, a carbon atom is neutral. It has 6 protons, so we know it must have 6 electrons.

> **a** What are the names of the three particles that make up an atom?
>
> **b** An oxygen atom has 8 protons – how many electrons does it have?

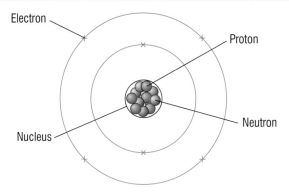

Figure 1 Understanding the structure of an atom gives us important clues to the way chemicals react together

Type of subatomic particle	Relative charge
Proton	+1
Neutron	0
Electron	−1

To help you remember the charge on the subatomic particles:

- **P**rotons are **P**ositive;
- **Neutr**ons are **Neutr**al;
- so that means Electrons must be Negative!

Atomic number and the periodic table

All the atoms of a particular element have the same number of protons. For example, hydrogen has 1 proton in its nucleus, carbon has 6 protons in its nucleus and sodium has 11 protons in its nucleus.

We call the **number of protons** in each atom of an element its **atomic number**.

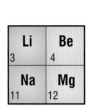
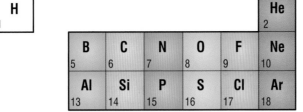

Figure 2 The elements in the periodic table are arranged in order of their atomic number. (As atoms are neutral, this is also the same order as their number of electrons.)

The elements in the periodic table are arranged in order of their atomic number (number of protons). If you are told that the atomic number of an element is 8, you can identify it using the periodic table. It will be the 8th element listed. In this case it is oxygen.

> **c** What is the 14th element in the periodic table?

You read the periodic table from left to right, and from the top down – just like reading a page of writing.

> **d** Look at the elements in the last group of the abbreviated periodic table in Figure 2. What pattern do you notice about the number of protons going from helium to neon to argon?

Mass number

The **number of protons plus neutrons** in the nucleus of an atom is called its **mass number**.

● So, if an atom has 4 protons and 5 neutrons, its mass number will be 4 + 5 = **9**.

● Given the atomic number and mass number, we can work out how many protons, electrons and neutrons are in an atom. For example, an argon atom has an atomic number of 18 and a mass number of 40.

Its atomic number is 18 so it has **18 protons**. Remember that atoms have an equal number of protons and electrons. So argon also has **18 electrons**. The mass number is 40, so we know that:

18 (the number of protons) + the number of neutrons = 40

Therefore argon must have **22 neutrons** (as 18 + 22 = 40).

We can summarise the last part of the calculation as:

number of neutrons = mass number – atomic number

Summary questions

1 Copy and complete using the words below:

electrons   atomic   negative   neutrons

In the nucleus of atoms there are protons and .... Around the nucleus there are .... which have a .... charge. In the periodic table, atoms are arranged in order of their .... number.

2 Atoms are always neutral. Explain why.

3 How many protons, electrons and neutrons do the following atoms contain?

  a A nitrogen atom whose atomic number is 7 and its mass number is 14.

  b A chlorine atom whose atomic number is 17 and its mass number is 35.

Study tip

In an atom, the number of protons is always equal to the number of electrons. You can find out the number of protons and electrons in an atom by looking up its atomic number in the periodic table.

links

For more information on the patterns in the periodic table, see C1 1.3 The arrangement of electrons in atoms.

Key points

● Atoms are made of protons, neutrons and electrons.

● Protons and electrons have equal and opposite electric charges. Protons are positively charged, and electrons are negatively charged.

● Neutrons have no electric charge. They are neutral.

● Atomic number
    = number of protons
    (= number of electrons)
  Mass number = number of protons + neutrons

● Atoms are arranged in the periodic table in order of their atomic number.

The arrangement of electrons in atoms

Learning objectives

- How are the electrons arranged inside an atom?

- How is the number of electrons in the highest energy level of an atom related to its group in the periodic table?

- How is the number of electrons in the highest energy level of an atom related to its chemical properties?

- Why are the atoms of Group 0 elements so unreactive?

One model of the atom which we use has electrons arranged around the nucleus in **shells**, rather like the layers of an onion. Each shell represents a different **energy level**. The lowest energy level is shown by the shell which is nearest to the nucleus. The electrons in an atom occupy the lowest available energy level (the shell closest to the nucleus).

a Where are the electrons in an atom?
b Which shell represents the lowest energy level in an atom?

Electron shell diagrams

We can draw diagrams to show the arrangement of electrons in an atom. A carbon atom has 6 protons, which means it has 6 electrons. Figure 1 shows how we represent an atom of carbon.

An energy level (or shell) can only hold a certain number of electrons.

- The first, and lowest, energy level holds 2 electrons.
- The second energy level can hold up to 8 electrons.
- Once there are 8 electrons in the third energy level, the fourth begins to fill up, and so on.

To save drawing atoms all the time, we can write down the numbers of electrons in each energy level. This is called the **electronic structure**. For example, the carbon atom in Figure 1 has an electronic structure of 2,4.

A silicon atom with 14 electrons has the electronic structure 2,8,4. This represents 2 electrons in the first, and lowest, energy level, then 8 in the next energy level. There are 4 in the highest energy level (its outermost shell).

The best way to understand these arrangements is to look at some examples.

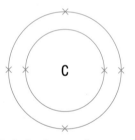

Figure 1 A simple way of representing the arrangement of electrons in the energy levels (shells) of a carbon atom

Study tip

Make sure that you can draw the electronic structure of the atoms for all of the first 20 elements. You will always be given their atomic number or their position in the periodic table (which tells you the number of electrons) – so you don't have to memorise these numbers.

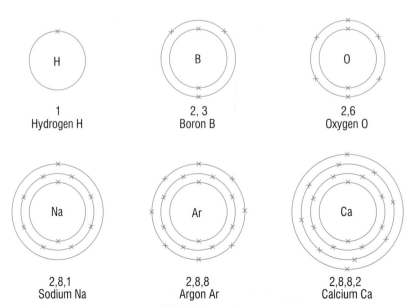

Figure 2 Once you know the pattern, you should be able to draw the energy levels (shells) and electrons in any of the first 20 atoms (given their atomic number)

c How many electrons can the first energy level hold?

d What is the electronic structure of sulfur (whose atoms contain 16 electrons)?

Electrons and the periodic table

Look at the elements in any one of the main groups of the periodic table. Their atoms will all have the same number of electrons in their highest energy level. These electrons are often called the outer electrons because they are in the outermost shell. Therefore, all the elements in Group 1 have one electron in their highest energy level.

Demonstration

Properties of the Group 1 elements

Your teacher will show you the Group 1 elements lithium, sodium and potassium. The elements in this group are called the **alkali metals**. Make sure you wear eye protection for all the demonstrations.

- In what ways are the elements similar?
- Watch their reactions with water and comment on the similarities.
- You might also be shown their reactions with oxygen.

Figure 3 The Group 1 metals are all reactive metals, stored under oil

The chemical properties of an element depend on how many electrons it has. The way an element reacts is determined by the number of electrons in its highest energy level (or outermost shell). So as the elements in a particular group all have the same number of electrons in their highest energy level, they all react in a similar way.

For example:

lithium + water → lithium hydroxide + hydrogen
sodium + water → sodium hydroxide + hydrogen
potassium + water → potassium hydroxide + hydrogen

The elements in Group 0 of the periodic table are called the noble gases because they are unreactive. Their atoms have a very stable arrangement of electrons. They all have 8 electrons in their outermost shell, except for helium, which has only 2 electrons.

Summary questions

1 Copy and complete using the words below:

electrons energy group nucleus shells

The electrons in an atom are arranged around the .... in .... (energy levels). The electrons further away from the nucleus have more .... than those close to the nucleus. All elements in the same .... of the periodic table have the same number of .... in their outermost shell.

2 Using the periodic table, draw the arrangement of electrons in the following atoms and label each one with its electronic structure.

a Li   **b** B   **c** P   **d** Ar

3 What is special about the electronic structure of neon and argon?

Key points

- The electrons in an atom are arranged in energy levels or shells.

- Atoms with the same number of electrons in their outermost shell belong in the same group of the periodic table.

- The number of electrons in the outermost shell of an element's atoms determines the way that element reacts.

- The atoms of the unreactive noble gases (in Group 0) all have very stable arrangements of electrons.

C1 1.4 Forming bonds

Learning objectives

- How do metals and non-metals bond to each other?
- How do non-metals bond to each other?
- How do we write the formula of a compound?

It is useful for us to know how atoms bond to each other in different substances. It helps us to predict and explain their properties.

How Science Works

Predicting what material to use

A team of research chemists and material scientists are working to make a new compound for the latest surfboard. Knowing about chemical bonding will make the process of designing a new compound a lot quicker.

Figure 1 Surfboards have to be very strong and have a relatively low density

The substances used to make a surfboard have to be very strong (to withstand large forces) and have a relatively low density (to float on water). Chemists help design materials with suitable properties. They will know before they start which combinations of atoms might prove useful to investigate.

Sometimes atoms react together by **transferring** electrons to form chemical bonds. This happens when metals react with non-metals. If the reacting atoms are all non-metals, then the atoms **share** electrons to form chemical bonds.

Forming ions

When a metal bonds with a non-metal, the metal atom gives one or more electrons to the non-metal atom. Both atoms become charged particles called **ions**.

- Metal atoms form positively charged ions (+).
- Non-metal atoms form negatively charged ions (−).

Opposite charges attract each other. There are strong attractions between the positive and negative ions in a compound of a metal and non-metal. These strong forces of attraction are the chemical bonds that form. They are called **ionic bonds**.

To see how ions are formed we can look at an example. Lithium metal will react with the non-metal fluorine. They make the compound lithium fluoride. Lithium atoms have 3 electrons, each negatively charged. As all atoms are neutral, we know it also has 3 positive protons in its nucleus. The charges on the negative electrons are balanced by the positive protons.

Study tip

When counting atoms, think of each symbol as a single atom and the formula of each ion as a single ion. Small numbers in a chemical formula only multiply the symbol they follow. Brackets are needed when there is more than one atom in the ion being multiplied. For example, a hydroxide ion has the formula OH^-. So calcium hydroxide, in which Ca^{2+} and OH^- combine, has the formula $Ca(OH)_2$.

Figure 2 The positive and negative charge on the ions in a compound balance each other, making the total charge zero

When lithium reacts with fluorine it loses 1 electron. This leaves it with only 2 electrons. However, there are still 3 protons in the nucleus. Therefore the lithium ion carries a 1+ charge.

3 protons	= 3+
2 electrons	= 2−
Charge on ion	= 1+

We show the formula of a lithium ion as **Li⁺**.

The electron lost from lithium is accepted by a fluorine atom. A fluorine atom has 9 electrons and 9 protons, making the atom neutral. However, with the extra electron from lithium, it has an extra 1− charge:

9 protons	= 9+
10 electrons	= 10−
Charge on ion	= 1−

We show the formula of a fluoride ion as **F⁻**.

Notice the spelling – we have a fluor**ine** atom which turns into a negatively charged fluor**ide** ion.

In compounds between metals and non-metals, the charges on the ions always cancel each other out. This means that their compounds have no overall charge. So the formula of lithium fluoride is written as **LiF**.

> **a** Potassium (K) is a metal. It loses one electron when it forms an ion. What is the formula of a potassium ion?

Forming molecules

Non-metal atoms bond to each other in a different way. The outermost shells of their atoms overlap and they share electrons. Each pair of shared electrons forms a chemical bond between the atoms. These are called **covalent bonds**. No ions are formed. They form molecules, such as hydrogen sulfide, H_2S, and methane, CH_4 (see Figure 3).

> **b** What do we call the bonds between nitrogen and hydrogen atoms in an ammonia molecule, NH_3?

Chemical formulae

The chemical formula of an ionic compound tells us the ratio of each type of ion in the compound. We use a ratio because when ions bond together they form structures made of many millions of ions. The ratio depends on the charge on each ion. The charges must cancel each other out.

An example is magnesium chloride. Magnesium forms Mg^{2+} ions and chlorine forms Cl^- ions. So the formula of magnesium chloride is $MgCl_2$. We have 2 chloride ions for every one magnesium ion in the compound (see Figure 4).

In covalent molecules we can just count the number of each type of atom in a molecule to get its formula. Figure 3 shows two examples.

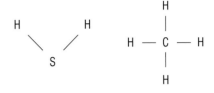

Hydrogen sulfide   Methane

Figure 3 There are strong covalent bonds between the non-metal atoms in each of these molecules. These are shown as lines between each atom or between the symbols of each atom in the molecule (H_2S and CH_4).

Figure 4 The 2+ positive charge on the magnesium ion balances the two 1− negative charges on the chloride ions in magnesium chloride ($MgCl_2$)

Key points

- When atoms from different elements react together they make compounds. The formula of a compound shows the number and type of atoms that have bonded together to make that compound.

- When metals react with non-metals, charged particles called ions are formed.

- Metal atoms form positively charged ions. Non-metal atoms form negatively charged ions. These oppositely charged ions attract each other in ionic bonding.

- Atoms of non-metals bond to each other by sharing electrons. This is called covalent bonding.

Summary questions

1 Copy and complete using the words below:

covalent lose gain ionic negative attract share positive

Metal atoms form .... ions because they .... one or more electrons when they combine with non-metals. Non-metal atoms .... electrons in the reaction, forming .... ions. The oppositely charged ions .... each other. This is called .... bonding.

When non-metals combine with each other, they form .... bonds. Their atoms .... electrons.

2 Sodium (Na) atoms lose one electron when they combine with fluorine (F). Each fluorine atom gains one electron in the reaction.
 a What is the name of the compound formed when sodium reacts with fluorine?
 b Write down the formula of a sodium ion and a fluoride ion.
 c What is the formula of the compound made when sodium reacts with fluorine.

C1 1.5 | Chemical equations

Learning objectives

- What happens to the atoms in a chemical reaction?
- How does the mass of reactants compare with the mass of products in a chemical reaction?
- How can we write balanced symbol equations to represent reactions? **[H]**

Chemical equations show the **reactants** (the substances we start with) and the **products** (the new substances made) of a reaction.

We can represent the test for hydrogen gas using a **word equation**:

$$\text{hydrogen} + \text{oxygen} \rightarrow \text{water}$$
$$\text{(reactants)} \qquad \text{(product)}$$

a State what happens in a positive test for hydrogen gas.

In chemical reactions the atoms get rearranged. You can think of them 'swapping partners'. Now you can investigate what happens to the mass of reactants compared with mass of products in a reaction.

Practical

Investigating the mass of reactants and products

You are given solutions of lead nitrate (toxic) and potassium iodide.

Wearing chemical splashproof eye protection, add a small volume of each solution together in a test tube.

- What do you see happen?

The formula of lead nitrate is $Pb(NO_3)_2$ and potassium iodide is KI.

The precipitate (a solid suspended in the solution) formed in the reaction is lead iodide, PbI_2 (toxic).

- Predict a word equation for the reaction.
- How do you think that the mass of reactants compare with the mass of the products?

Now plan an experiment to test your answer to this question.

Using **symbol equations** helps us to see how much of each substance is reacting. Representing reactions in this way is better than using word equations, for three reasons.

- Word equations are only useful if everyone who reads them speaks the same language.
- Word equations do not tell us how much of each substance is involved in the reaction.
- Word equations can get very complicated when lots of chemicals are involved.

For example, calcium carbonate decomposes (breaks down) on heating. We can show the reaction using a symbol equation like this:

$$CaCO_3 \rightarrow CaO + CO_2$$

This equation is **balanced** – there is the same number of each type of atom on both sides of the equation. This is very important, because atoms cannot be created or destroyed in a chemical reaction. This also means that:

The total mass of the products formed in a reaction is equal to the total mass of the reactants.

links

For more information on the decomposition of calcium carbonate, see C1 2.1 Limestone and its uses.

b Name the reactant and products in the decomposition of calcium carbonate.

We can check if an equation is balanced by counting the number of each type of atom on either side of the equation. If the numbers are equal, then the equation is balanced.

Maths skills

Look at the chemical equation on the previous page. We can work out the mass of $CaCO_3$, CaO or CO_2 given the masses of the other two compounds.

Because the total mass of the products formed in a reaction is equal to the total mass of the reactants we can write:

$$CaCO_3 \rightarrow CaO + CO_2$$
$$Mass = \quad a \quad = \quad b \quad + \quad c$$

So if the mass of CaO formed is 2.8 g (b above) and the mass of CO_2 is 2.2 g (c above); the mass of $CaCO_3$

(a) that we start with must be 2.8 + 2.2 (b + c) which equals **5.0 g**.

Rearranging the equation for a, b and c we get a – c = b.

So if the reaction started with 100 tonnes of $CaCO_3$ (a) and it gave off 44 tonnes of CO_2 (c),

Then the mass of CaO (b) made is 100 – 44 (a – c) = **56 tonnes**.

 Higher

Making an equation balance

In the case of hydrogen reacting with oxygen it is not so easy to balance the equation. First of all we write the formula of each reactant and product:

$$H_2 + O_2 \rightarrow H_2O$$

Counting the atoms on either side of the equation we see that we have:

Reactants
2 H atoms, 2 O atoms

Products
2 H atoms, 1 O atom

So we need another oxygen atom on the product side of the equation. We can't simply change the formula of H_2O to H_2O_2. (H_2O_2 – hydrogen peroxide is a bleaching agent which is certainly not suitable to drink!) But we can have **2 water molecules** in the reaction – this is shown in a symbol equation as:

$$H_2 + O_2 \rightarrow 2H_2O$$

Counting the atoms on either side of the equation again we get:

Reactants
2 H atoms, 2 O atoms

Products
4 H atoms, 2 O atom

Although the oxygen atoms are balanced, we now need two more hydrogen atoms on the reactant side. We do this by putting 2 in front of H_2:

$$\mathbf{2H_2 + O_2 \rightarrow 2H_2O}$$

Now we have:

Reactants
4 H atoms, 2 O atoms

Products
4 H atoms, 2 O atom

…. and the equation is balanced.

c Balance the following equation: $H_2 + Cl_2 \rightarrow HCl$

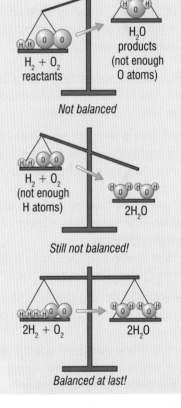

$H_2 + O_2$
reactants

H_2O
products
(not enough
O atoms)

Not balanced

$H_2 + O_2$
(not enough
H atoms)

$2H_2O$

Still not balanced!

$2H_2 + O_2$

$2H_2O$

Balanced at last!

Summary questions

1 Why must all symbol equations be balanced?

2 **a** A mass of 8.4 g of magnesium carbonate ($MgCO_3$) completely decomposes when it is heated. It made 4.0 g of magnesium oxide (MgO). What is the total mass of carbon dioxide (CO_2) produced in this reaction?

 b Write a word equation to show the reaction in part **a**.

3 Balance these symbol equations:
 a $Ca + O_2 \rightarrow CaO$ **b** $Al + O_2 \rightarrow Al_2O_3$ **c** $Na + H_2O \rightarrow NaOH + H_2$ **[H]**

Key points

● As no new atoms are ever created or destroyed in a chemical reaction:
The total mass of reactants = the total mass of products

● There is the same number of each type of atom on each side of a balanced symbol equation.

Summary questions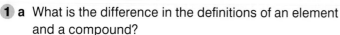

1 a What is the difference in the definitions of an element and a compound?

b The chemical formula of ethanol is written as C_2H_5OH.
  i How many atoms of hydrogen are there in an ethanol molecule?
  ii How many different elements are there in ethanol?
  iii What is the total number of atoms in an ethanol molecule?

2 a Draw a table to show the relative charge on protons, neutrons and electrons.

b In which part of an atom do we find:
  i protons
  ii neutrons
  iii electrons.

c i What is the overall charge on any atom?
  ii A nitrogen atom has 7 protons. How many electrons does it have?

3 This question is about the periodic table of elements. You will need to use the periodic table at the back of this book to help you answer some parts of the question.

a Argon (Ar) is the 18th element in the periodic table.
  i Is argon a metal or a non-metal?
  ii Are there more metals or non-metals in the periodic table?
  iii How many protons does an argon atom contain?
  iv State the name and number of the group to which argon belongs.
  v How many electrons does argon have in its highest energy level (outermost shell)?

b The element barium (Ba) has 56 electrons.
  i How many protons are in the nucleus of each barium atom?
  ii How many electrons does a barium atom have in its highest energy level (outermost shell)? How did you decide on your answer?
  iii Is barium a metal or a non-metal?

4 The diagram below shows the arrangement of electrons in an atom.

a How many protons are in the nucleus of this atom?

b Use the periodic table at the back of this book to give the name and symbol of the element whose atom is shown here.

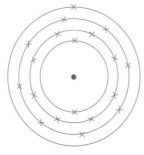

c This element forms ions with a 1+ charge.
  i What is an ion?
  ii How does the charge on the ion tell us whether the element above is a metal or non-metal?
  iii Describe what happens to the number of electrons when the atom forms a 1+ ion.
  iv Write the chemical formula of the ion.
  v This ion can form compounds with negatively charged ions. What type of bonding will we find in these compounds?

d A compound is formed when this element reacts with chlorine gas.
  i What is the name of the compound formed?
  ii Chloride ions carry a 1− charge. Write the chemical formula of the compound formed.

5 What is the missing number needed to balance the following symbol equations?

a $2Na + Cl_2 \rightarrow \ldots. \; NaCl$

b $2Zn + O_2 \rightarrow \ldots. \; ZnO$

c $\ldots. \; Cr + 3O_2 \rightarrow 2Cr_2O_3$

d $C_3H_8 + \ldots. \; O_2 \rightarrow 3CO_2 + 4H_2O$ [H]

6 Balance the following symbol equations:

a $H_2 + Br_2 \rightarrow HBr$

b $Mg + O_2 \rightarrow MgO$

c $H_2O_2 \rightarrow H_2O + O_2$

d $Li + H_2O \rightarrow LiOH + H_2$

e $NaNO_3 \rightarrow NaNO_2 + O_2$

f $Fe + O_2 \rightarrow Fe_2O_3$ [H]

7 When a mixture of iron and sulfur is heated, a compound called iron sulfide is made.

In an experiment 2.8 g of iron made 4.4 g of iron sulfide.

a What mass of sulfur reacted with the 2.8 g of iron?

b Explain how you worked out your answer to part **a**.

Practice questions (k)

1 Use numbers from the list to complete the table to show the charge on each subatomic particle.

+2   +1   0   –1   –2

Subatomic particle	Charge
electron	
neutron	
proton	

(3)

2 Use the periodic table at the back of your book to help you to answer this question.

a How many protons are in an atom of fluorine? (1)

b How many electrons are in an atom of carbon? (1)

c Complete the electronic structure of aluminium:
2,8, ... (1)

d What is the electronic structure of potassium? (1)

3 Neon is a noble gas.

a What does this tell you about its electronic structure? (1)

b Draw a diagram to show the electronic structure of neon. (2)

4 a Magnesium has the electronic structure 2,8,2. Explain, in terms of its electronic structure, why magnesium is in Group 2 of the periodic table. (1)

b Give **one** way in which the electronic structures of the atoms of Group 2 elements are:
i the same (1)
ii different. (1)

c When magnesium is heated in air it burns with a bright flame and produces magnesium oxide.

Calcium is also in Group 2. Describe what you expect to happen and what would be produced when calcium is heated in air. (2)

5 Sodium reacts with water to produce sodium hydroxide and hydrogen.

The word equation for this reaction is:

sodium + water → sodium hydroxide + hydrogen

a Name one substance in this equation that is:
i an element (1)
ii a compound (1)
iii has ionic bonds (1)
iv has covalent bonds (1)

b If 2.3 g of sodium reacted with 1.8 g of water, what would be the total mass of sodium hydroxide and hydrogen produced?
Explain your answer. (2)

c Balance the symbol equation for this reaction.

.... Na + ... H_2O → .... $NaOH$ + H_2 **[H]** (1)

d Lithium is in the same group of the periodic table as sodium.
i Write a word equation for the reaction of lithium with water. (1)
ii What is the formula of lithium hydroxide? (1)
iii How many atoms are shown in the formula of lithium hydroxide you have written? (1)

C1 2.1

Limestone and its uses

Learning objectives

- What are the uses of limestone?

- What happens when we heat limestone?

Uses of limestone

Limestone is a rock that is made mainly of **calcium carbonate**. Some types of limestone were formed from the remains of tiny animals and plants that lived in the sea millions of years ago. We dig limestone out of the ground in quarries all around the world. It has many uses, including its use as a building material.

Many important buildings around the world are made of limestone. We can cut and shape the stone taken from the ground into blocks. These can be placed one on top of the other, like bricks in a wall. We have used limestone in this way to make buildings for hundreds of years.

Powdered limestone can also be heated with powdered clay to make **cement**. When we mix cement powder with water, sand and crushed rock, a slow chemical reaction takes place. The reaction produces a hard, stone-like building material called **concrete**.

Figure 1 St Paul's Cathedral in London is built from limestone blocks

Figure 2 These white cliffs are made of chalk. This is one type of limestone, formed from the shells of tiny sea plants.

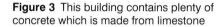

Figure 3 This building contains plenty of concrete which is made from limestone

a What is the main compound found in limestone?
b How do we use limestone to make buildings?

Heating limestone

The chemical formula for calcium carbonate is $CaCO_3$. It is made up of calcium ions, Ca^{2+}, and carbonate ions, CO_3^{2-}. The 2+ and 2− charges tell us that there are the same number of calcium ions and carbonate ions in calcium carbonate. Remember that the charges on the ions cancel out in compounds.

When we heat limestone strongly, the calcium carbonate breaks down to form calcium oxide. Carbon dioxide is also produced in this reaction. Breaking down a chemical by heating is called **thermal decomposition**.

We can show the thermal decomposition reaction using the following equations:

Word equation: calcium carbonate $\xrightarrow{\text{heat}}$ calcium oxide + carbon dioxide

Balanced symbol equation: $CaCO_3 \rightarrow CaO + CO_2$

The calcium oxide made is also a very useful substance in the building and farming industries.

Did you know ...?

Chalk is a form of limestone. It was formed millions of years ago from the skeletal remains of tiny sea plants called coccoliths. They were deposited on the seabed between 65 and 130 million years ago. It has been estimated that it took almost 100 000 years to lay down each metre of chalk in a cliff face.

links

For information on the formulae of compounds made up of ions, look back at C1 1.4 Forming bonds.

Practical

Thermal decomposition

In this experiment you can carry out the reaction that takes place in a lime kiln.

Safety: Make sure the rubber tube is tightly secured to the gas tap and the Bunsen burner before starting the experiment. Do not overstretch the tubing. Do not touch the decomposed carbonate as it is corrosive. Wash your hands if you get any chemicals on them. Wear eye protection.

Place a limestone chip on a tripod and gauze. Using a roaring flame, hold the base of the Bunsen burner and heat a limestone chip strongly from the side. It is best if the tip of the blue cone of the flame heats the limestone directly. You will see signs of a reaction happening on the surface of the limestone.

- What do you see happen as the limestone is heated strongly?

A rotary lime kiln

To make lots of calcium oxide this reaction is done in a furnace called a **lime kiln**. We fill the kiln with crushed limestone and heat it strongly using a supply of hot air. Calcium oxide comes out of the bottom of the kiln. Waste gases, including the carbon dioxide made, leave the kiln at the top.

Calcium oxide is often produced in a **rotary kiln**, where the limestone is heated in a rotating drum. This makes sure that the limestone is thoroughly mixed with the stream of hot air. This helps the calcium carbonate to decompose completely.

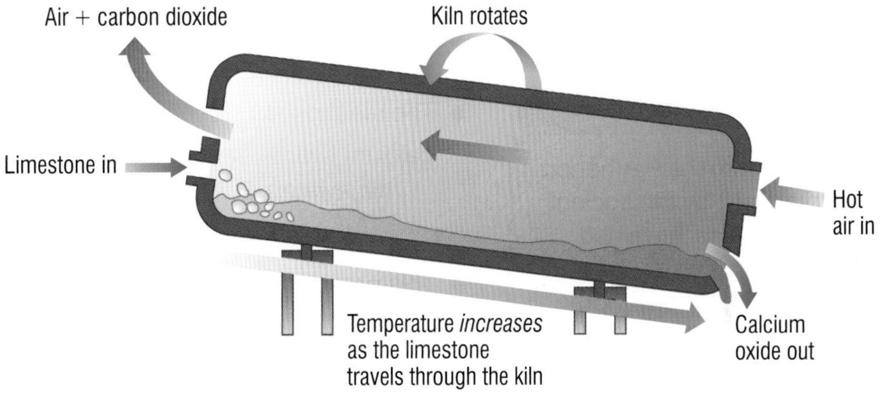

Air + carbon dioxide   Kiln rotates

Limestone in

Hot air in

Temperature *increases* as the limestone travels through the kiln

Calcium oxide out

Figure 4 Calcium oxide is produced in a rotary lime kiln

⬭ links

For more information on the uses of calcium oxide, see C1 2.3 The 'limestone reaction cycle'.

Summary questions

1 Copy and complete using the words below:

   building   calcium   cement   concrete

   Limestone is mostly made of .... carbonate (whose chemical formula is $CaCO_3$). As well as making blocks of building material, limestone can be used to produce .... and .... that are also used in the .... industry.

2 Produce a poster or PowerPoint presentation to show how limestone is used in building.

3 The stone roof of a building is supported by columns made of limestone. Why might this be unsafe after a fire in the building? Explain the chemical reaction involved in weakening the structure.

Key points

- Limestone is made mainly of calcium carbonate.

- Limestone is widely used in the building industry.

- The calcium carbonate in limestone breaks down when we heat it strongly to make calcium oxide and carbon dioxide. The reaction is called thermal decomposition.

C1 2.2   Reactions of carbonates

Learning objectives

- Do other carbonates behave in the same way as calcium carbonate?

- What happens when dilute acid is added to a carbonate?

- What is the test for carbon dioxide gas?

Buildings and statues made of limestone suffer badly from damage by acid rain. You might have noticed statues where the fine features have been lost. Limestone is mostly calcium carbonate, which reacts with acid. A gas is given off in the reaction.

Testing for carbon dioxide

You can use a simple test to find out if the gas given off is carbon dioxide. Carbon dioxide turns **limewater** solution cloudy. The test works as follows:

- Limewater is a solution of calcium hydroxide. It is alkaline.

- Carbon dioxide is a weakly acidic gas so it reacts with the alkaline limewater.

- In this reaction tiny solid particles of insoluble calcium carbonate are formed as a precipitate.

- The reaction is:

Figure 1 Limestone is attacked and damaged by acids

calcium hydroxide + carbon dioxide →   calcium carbonate   + water
    (limewater)   (an insoluble precipitate)
    $Ca(OH)_2$ + CO_2 → $CaCO_3$ + H_2O

- This precipitate of calcium carbonate makes the limewater turn cloudy. That's because light can no longer pass through the solution with tiny bits of white solid suspended in it.

> **a** What is a precipitate?

Carbonates react with acids to give a salt, water and carbon dioxide. For calcium carbonate the reaction with hydrochloric acid is:

calcium carbonate + hydrochloric acid → calcium chloride + water + carbon dioxide

The balanced symbol equation is:

$$CaCO_3 + 2HCl → CaCl_2 + H_2O + CO_2$$

> **b** Write a word equation for the reaction of magnesium carbonate with hydrochloric acid.

??? Did you know ...?

Sculptures from the Parthenon (a temple), built by the ancient Greeks in Athens, have had to be removed and replaced by copies to avoid any more damage from acid pollution from vehicle exhausts.

Figure 2 The Parthenon in Greece

Practical

Acid plus carbonates

Set up the apparatus as shown.

Try the test with some other carbonates, such as those of magnesium, copper, zinc and sodium.

Record your observations.

- What conclusion can you draw?

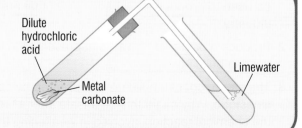

Dilute hydrochloric acid

Metal carbonate

Limewater

Decomposing carbonates

In C1 2.1 we saw that limestone is made up mainly of calcium carbonate. This decomposes when we heat it. The reaction produces calcium oxide and carbon dioxide. Calcium is an element in Group 2 of the periodic table. As we have already seen, the elements in a group tend to behave in the same way. So, does magnesium carbonate also decompose when you heat it? And what about other carbonates too?

c Why might you expect magnesium carbonate to behave in a similar way to calcium carbonate?

Practical

Investigating carbonates

You can investigate the thermal decomposition of carbonates by heating samples in a Bunsen flame. You will have samples of the carbonates listed below.

Powdered carbonate samples: sodium carbonate, potassium carbonate, magnesium carbonate, zinc carbonate, copper carbonate

- What observations might tell you if a sample decomposes when you heat it?
- How could you test any gas given off?

Plan an investigation to find out how easily different carbonates decompose.

- How will you try to make it a fair test?
- How will you make your investigation safe?

Before you start any practical work, your teacher must check your plan.

Safety: It is important to remove the delivery tube from the limewater before you stop heating the carbonate. If you don't, the cold limewater will be 'sucked back' into the hot boiling tube causing it to smash. You must wear eye protection when doing this practical.

Figure 3 Investigating the thermal decomposition of a solid

Investigations like this show that many metal carbonates decompose when they are heated in a Bunsen flame. They form the metal oxide and carbon dioxide – just as calcium carbonate does. Sodium and potassium carbonate do not decompose at the temperature of the Bunsen flame. They need a higher temperature.

Magnesium carbonate decomposes like this:

$$\text{magnesium carbonate} \rightarrow \text{magnesium oxide} + \text{carbon dioxide}$$
$$MgCO_3 \rightarrow MgO + CO_2$$

Summary questions

1 Give a general word equation for:
   a the reaction of a carbonate plus an acid
   b the thermal decomposition of a carbonate.

2 Write a word equation for the reaction of sodium carbonate with dilute hydrochloric acid.

3 The formula of zinc carbonate is $ZnCO_3$.
   a Zinc carbonate decomposes when heated, giving zinc oxide and carbon dioxide. Write the balanced equation for this reaction. **[H]**
   b Write the balanced symbol equation for the reaction of zinc carbonate with dilute hydrochloric acid. **[H]**

Key points

- Carbonates react with dilute acid to form a salt, water and carbon dioxide.
- Limewater turns cloudy in the test for carbon dioxide gas. A precipitate of insoluble calcium carbonate causes the cloudiness.
- Metal carbonates decompose on heating to form the metal oxide and carbon dioxide.

C1 2.3 The 'limestone reaction cycle' (k)

Learning objectives

- How can we make calcium hydroxide from calcium oxide?
- Why is calcium hydroxide a useful substance?
- What is the 'limestone reaction cycle'?

Limestone is used very widely as a building material. We can also use it to make other materials for the construction industry.

As we saw in C1 2.1 **calcium oxide** is made when we heat limestone strongly. The calcium carbonate in the limestone undergoes thermal decomposition.

When we add water to calcium oxide it reacts to produce **calcium hydroxide**. This reaction gives out a lot of heat.

$$\text{calcium oxide} + \text{water} \rightarrow \text{calcium hydroxide}$$
$$\text{CaO} + \text{H}_2\text{O} \rightarrow \text{Ca(OH)}_2$$

Although it is not very soluble, we can dissolve a little calcium hydroxide in water. After filtering, this produces a colourless solution called limewater. We can use limewater to test for carbon dioxide.

> **a** What substance do we get when calcium oxide reacts with water?
> **b** Describe how we can make limewater from calcium hydroxide.

Study tip

Make sure that you know the limestone reaction cycle and the equations for each reaction.

Practical

Investigating the 'limestone reaction cycle' (k)

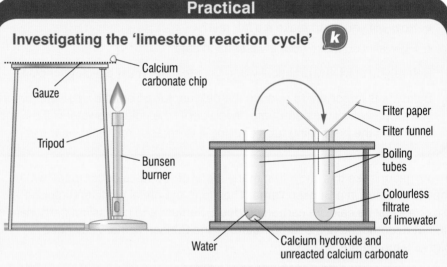

Gauze
Calcium carbonate chip
Tripod
Bunsen burner
Filter paper
Filter funnel
Boiling tubes
Colourless filtrate of limewater
Water
Calcium hydroxide and unreacted calcium carbonate

Heat the calcium carbonate chip very strongly, making it glow. Make sure you are wearing eye protection. The greater the area of the chip that glows, the better the rest of the experiment will be. This reaction produces calcium oxide (corrosive). Let the calcium oxide cool down. Then, using tongs, add it to the empty boiling tube.

Then you add a few drops of water to the calcium oxide, one drop at a time. This reaction produces calcium hydroxide.

When you dissolve this calcium hydroxide in more water and filter, it produces limewater.

Carbon dioxide bubbled through the limewater produces calcium carbonate. This turns the solution cloudy.

- The reaction between calcium oxide and water gives out a lot of energy. What do you observe during the reaction?
- Why does bubbling carbon dioxide through limewater make the solution go cloudy?

∞ links

For information on the test for carbon dioxide, look back at C1 2.2 Reactions of carbonates.

The reactions in the experiment can be shown on a flow diagram:

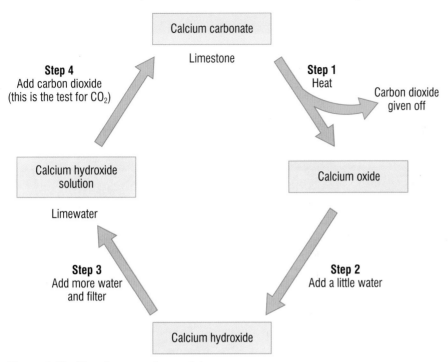

Figure 1 The 'limestone reaction cycle'

Did you know ...?

The saying 'to be in the limelight' originated from the theatre, well before the days of electricity. Stages were lit up by heated limestone before electric or gas lamps were invented.

Neutralising acids

Calcium hydroxide is an alkali. It reacts with acids in a neutralisation reaction. The products of the reaction are a calcium salt and water.

Calcium hydroxide is used by farmers to improve soil that is acidic. Because it is an alkali, it will raise the pH of acidic soil. It is also used to neutralise acidic waste gases in industry before releasing gases into the air.

Summary questions

1 Copy and complete using the words below:

carbon limewater hydroxide carbonate water oxide

When limestone is heated, the calcium .... in it decomposes to produce calcium .... and .... dioxide gas. If calcium oxide is reacted with water, calcium .... is produced. When we add more .... and filter we make a solution of calcium hydroxide called ....

2 Describe and explain the positive test for carbon dioxide gas. Include a word equation in your answer. (See C1 2.2).

3 a When calcium oxide reacts with water, calcium hydroxide is produced. Write a word equation and a balanced symbol equation to show the reaction. [H]

   b Calcium hydroxide is an alkali so it reacts with acids.
   i Give one use of calcium hydroxide that relies on this reaction.
   ii What do we call this type of reaction?

4 Why do we refer to the series of reactions in the practical box on the previous page as the 'limestone reaction cycle'?

Key points

- When water is added to calcium oxide it produces calcium hydroxide.

- Calcium hydroxide is alkaline so it can be used to neutralise acids.

- The reactions of limestone and its products that you need to know are shown in the 'limestone reaction cycle'.

C1 2.4

Cement and concrete

Learning objectives

- How has mortar developed over time?
- How do we make cement?
- What is concrete?
- How can you improve the quality of data collected in an investigation?

Figure 2 The original lime mortar has flaked away from the surface of the Sphinx in Egypt, and many of the stones are now missing

Figure 1 Lime mortar is not suitable for building pools as it will not harden when in contact with water

How Science Works

Development of lime mortar

About 6000 years ago the Egyptians heated limestone strongly in a fire and then combined it with water. This produced a material that hardened with age. They used this material to plaster the pyramids. Nearly 4000 years later, the Romans mixed calcium hydroxide with sand and water to produce **mortar**.

Mortar holds other building materials together – for example, stone blocks or bricks. It works because the lime in the mortar reacts with carbon dioxide in the air, producing calcium carbonate again. This means that the bricks or stone blocks are effectively held together by rock.

$$\text{calcium hydroxide} + \text{carbon dioxide} \rightarrow \text{calcium carbonate} + \text{water}$$
$$Ca(OH)_2 + CO_2 \rightarrow CaCO_3 + H_2O$$

The amount of sand in the mixture is very important. Too little sand and the mortar shrinks as it dries. Too much sand makes it too weak.

Even today, mortar is still used widely as a building material. However, modern mortars, made with cement in place of calcium hydroxide, can be used in a much wider range of ways than lime mortar.

Cement

Although lime mortar holds bricks and stone together very strongly, it does have some disadvantages. For example, lime mortar does not harden very quickly. It will not set at all where water prevents it from reacting with carbon dioxide.

Then people found that heating limestone with clay in a kiln produced cement. Much experimenting led to the invention of Portland cement. This is manufactured from a mixture of limestone, clay and other minerals. They are heated and then ground up into a fine powder.

This type of cement is still in use today. The mortar used to build a modern house is made by mixing Portland cement and sand. This sets when it is mixed thoroughly with water and left for a few days.

a What does lime mortar need in order to set hard?

b Why will lime mortar not set under water?

Concrete

Sometimes builders add small stones or crushed rocks, called aggregate, to the mixture of water, cement and sand. When this sets, it forms a hard, rock-like building material called concrete.

This material is very strong. It is especially good at resisting forces which tend to squash or crush it. We can make concrete even stronger by pouring the wet mixture around steel rods or bars and then allowing it to set. This makes reinforced concrete, which is also good at resisting forces that tend to pull it apart.

Did you know ...?

The Romans realised that they needed to add something to lime mortar to make it set in wet conditions. They found that adding brick dust or volcanic ash improved its setting. The modified mortar mixture could harden even under water. This method remained in use until the 18th century.

Practical

Which mixture makes the strongest concrete?

Try mixing different proportions of cement, gravel and sand, then adding water, to find out how to make the strongest concrete.

● How can you test the concrete's strength?

● How could you improve the quality of the data you collect?

Figure 3 Portland cement was invented nearly 200 years ago. It is still in use all around the world today.

Summary questions

1 Copy and complete using the words below:

mortar   concrete   clay   sand   bricks

Cement is made in industry by heating limestone with .... It can be mixed with sand to produce ...., used to hold building materials like .... in place. An even stronger material is made by mixing cement, ... and aggregate to make ....

2 List the different ways in which limestone has been used to build your home or school.

3 Concrete and mortar are commonly used building materials. Evaluate the use of:

a concrete to make a path rather than using mortar

b mortar to bind bricks to each other rather than using concrete.

Key points

● Cement is made by heating limestone with clay in a kiln.

● Mortar is made by mixing cement and sand with water.

● Concrete is made by mixing crushed rocks or small stones called aggregate, cement and sand with water.

C1 2.5　Limestone issues

Learning objectives

● What are the environmental, social and economic effects of quarrying limestone?

● What are the advantages and disadvantages of using limestone, cement and concrete as building materials?

Limestone is a very useful raw material, but mining for limestone can affect the local community and environment.

Limestone quarrying

Limestone is quarried from the ground. A quarry forms a huge hole in the ground. The limestone is usually blasted from a quarry by explosives. Then it is taken in giant lorries to be processed. Much of the limestone goes to cement factories which are often found near the quarry.

Figure 1 Limestone is often found in beautiful countryside. Quarrying the limestone scars the landscape.

Explosive charges are used to dislodge limestone from the rock face. This is known as blasting. As well as scarring the landscape the blasting noise scares off wildlife and can disturb local residents. Eventually a huge crater is formed. These can later be filled with water and can be used as a reservoir or for leisure activities. There is also the possibility of use as landfill sites for household rubbish before covering with soil and replanting.

Figure 2 Explosive charges are used to dislodge limestone from the rock face

Activity

Limestone debate

A large mining company wants to open a new limestone quarry on the edge of a National Park. Look at the views of different people affected by the planning decision to allow the quarry or not.

Take the role of one of the people shown and debate the issues involved. Assign a chairperson to make sure each person gets their say.

● Write your own response to the planning application in a letter to the mining company's managing director after your debate.

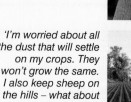
'Now we might get that by-pass we've been asking for.'

'This quarry will obviously destroy the habitats of birds and animals. A rare species of toad is found near the proposed site.'

'At last I might be able to get a job around here! I was born here and I really don't want to leave.'

'I'm worried about all the dust that will settle on my crops. They won't grow the same. I also keep sheep on the hills – what about the noise from the blasting?'

'We'll be able to supply limestone for the glass, steel and cement industries in this region now. We predict we'll be quarrying here for 10 years – then we'll landscape the crater before moving on.'

'The lorries carrying limestone will have to go straight through our village. My daughter's primary school is on the main road.'

'I think I'll get a lot more business from the workers at the quarry. I might start selling sandwiches and employ someone to make them freshly each day.'

Developments in limestone, cement and concrete

Bathroom tiles have traditionally been made from ceramics with a glazed finish to make them waterproof. They are very hard wearing. Nowadays more tiles are made from natural stone, such as travertine. These look very attractive with each tile having unique markings. However, travertine tiles are porous and can be easily scratched. They need to be sealed with a waterproof coating.

Cement is used to make mortar and concrete on building sites. Before cement mortar was invented, builders used lime mortar. However, this takes much longer to set fully than cement mortar, especially in wet conditions. The restoration of old buildings still needs lime mortar to repair brickwork. Often the old buildings have shallow, if any, foundations. Their brick walls are much more likely to move than modern buildings. With hard cement mortar this results in cracking along weak points in the walls. However, lime mortar offers more flexibility and will not crack as easily.

Carbon dioxide is a greenhouse gas. The manufacture of cement contributes about 5% of the CO_2 gas produced by humans emitted into the air. About half of this comes from burning fuels used to heat the kilns that decompose limestone. The rest comes from the reaction itself:

$$\text{calcium carbonate} \rightarrow \text{calcium oxide} + \text{carbon dioxide}$$

Using lime mortar would contribute less to carbon dioxide emissions as it absorbs CO_2 as it sets.

Concrete is the world's most widely used building material. Concrete was first reinforced using a wire mesh to strengthen it. Nowadays we can also use:

- glass fibres
- carbon fibres
- steel rods
- poly(propene), nylon, polyesters and Kevlar.

Some of the latest research uses pulp from wood, plants and recycled paper. A little recycled paper can improve concrete's resistance to cracking, impact (making it tougher) and scratching. These reinforcing materials are shredded into small pieces before adding them to the concrete mixture.

It is much cheaper to use reinforced concrete to make a bridge than to make it from iron or steel. However, steel is much stronger (harder to snap) than concrete. Over long spans, suspension bridges can use steel's high-tensile strength in cables between concrete towers. This will support the cheap reinforced concrete sections of bridges on which cars travel. Short span bridges will always be made from reinforced concrete because of its low cost.

Figure 3 Travertine is a form of limestone. Because travertine is made up mainly of calcium carbonate, tiles and worktops can be damaged by acidic solutions.

⚭ links

For information on how lime mortar reacts with CO_2 when setting, look back at C1 2.4 Cement and concrete.

Figure 4 The latest high performance concretes give architects new opportunities when designing buildings

Summary questions

1 Give one effect of starting up a new limestone quarry in a National Park in each of the following:
 a an environmental effect
 b a social effect
 c an economic effect.

2 A new material has been developed called ConGlassCrete. It has large pieces of recycled glass embedded into concrete. Its surface is polished smooth which gives a very attractive finish. Give one environmental advantage and one disadvantage of using ConGlassCrete instead of slate as a building material.

Key points

- There are good and bad points about quarrying for limestone. For example, more jobs will be created but there will be a large scar on the landscape.

- Limestone, cement and concrete all have useful properties for use as building materials but the mining and processing of limestone and its products has a major effect on our environment.

Summary questions

1 In the process of manufacturing cement, calcium carbonate is broken down by heat.

  a i Write a word equation to show the reaction that happens inside a lime kiln.

    ii What do we call this type of reaction?

  b Draw a diagram to show how you could test for the gas given off in the reaction described in part a.

  c Write a word equation to show the reaction between calcium oxide and water.

2 Write balanced symbol equations for the reactions in Question 1 parts a and c.   **[H]**

3 a How is limestone turned into cement?

  b Given cement powder, how would you make:

    i mortar

    ii concrete?

4 Potassium carbonate reacts with dilute hydrochloric acid. The gas given off gives a positive test for carbon dioxide.

  a Write a word equation and a balanced symbol equation to show the reaction between potassium carbonate, K_2CO_3, and dilute hydrochloric acid.   **[H]**

  b Describe what you see in a positive test for carbon dioxide.

  c Explain your observations made in part **b**. Include a word equation in your answer.

  d Write a balanced symbol equation for the reaction in part **c**.   **[H]**

5 a Here is a set of instructions for making concrete:

  'To make good, strong concrete, thoroughly mix together

  • 4 buckets of gravel

  • 3 buckets of sand

  • 1 bucket of cement

  When you have done this, add half a bucket of water.'

  Design and fill in a table to show the percentage of each substance in the concrete mixture. Give your values to the nearest whole number.

  b Describe an investigation you could use to find out which particular mixture of gravel, sand and cement makes the strongest concrete. What would you vary, what would you keep the same and how would you test the 'strength' of the concrete?

6 In an investigation into the behaviour of carbonates, a student draws the following conclusions when he heats samples of carbonates with a Bunsen burner:

Calcium carbonate	✓
Sodium carbonate	✗
Potassium carbonate	✗
Magnesium carbonate	✓
Zinc carbonate	✓
Copper carbonate	✓

(✓ = decomposes, ✗ = does not decompose

  a What was the independent variable in the investigation?

  b To which group in the periodic table do sodium and potassium belong?

  c To which group in the periodic table do magnesium and calcium belong?

  d What do these conclusions suggest about the behaviour of the carbonates of elements in Group 1 and Group 2?

  e Can you be certain about your answer to question d? Give reasons.

  f Write a word equation for the thermal decomposition of copper carbonate.

  g Write a balanced symbol equation for the thermal decomposition of magnesium carbonate.   **[H]**

Practice questions

1 Use words from the list to complete the sentences.
*calcium carbonate   calcium hydroxide
calcium oxide   carbon dioxide*

Limestone is mainly made of the compound ....
When limestone is heated strongly it decomposes
producing the gas .... and solid .... When the
solid reacts with water it produces ....   (4)

2 Match the compounds in the list with the descriptions.
*calcium carbonate   copper carbonate
sodium carbonate   zinc carbonate*

a When heated with a Bunsen burner it does not
decompose.   (1)

b It decomposes when heated to give zinc oxide.   (1)

c It is a blue solid that produces a black solid when
heated.   (1)

d It can be heated with clay to make cement.   (1)

3 Limestone blocks are damaged by acid rain.

Use words from the list to complete the sentences.
dissolves   escapes   produces   reacts

Calcium carbonate in the limestone .... with
acids in the rain. With sulfuric acid it ....
calcium sulfate, carbon dioxide and water. The carbon
dioxide .... into the air. The calcium sulfate
.... in the rainwater.   (4)

4 A student wanted to make calcium oxide from limestone.
The student heated a piece of limestone strongly in a
Bunsen burner flame.

a Complete the word equation for the reaction that
happened:
calcium carbonate → calcium oxide + ...   (1)

The student wanted to be sure he had made calcium
oxide. He crushed the heated limestone and added
water. The mixture got hot. The student cooled the
mixture and filtered it. This gave a colourless solution
and a white solid that was left in the filter paper.

b The student added universal indicator to the
colourless solution and it turned purple.

i Name the compound in the solution that causes
the indicator to turn purple.   (1)

ii Explain how the student's observations show
that he had made some calcium oxide by heating
limestone.   (1)

c The student added dilute hydrochloric acid to the
white solid from the filter paper.

The mixture fizzed and produced a gas that turned
limewater cloudy.

i What does this tell you about the white solid?   (1)

ii Was the student successful in changing all of the
limestone into calcium oxide? Explain your answer.   (1)

d Write balanced equations for the three chemical
reactions that the student did.   [H] (3)

5 Residents living near a cement works are concerned
because more children are suffering asthma attacks.
Residents have also noticed that parked cars are
becoming dirty because of smoke particles from the
chimney.

The table shows the possible medical risk from smoke
particles.

Particle size in mm	Medical effect
Larger than 0.4	No medical risks known
0.3 and smaller	Causes asthma attacks
0.2 and smaller	May cause cancer

It is also recommended that to avoid damage to health,
the concentration of any particles should be no higher
than 2 parts per million (ppm).

Scientists were brought in to monitor the emissions
from the cement works' chimney. They positioned four
sensors around the cement works to monitor airborne
smoke particles.

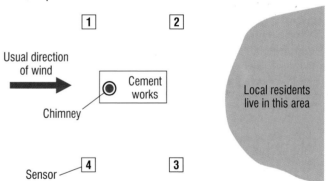

These four sensors only detect particle sizes larger than
0.5 mm and measure the concentration of particles in
ppm. The scientists reported that the particle sensors
showed that the average concentration of particles was
1.8 ppm. The scientists concluded that there was no risk
to health.

a Suggest **two** reasons why the local residents objected
to the positions of the four sensors.   (2)

b What evidence did the scientists use to conclude that
there was no risk to health?   (1)

c The local residents were still concerned that there was
a risk to health. Suggest **three** reasons why.   (3)

AQA, 2009

C1 3.1 — Extracting metals

Learning objectives

- Where do metals come from?
- How can we extract metals from their ores?

Figure 1 The Angel of the North stands 20 metres tall. It is made of steel which contains a small amount of copper.

??? Did you know ...?

Gold in Wales is found in seams, just like coal – although not as thick, unfortunately! Gold jewellery was worn by early Welsh princes as a badge of rank. Welsh gold has been used in modern times to make the wedding rings of royal brides.

Figure 3 Panning for gold. Mud and stones are washed away while the dense gold remains in the pan.

Metals have been important to people for thousands of years. You can follow the course of history by the materials people used. Starting from the Stone Age, we go to the Bronze Age (copper/tin) and then on to the Iron Age.

Where do metals come from?

Metals are found in the Earth's crust. We find most metals combined chemically with other chemical elements, often with oxygen. This means that the metal must be chemically separated from its compounds before you can use it.

In some places there is enough of a metal or metal compound in a rock to make it worth extracting the metal. Then we call the rock a metal **ore**. Ores are mined from the ground. Some need to be concentrated before the metal is extracted and purified. For example, copper ores are ground up into a powder. Then they are mixed with water and a chemical that makes the copper compound repel water. Air is then bubbled through the mixture and the copper compound floats on top as a froth. The rocky bits sink and the concentrated copper compound is scraped off the top. It is then ready to have its copper extracted.

Whether it is worth extracting a particular metal depends on:

- how easy it is to extract it from its ore
- how much metal the ore contains.

These two factors can change over time. For example, a new, cheaper method might be discovered for extracting a metal. We might also discover a new way to extract a metal efficiently from rock which contains only small amounts of a metal ore. An ore that was once thought of as 'low grade' could then become an economic source of a metal.

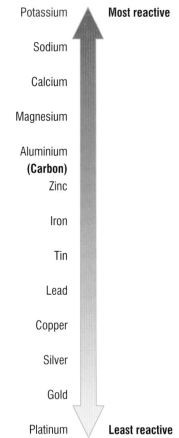

Figure 2 This reactivity series shows how reactive each element is compared to the other elements

Potassium — **Most reactive**
Sodium
Calcium
Magnesium
Aluminium
(Carbon)
Zinc
Iron
Tin
Lead
Copper
Silver
Gold
Platinum — **Least reactive**

A few metals, such as gold and silver, are so unreactive that they are found in the Earth as the metals (elements) themselves. We say that they exist in their native state.

Sometimes a nugget of gold is so large it can simply be picked up. At other times tiny flakes have to be physically separated from sand and rocks by panning.

a If there is enough metal in a rock to make it economic to extract it, what do we call the rock?
b Why is gold found as the metal rather than combined with other elements in compounds?

How do we extract metals?

The way that we extract a metal depends on its place in the **reactivity series**. The reactivity series lists the metals in order of their reactivity (see Figure 2). The most reactive are placed at the top and the least reactive at the bottom.

A more reactive metal will displace a less reactive metal from its compounds. Carbon (a non-metal) will also displace less reactive metals from their oxides. We use carbon to extract some metals from their **ores** in industry.

> **c** A metal cannot be extracted from its ore using carbon. Where is this metal in the reactivity series?

We can find many metals, such as copper, lead, iron and zinc, combined with oxygen. The compounds are called metal oxides. Because carbon is more reactive than each of these metals, we can use carbon to extract the metals from their oxides.

We must heat the metal oxide with carbon. The carbon removes the oxygen from the metal oxide to form carbon dioxide. The metal is also formed, as the element:

$$\text{metal oxide} + \text{carbon} \rightarrow \text{metal} + \text{carbon dioxide}$$

For example:

$$\text{lead oxide} + \text{carbon} \rightarrow \text{lead} + \text{carbon dioxide}$$

$$2PbO + C \rightarrow 2Pb + CO_2$$

We call the removal of oxygen from a compound chemical **reduction**.

> **d** What do chemists mean when they say that a metal oxide is reduced?

Metals that are more reactive than carbon are not extracted from their ores by reduction. Instead they are extracted using **electrolysis**.

Practical

Reduction by carbon

Heat some copper oxide with carbon powder in a test tube, gently at first then more strongly.

Empty the contents into an evaporating dish.

You can repeat the experiment with lead oxide and carbon if you have a fume cupboard to work in.

- Explain your observations. Include a word equation or a balanced symbol equation.

Key points

- A metal ore contains enough of the metal to make it economic to extract the metal. Ores are mined and might need to be concentrated before the metal is extracted and purified.

- We can find gold and other unreactive metals in their native state.

- The reactivity series helps us decide the best way to extract a metal from its ore. The oxides of metals below carbon in the series can be reduced by carbon to give the metal element.

- Metals more reactive than carbon *cannot* be extracted from their ores using carbon.

Summary questions

1 Copy and complete using the words below:

   crust   lead   extracted   native   elements   reduced

   Metals come from the Earth's .... Some metals are very unreactive and are found as ...., in their .... state. Metals, such as zinc, iron and ...., are found combined with oxygen in compounds. These metals can be .... using chemical reactions. The metal oxides are .... as oxygen is removed from the compound.

2 Define the word 'ore'.

3 Platinum is never found combined with oxygen. What does this tell you about its reactivity? Give a use of platinum that depends on this property.

4 Zinc oxide (ZnO) can be reduced to zinc by heating it in a furnace with carbon. Carbon monoxide (CO) is given off in the reaction.
   a Write a word equation for the reduction of zinc oxide.
   b Now write a balanced symbol equation for the reaction in part a.   **[H]**

C1 3.2 | Iron and steels

Learning objectives

- How is iron ore reduced?
- Why is iron from a blast furnace not very useful?
- How is iron changed to make it more useful?
- What are the main types of steel?

Iron ore contains iron combined with oxygen in iron oxide. Iron is less reactive than carbon. So we can extract iron by using carbon to remove oxygen from the iron(III) oxide in the ore. We extract iron in a **blast furnace**.

Some of the iron(III) oxide reacts with carbon. The carbon reduces iron(III) oxide, forming molten iron and carbon dioxide gas. This is one of the reduction reactions which takes place in a blast furnace:

$$\text{iron(III) oxide} + \text{carbon} \rightarrow \text{iron} + \text{carbon dioxide}$$

Iron straight from the blast furnace has limited uses. It contains about 96% iron and contains impurities, mainly carbon. This makes it very brittle, although it is very hard and can't be easily compressed. When molten it can be run into moulds and cast into different shapes. This **cast iron** is used to make wood-burning stoves, man-hole covers on roads, and engines.

We can treat the iron from the blast furnace to remove some of the carbon.

Removing all the carbon and other impurities from cast iron gives us pure iron. This is very soft and easily-shaped. However, it is too soft for most uses. If we want to make iron really useful we have to make sure that it contains tiny amounts of other elements. These include carbon and metals, such as nickel and chromium.

We call a metal that is mixed with other elements an **alloy**.

Steel is an alloy of iron. By adding elements in carefully controlled amounts, we can change the properties of the steel.

> **a** Why is iron from a blast furnace very brittle?
> **b** Why is pure iron not very useful?
> **c** How do we control the properties of steel?

Figure 1 The iron which has just come out of a blast furnace contains about 96% iron. The main impurity is carbon.

Steels

Steel is not a single substance. Like all alloys, it is a mixture. There are lots of different types of steel. All of them are alloys of iron with carbon and/or other elements.

Carbon steels

The simplest steels are the **carbon steels**. We make these by removing most of the carbon from cast iron, just leaving small amounts of carbon (from 0.03% to 1.5%). These are the cheapest steels to make. We use them in many products, such as the bodies of cars, knives, machinery, ships, containers and structural steel for buildings.

Often these carbon steels have small amounts of other elements in them as well. High carbon steel, with a relatively high carbon content, is very strong but brittle. On the other hand, low carbon steel is soft and easily shaped. It is not as strong, but is much less likely to shatter on impact with a hard object.

Mild steel is one type of low carbon steel. It contains less than 0.1% carbon. It is very easily pressed into shape. This makes it particularly useful in mass production, such as making car bodies.

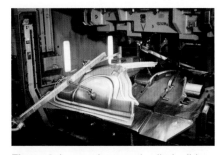

Figure 2 Low carbon steel called mild steel is easily pressed into shapes

Alloy steels

Low-alloy steels are more expensive than carbon steels because they contain between 1% and 5% of other metals. Each of these metals produces a steel that is well-suited for a particular use.

Figure 3 The properties of steel alloys make them ideal for use in suspension bridges

Even more expensive are the **high-alloy steels**. These contain a much higher percentage of other metals. The chromium–nickel steels are known as **stainless steels**. We use them to make cooking utensils and cutlery. They are also used to make chemical reaction vessels. That's because they combine hardness and strength with great resistance to corrosion. Unlike most other steels, they do not rust!

Figure 4 The properties of stainless steels make them ideal for making utensils and cutlery

?? Did you know …?

We use nickel–steel alloys to build long-span bridges, bicycle chains and military armour-plating. That's because they are very resistant to stretching forces. Tungsten steel operates well under very hot conditions so it is used to make high-speed tools such as drill bits.

Study tip

Know how the hardness of steels is related to their carbon content.

Summary questions

1 Copy and complete the following sentences using the terms below:

    carbon   pure   steel   cast   reduced

    Iron(III) oxide is .... (has its oxygen removed) in a blast furnace.

    Iron from the blast furnace, poured into moulds and left to solidify is called .... iron.

    If all the carbon and other impurities are removed from cast iron we get .... iron.

    Iron that has been alloyed with carbon and other elements is called ....

    Iron that contains just a small percentage of carbon is called .... steel.

2 How does cast iron differ from pure iron?

3 **a** Make a table to summarise the properties and some uses of low carbon steel, high carbon steel and chromium–nickel steel.

    b Why are surgical instruments made from steel containing chromium and nickel?

Key points

- We extract iron from iron ore by reducing it using carbon in a blast furnace.

- Pure iron is too soft for it to be very useful.

- Carefully controlled quantities of carbon and other elements are added to iron to make alloys of steel with different properties.

- Important examples of steels are:
    – low carbon steels which are easily shaped,
    – high carbon steels which are very hard,
    – stainless steels which are resistant to corrosion.

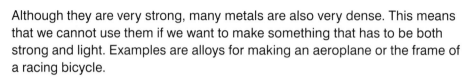

C1 3.3 Aluminium and titanium

Learning objectives

- Why are aluminium and titanium so useful?
- What method is used to extract metals that are more reactive than carbon?
- Why does it cost so much to extract aluminium and titanium?

Although they are very strong, many metals are also very dense. This means that we cannot use them if we want to make something that has to be both strong and light. Examples are alloys for making an aeroplane or the frame of a racing bicycle.

Where we need metals which are both strong and have a low density, **aluminium** and **titanium** are often chosen. These are also metals which do not corrode.

Properties and uses of aluminium

Aluminium is a silvery, shiny metal. It is surprisingly light for a metal as it has a relatively low density. It is an excellent conductor of energy and electricity. We can also shape it and draw it into wires very easily.

Although aluminium is a relatively reactive metal, it does not corrode easily. This is because the aluminium atoms at its surface react with oxygen in air. They form a thin layer of aluminium oxide. This layer stops any further corrosion taking place.

Aluminium is not a particularly strong metal, but we can use it to form alloys. These alloys are harder, more rigid and stronger than pure aluminium.

Because of these properties, we use aluminium to make a whole range of goods. These include:

- drinks cans
- cooking foil
- saucepans
- high-voltage electricity cables
- aeroplanes and space vehicles
- bicycles.

Figure 1 We use aluminium alloys to make bicycles because of their combination of low density and strength

a Why does aluminium resist corrosion?
b How do we make aluminium stronger?

Extracting aluminium

Because aluminium is a reactive metal we cannot use carbon to displace it from its oxide. Instead we extract aluminium using electrolysis. An electric current is passed through molten aluminium oxide at high temperatures to break it down.

First we must mine the aluminium ore. This contains aluminium oxide mixed with impurities. Then the aluminium oxide is separated from the impurities. The oxide must then be melted before electrolysis can take place.

The problem with using electrolysis to extract metals is that it is a very expensive process. That's because we need to use high temperatures to melt the metal compound. Then we also need a great deal of electricity to extract the metal from its molten compound. There are also environmental issues to consider when using so much energy.

Figure 2 We use aluminium alloys to make aircraft. The alloys are strong yet have a low density so the plane can carry more passengers and cargo.

Properties and uses of titanium

Titanium is a silvery-white metal. It is very strong and very resistant to corrosion. Like aluminium it has an oxide layer on its surface that protects it. Although it is denser than aluminium, it is less dense than most other metals.

Titanium has a very high melting point – about 1660 °C – so we can use it at very high temperatures.

We use titanium for:

● the bodies of high-performance aircraft and racing bikes (because of its combination of strength and relatively low density)
● parts of jet engines (because it keeps its strength even at high temperatures)
● parts of nuclear reactors (where it can stand up to high temperatures and its tough oxide layer means that it resists corrosion)
● replacement hip joints (because of its low density, strength and resistance to corrosion).

   c What properties make titanium ideal to use in jet engines and nuclear reactors?

Extracting titanium

Titanium is not particularly reactive, so we could produce it by displacing it from its oxide with carbon. But unfortunately carbon reacts with titanium metal making it very brittle. So we have to use a more reactive metal to displace titanium. We use sodium or magnesium. However, both sodium and magnesium have to be extracted by electrolysis themselves in the first place.

Before displacement of titanium can take place, the titanium ore must be processed. This involves separating the titanium oxide and converting it to a chloride. Then the chloride is distilled to purify it. Only then is it ready for the titanium to be displaced by the sodium or magnesium. Each one of these steps takes time and costs money.

   d Why do we need electricity to make:
      i aluminium and **ii** titanium?

Figure 3 We can use titanium inside the body as well as outside. This is an artificial hip joint, used to replace a natural joint damaged by disease or wear and tear.

∞ links

For more information on the environmental impact of extracting metals, see C1 3.6 Metallic issues.

Summary questions

1 Copy and complete using the words below:

   corrode   energy   expensive   high   low   oxide   reactive   strong

   Aluminium and titanium alloys are useful as they are .... and have a .... density. Although aluminium is reactive, it does not .... because its surface is coated with a thin, tough layer of aluminium .... Titanium does not corrode because it is not very .... and also has its oxide layer to protect it. We use large amounts of .... in the extraction of both metals from their ores which makes them .... The large number of steps involved in the extraction of the metals also contributes to their .... cost.

2 Why is titanium used to make artificial hip joints?

3 **a** Explain the different reasons why carbon cannot be used to extract:
      i aluminium, or **ii** titanium.
   b Name two processes in the extraction of aluminium that require large amounts of energy.

Key points

● Aluminium and titanium are useful because they resist corrosion.

● Aluminium requires the electrolysis of molten aluminium oxide to extract it as it is too reactive to reduce using carbon.

● Aluminium and titanium are expensive because extracting them from their ores involves many stages and requires large amounts of energy.

C1 3.4 Extracting copper

Learning objectives

- How is copper obtained from copper-rich ores?

- What methods can be used to obtain copper from low-grade ores?

- How is copper purified?

Figure 1 Mining copper ores can leave huge scars on the landscape. This is called open-cast mining. About 90% of copper comes from open-cast mines. Our supplies of copper-rich ores are a limited resource.

⚭ links

For information on the charges on metal ions, look back at C1 1.4 Forming bonds.

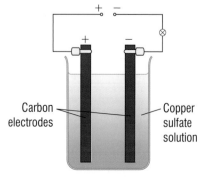

Figure 2 Extracting copper metal using electricity

Extracting copper from copper-rich ores

We extract most of our copper from **copper-rich ores**. These are a limited resource and are in danger of running out.

There are two main methods used to remove the copper from the ore.

- In one method we use sulfuric acid to produce copper sulfate solution, before extracting the copper.

- The other process is called **smelting** (roasting). We heat copper ore very strongly in a furnace with air to produce crude copper.

For example, copper can be found in an ore called chalcocite. This contains copper(I) sulfide, Cu_2S. If we heat the copper(I) sulfide in air, it decomposes to give copper metal:

$$\text{copper(I) sulfide} + \text{oxygen} \rightarrow \text{copper} + \text{sulfur dioxide}$$

Care has to be taken to avoid letting sulfur dioxide gas into the air. This gas causes acid rain. So chimneys are fitted with basic 'scrubbers' to neutralise the acidic gas.

Then we use the impure copper as the positive electrode in electrolysis cells to make pure copper. About 80% of copper is still produced by smelting.

a What chemical do we use to treat copper ore in order to form copper sulfate?

Smelting and purifying copper ore uses huge amounts of heat and electricity. This costs a lot of money and will have an impact on the environment.

Practical

Extracting copper from malachite

Malachite is a copper ore containing copper carbonate. To extract the copper we first heat the copper carbonate in a boiling tube. Thermal decomposition takes place. Copper oxide is left in the tube.

- Which gas is given off?

We then add dilute sulfuric acid to the copper oxide. Stopper and shake the tube. This makes copper sulfate solution. Filter off any excess black copper oxide in the solution.

To extract the copper metal, either

1. Put an iron nail into the copper sulfate solution

- What happens to the iron nail?

Or

2. Collect some extra copper sulfate solution and place it in a small beaker. Set up the circuit as shown in Figure 2. Turn the power on until you see copper metal collecting.

- Which electrode – the positive or the negative – does the copper form on?

Metal ions are always positively charged. Therefore, in electrolysis they are attracted to the negative electrode. So metals are always deposited at the negative electrode. In industry the electrolysis is carried out in many cells running at once. This method gives the very pure copper needed to make electrical wiring. Electrolysis is also used to purify the impure copper extracted by smelting. In the industrial process, the electrolysis cells use copper electrodes.

The copper can also be extracted from copper sulfate solution in industry by adding scrap iron. Iron is more reactive than copper, so it can **displace** copper from its solutions:

$$\text{iron} + \text{copper sulfate} \rightarrow \text{iron sulfate} + \text{copper}$$

Extracting copper from low-grade copper ores

Instead of extracting copper from our limited copper-rich ores, scientists are developing new ways to get copper from low-grade ores. This would be uneconomical using traditional methods. We can now use bacteria (**bioleaching**) and even plants (**phytomining**) to help extract copper.

In phytomining, plants can absorb copper ions from low-grade copper ore as they grow. This could be on slag heaps of previously discarded waste from the processing of copper-rich ores. Then the plants are burned and the metals can be extracted from the ash. The copper ions can be 'leached' (dissolved) from the ash by adding sulfuric acid. This makes a solution of copper sulfate. Then we can use displacement by scrap iron and electrolysis to extract pure copper metal.

In bioleaching, bacteria feed on low-grade metal ores. By a combination of biological and chemical processes, we can get a solution of copper ions (called a 'leachate') from waste copper ore. Once again, we use scrap iron and electrolysis to extract the copper from the leachate.

About 20% of our copper comes from bioleaching. This is likely to increase as sources of copper-rich ores run out.

Bioleaching is a slow process so scientists are researching ways to speed it up. At present it can take years to extract 50% of the metal from a low-grade ore.

??? Did you know ...?

Copper metal is so unreactive that some samples of copper exist in nature as the element itself. It is found native. A huge copper boulder was discovered by a diver at the bottom of Lake Superior in North America. It was raised to the surface in 2001. It has a mass of about 15 000 kg.

∞ links

For more information on the environmental impact of extracting metals and phytomining, see C1 3.6 Metallic issues.

Figure 3 In Australia Dr Jason Plumb looks for bacteria that can extract metals from ores. His search takes him to some exciting places – including volcanoes!

Summary questions

1 Copy and complete using the words below:

bacteria   smelting   electricity   phytomining   iron   low   sulfuric

Traditionally, copper can be extracted from some of its ores by heating (............). If copper ore is treated with .... acid, we get a solution of copper sulfate. We can obtain copper metal from this solution either by adding .... metal or by passing .... through the solution. Now new ways are being developed to extract copper using .... (bioleaching) or plants (............). These can extract the copper from ....-grade ores.

2 a Explain briefly two traditional ways of extracting copper metal.
   b State an advantage of extracting copper using bacteria rather than traditional methods.
   c Why can copper sometimes be found native (as the element itself)?
   d When copper is purified by electrolysis, which electrode do you think that the pure copper collects at? Why?

3 Write a balanced chemical equation for the extraction of copper:
   a from copper(I) sufide   [H]
   b from copper sulfate solution using scrap iron.   [H]

Key points

● Most copper is extracted by smelting (roasting) copper-rich ores, although our limited supplies of ores are becoming more scarce.

● Copper can be extracted from copper solutions by electrolysis or by displacement using scrap iron. Electrolysis is also used to purify impure copper, e.g. from smelting.

● Scientists are developing ways to extract copper that use low-grade copper ores. Bacteria are used in bioleaching and plants in phytomining.

C1 3.5 Useful metals

- What are transition metals and why are they so useful?

- Why is copper such a useful metal?

- Why are alloys more useful than pure metals?

Transition metals

In the centre of the periodic table there is a large block of metallic elements. They are called the **transition metals**. Many of them have similar properties. Like all metals, the transition metals are very good conductors of electricity and energy. They are strong but can also be bent or hammered into useful shapes.

Transition metals

Figure 1 The position of the transition metals in the periodic table

a In which part of the periodic table do we find the transition metals?

b Name three properties of these elements.

The properties of the transition metals mean that we can use them in many different ways. You will find them in buildings and in cars, trains and other types of transport. Their strength makes them useful as building materials. We use them in heating systems and for electrical wiring because energy and electricity pass through them easily.

Copper is a very useful transition metal. It can be bent but is still hard enough for plumbers to use as water tanks or pipes. Fortunately, it does not react with water. Copper also conducts electricity and energy very well. So it is ideal where we need:

- pipes that will carry water, or

- wires that will conduct electricity.

c What makes copper so useful for a plumber?

Figure 2 Copper is particularly useful because it is such a good conductor of electricity

Figure 3 Transition metals are used in many different ways because of their useful properties

Copper alloys

Bronze was probably the first alloy made by humans, about 5500 years ago. It is usually made by mixing copper with tin. We use it to make ship's propellers because of its toughness and resistance to corrosion.

We make brass by alloying copper with zinc. Brass is much harder than copper but it is workable. It can be hammered into sheets and pressed into intricate shapes. This property is used to make musical instruments.

> **d** Why are copper alloys more suitable for some uses than pure copper metal?

Aluminium alloys

Aluminium has a low density for a metal. It can be alloyed with a wide range of other elements. There are over 300 alloys of aluminium available. These alloys have very different properties. We can use some to build aircraft while others can be used as armour plating on tanks and other military vehicles.

Gold alloys

As with copper and iron, we can make gold and aluminium harder by adding other elements. We usually alloy gold with copper when we use it in jewellery. Pure gold wears away more easily than its alloy with copper. By varying the proportions of the two metals we also get different shades of 'gold' objects.

Figure 4 The Statue of Liberty in New York contains over 80 tonnes of copper

Figure 5 Alloying with copper makes gold more hardwearing. This is especially important in wedding rings, which many people wear most of the time.

?? Did you know …?

The purity of gold is often expressed in 'carats', where 24-carat gold is almost pure gold (99.9%). If you divide the carat number by 24, you get the fraction of gold in your jewellery. So an 18-carat gold ring will contain ¾ (75%) gold.

> **e** What property of aluminium makes it useful for making alloys in the aircraft industry?
>
> **f** Apart from making gold harder, what else can alloying change?

Summary questions

1 Copy and complete using the words below:

   aluminium   brass   aircraft   bronze   soft   transition

   The .... metals are found in the central block of the periodic table. Like pure iron, pure copper is too .... to be very useful. We can make copper harder by alloying it with tin to make ...., and with zinc to make ....

   There are over 300 alloys of the low-density metal .... Many of these are used to make .... where strength is also an important property.

2 **a** Write a list of the properties of a typical transition metal.

   b Why is copper metal used so much in plumbing?

3 Silver and gold are transition metals that conduct electricity even better than copper. Why do we use copper to make electric cables instead of either of these metals?

4 Why can aluminium alloys be used in so many different ways?

Key points

- The transition metals are found in the central block of elements in the periodic table.

- Transition metals have properties that make them useful for building and making things. For example, copper is used in wiring because of its high electrical conductivity.

- Copper, gold and aluminium are all alloyed with other metals to make them harder.

C1 3.6 Metallic issues

Learning objectives

- What issues arise in exploiting metal ores?
- How can plants help to exploit low-grade metal ores?
- Why should we recycle metals?
- What are the good and bad points in using metals to build structures?

Exploiting metal ores

It is difficult to imagine our lives without metals. They play a vital role in our technological society. Just think of all those electrical devices we depend on! However, whenever we mine metal ores from the Earth's crust there are consequences for our environment.

You have seen that open cast mining is often used to get copper ore from the ground. The ores of iron and aluminium are also mainly mined like this. Huge pits that scar the landscape are made, creating noise and dust and destroying the habitats of plants and animals. The mines also leave large heaps of waste rock.

The water in an area can also be affected by mining. As rain drains through exposed ores and slag heaps of waste, the groundwater can become acidic.

Then the ores must be processed to extract the metals. For example, sulfide ores are heated strongly in smelting. Any sulfur dioxide gas that escapes into the air will cause acid rain.

Phytomining

As plants grow, they absorb dissolved ions in the soil through their roots. Some plants are very effective at absorbing metal ions. Once harvested, we can extract the metals from ash left after burning the plants. This can be used in the phytomining of low-grade metal ores, such as copper ores.

Copper metal is extracted from the plant by dissolving the ash in sulfuric acid first of all. Then the solution made can be electrolysed to get the copper. The copper collects at the negative electrode. Alternatively, scrap iron can be added to the solution to displace copper:

iron + copper sulfate → iron sulfate + copper

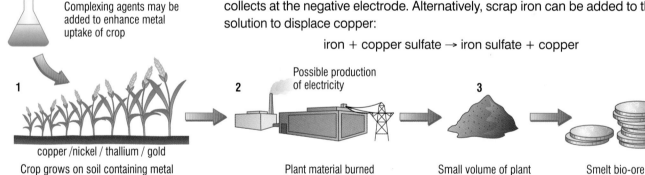

Complexing agents may be added to enhance metal uptake of crop

1 copper /nickel / thallium / gold
Crop grows on soil containing metal concentration too low for conventional exploitation

2 Possible production of electricity
Plant material burned

3 Small volume of plant ash (bio-ore) containing high concentration of target metal

Smelt bio-ore to yield metal

Figure 1 Metal ions are absorbed by plants which can then be processed to extract the metals

∞ links

For more information on open cast copper mining and phytomining, look back at C1 3.4 Extracting copper.

Recycling metals

In the UK each person uses around 8 kg of aluminium every year. This is why it is important to **recycle** aluminium. It saves energy, and therefore money, since recycling aluminium does not involve electrolysis. Comparing recycled aluminium with aluminium extracted from its ore, there is a 95% energy saving.

We also recycle iron and steel. 'Tin cans' are usually steel cans with a very thin coating of tin to prevent rusting. These cans are easy to separate from rubbish as they are magnetic. Using recycled steel saves about 50% of the energy used to extract iron and turn it into steel. Much of this energy is supplied by burning fossil fuels so recycling helps save our dwindling supplies of the fuels.

Copper is also recycled but this is tricky as it is often alloyed with other metals. So it would need to be purified for use in electrical wiring.

Recycling metals reduces the need to mine the metal ore and conserves the Earth's reserves of metal ores. It also prevents any pollution problems that arise from extracting the metal from its ore.

Metallic structures

Steel is the most commonly used metal. It is often used in the construction industry where strength is needed. For example:

- skyscrapers have steel girders supporting them
- suspension bridges use thick steel cables
- concrete bridges over motorways are made from concrete, reinforced with steel rods.

However, steel does have some drawbacks. Unfortunately the iron in it tends to rust. Stainless steel could be used but only for small specialist jobs. That's because it is much more expensive than ordinary steel. Even so, protecting the steel from rusting also costs money. Coatings, such as paint or grease, also have to be reapplied regularly. Rusting will affect the strength of steel and can be dangerous.

Figure 2 Recycling cans saves energy and reduces pollution

Figure 3 The Golden Gate Bridge in San Francisco uses thick steel cables to support it

Some benefits of using metals in construction	Some drawbacks of using metals in construction
<ul><li>Copper is a good electrical conductor for wiring; it is not reactive so can be made into water pipes</li><li>Lead can be bent easily so is used to seal joints on roofs</li><li>Steel is strong for girders and scaffolding</li><li>Aluminium alloys are corrosion-resistant.</li></ul>	<ul><li>Iron and steel can rust, severely weakening structures, e.g. if the iron rods used inside reinforced concrete rust, structures can collapse</li><li>The exploitation of metal ores to extract metals causes pollution and uses up the Earth's limited resources</li><li>Metals are more expensive than other materials such as concrete.</li></ul>

Activity

Saving energy

Make a list of the processes required in one of the following:

- extracting iron from its ore and then making steel
- extracting aluminium from its ore
- recycling steel or aluminium.

Highlight those processes that use a lot of energy. Then write a paragraph justifying the claims that recycling metal saves energy.

Summary questions

1 What can a mining company do to help the environment when an open-cast mine is no longer economic?

2 Each person in the UK uses about 8 kg of aluminium each year.
  a Recycling 1 kg of aluminium saves about enough energy to run a small electric fire for 14 hours. If you recycle 50% of the aluminium you use in one year, how long could you run a small electric fire on the energy you have saved?
  b Explain the benefits of recycling aluminium.

Key points

- There are social, economic and environmental issues associated with exploiting metal ores.
- Plants can remove metals from low-grade ores. The metals can be recovered by processing the ash from burning the plants.
- Recycling metals saves energy and our limited metal ores (and fossil fuels). The pollution from extracting metals is also reduced.
- There are drawbacks as well as benefits from the use of metals in structures.

Summary questions

1 Write simple definitions for the following terms:
  a metal ore
  b native state
  c chemical reduction.

2 We can change the properties of metals by alloying them with other elements.

Write down **three** ways that a metal alloy may be different from the pure metal.

3 a What name is given to the method of extracting copper from an ore:
  i using bacteria
  ii using plants
  iii using heat
  iv using electricity?

  b Which methods in part **a** are being developed to extract copper from low-grade copper ores?

4 Describe how brassicas can be used to decontaminate 'brown-field' sites and recover the polluting metals.　**[H]**

5 Carry out some research to find the advantages and disadvantages of using bioleaching to extract copper metal.

6 By the middle of the decade scrap car dealers are required to recover 95% of all materials used to make a car. The following table shows the metals we find in an average car:

Material	Average mass (kg)	% mass
Ferrous metal (steels)	780	68.3
Light non-ferrous metal (mainly aluminium)	72	6.3
Heavy non-ferrous metal (for example lead)	17	1.5

Other materials used include plastics, rubber and glass.

  a What is the average mass of metal in a car?

  b What percentage of a car's mass is made up of **non-metallic** materials?

  c i What is the main metal found in most cars?
  ii Which of this metal's properties allows it to be separated easily from other materials in the scrap from a car?

7 The following was overheard in a jeweller's shop:

"I would like to buy a 24-carat gold ring for my husband."

"Well madam, we would advise that you buy one which is a lower carat gold. It looks much the same but the more gold there is, the softer it is."

Is this actually the case? Let's have a look scientifically at the data.

Pure gold is said to be 24 carats. A carat is a twenty-fourth, so $24 \times \frac{1}{24} = 1$ or pure gold. So a 9-carat gold ring will have $\frac{9}{24}$ gold and $\frac{15}{24}$ of another metal, probably copper or sometimes silver. Most 'gold' sold in shops is therefore an alloy.

How hard the 'gold' is will depend on the amount of gold and on the type of metal used to make the alloy.

Here are some data on the alloys and the maximum hardness of 'gold'.

Gold alloy (carat)	Maximum hardness (BHN)
9	170
14	180
18	230
22	90
24	70

  a Which metals are used to alloy gold in jewellery?

  b The shop assistant said that 'the more gold there is, the less hard it is.' Was this based on science or was it hearsay? Explain your answer.

  c In this investigation which is the independent variable?

  d Which type of variable is 'the maximum hardness of the alloy' – continuous or categoric?

  e Plot a graph of the results.

  f What is the pattern in the results?

Practice questions ⓚ

1 Bicycle frames are often made from metal tubes. The metal tubes are produced using the steps in this list:
mining → concentrating → extracting → purifying → alloying → shaping
Match each of the following statements with the correct word from the list.
a The metal is produced using chemical reduction. (1)
b The metal is mixed with other metals to make it harder and stronger. (1)
c The metal ore is dug from the ground. (1)
d Waste rock is removed from the metal ore. (1)
e Other elements are removed from the metal. (1)

2 Choose the correct words from those shown to complete each sentence.
a Gold is found in the Earth as .... (1)
gold chloride   gold metal   gold oxide
b Iron is extracted by reacting iron oxide with .... (1)
carbon   copper   nitrogen
c Aluminium is extracted from aluminium oxide using .... (1)
combustion   distillation   electrolysis

3 Copper metal is used for electric wires. An alloy of copper, called brass, is used for pins and terminals of electric plugs.
a Copper metal is relatively soft and flexible. Give another reason why copper is used for electric wires. (1)
b Brass is an *alloy*. What is an *alloy*? (1)
c Open-cast mining of copper ore makes a very large hole.
  i Suggest **one** environmental problem that is caused by open-cast mining of copper ore. (1)
  ii Some copper ores contain copper sulfide, CuS. Copper sulfide in heated in air to produce copper and sulfur dioxide.
  $CuS + O_2 \rightarrow Cu + SO_2$
  Suggest **one** environmental problem caused by heating copper sulfide in air. (1)
d The amount of copper-rich ores is estimated to last only a few more years. New houses need several kilometres of copper wire.
  i Explain why the need to use so much copper will cause a problem in the future. (1)
  ii Suggest **two** ways in which society could overcome this problem. (2)
  AQA, 2008

4 *In this question you will be assessed on using good English, organising information clearly and using specialist terms where appropriate.*
Most of the iron we use is converted into steels.
Describe and explain how the differences in the properties of the three main types of steel allow them to be used in different ways. (6)

5 Titanium is used in aircraft, ships and hip replacement joints. Titanium is as strong as steel but 45% lighter, and is more resistant to acids and alkalis.
Most titanium is produced from its ore, rutile (titanium oxide), by a batch process that takes up to 17 days.
  Titanium oxide is reacted with chlorine to produce titanium chloride →
  Titanium chloride is reacted with magnesium at 900 °C in a sealed reactor for 3 days →
  The reactor is allowed to cool, then opened and the titanium is separated from the magnesium chloride by hand.
Titanium reactors produce about 1 tonne of the metal per day.
Iron blast furnaces produce about 20 000 tonnes of the metal per hour.
a Give **one** property of titanium that makes it more useful than steel for hip replacement joints. (1)
b Suggest **three** reasons why titanium costs more than steel. (3)
  AQA, 2008

6 Phytomining uses plants to absorb metal compounds from the ground. It is often used on land that has been contaminated by normal mining. It involves these stages:
Sow seeds → grow plants → harvest plants → dry plants → burn plants → collect ash
The ash is then treated like a metal ore obtained by normal mining.
a Suggest **one** environmental advantage of phytomining compared with normal mining. (1)
The table shows information about some metals that are absorbed by plants used for phytomining.

Metal	Value of metal in £ per kg	Maximum mass of metal in plants in g per kg	Percentage (%) of metal in normal ore
Gold	25 000	0.10	0.002
Nickel	17	38	2
Copper	4.9	14	0.5
Zinc	3.2	40	5
Lead	1.5	10	3

b The plants used for gold phytomining give a maximum yield of 20 tonnes of plants per hectare. Calculate the maximum value of the gold that can be recovered from 1 hectare. (2)
c One kilogram of plants used for nickel phytomining produces 150 g of ash.
What is the percentage of nickel in the ash? (2)
d Suggest reasons why phytomining has been used to produce gold, nickel and copper, but is only rarely used to produce zinc and lead. (4)

C1 4.1 Fuels from crude oil

Learning objectives

- What is in crude oil?
- What are alkanes?
- How do we represent alkanes?

Figure 1 The price of nearly everything we buy is affected by oil because the cost of moving goods to the shops affects the price we pay for them

Some of the 21st century's most important chemicals come from crude oil. These chemicals play a major part in our lives. We use them as fuels to run our cars, to warm our homes and to make electricity.

Fuels are important because they keep us warm and on the move. So when oil prices rise, it affects us all. Countries that produce crude oil can affect the whole world economy by the price they charge for their oil.

> **a** Why is oil so important?

Crude oil

Crude oil is a dark, smelly liquid. It is a **mixture** of lots of different chemical compounds. A mixture contains two or more elements or compounds that are not chemically combined together.

Crude oil straight from the ground is not much use. There are too many substances in it, all with different boiling points. Before we can use crude oil, we must separate it into different substances with similar boiling points. These are known as **fractions**. Because the properties of substances do not change when they are mixed, we can separate mixtures of substances in crude oil by using **distillation**. Distillation separates liquids with different boiling points.

> **b** What is crude oil?
> **c** Why can we separate crude oil using distillation?

Demonstration

Distillation of crude oil

Mixtures of liquids can be separated using distillation. This can be done in the lab on a small scale. We heat the crude oil mixture so that it boils. The different fractions vaporise between different ranges of temperature. We can collect the vapours by cooling and condensing them.

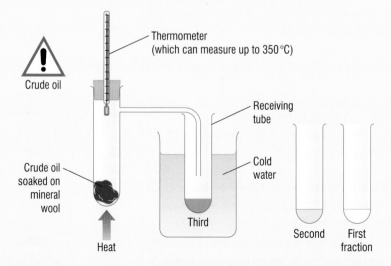

- What colour are the first few drops of liquid collected?

Hydrocarbons k

Nearly all of the compounds in crude oil are compounds containing only hydrogen and carbon. We call these compounds **hydrocarbons**. Most of the hydrocarbons in crude oil are **alkanes**. You can see some examples of alkane molecules in Figure 2.

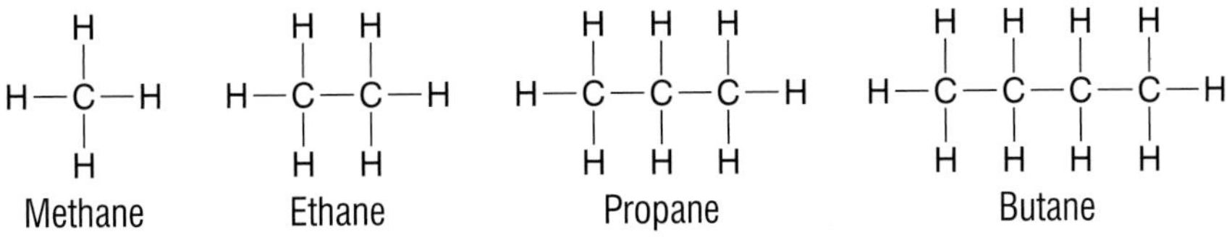

Figure 2 We can represent alkanes like this, showing all of the atoms in the molecule. They are called displayed formulae. The line drawn between two atoms in a molecule represents the covalent bond holding them together.

Look at the formulae of the first five alkane molecules:

CH$_4$ (methane)

C$_2$H$_6$ (ethane)

C$_3$H$_8$ (propane)

C$_4$H$_{10}$ (butane)

C$_5$H$_{12}$ (pentane).

Can you see a pattern in the formulae of the alkanes? We can write the general formula for alkane molecules like this:

$$C_nH_{(2n + 2)}$$

which means that 'for every n carbon atoms there are (2n + 2) hydrogen atoms'. For example, if an alkane contains 12 carbon atoms its formula will be **C$_{12}$H$_{26}$**.

We describe alkanes as **saturated hydrocarbons**. This means they have single carbon–carbon bonds only (do not contain any C=C).

links

For information on covalent bonding, look back at C1 1.4 Forming bonds.

Summary questions

1 Copy and complete using the words below:

carbon   distillation   hydrocarbons   hydrogen   mixture

Crude oil is a .... of compounds. Many of these only contain atoms of .... and .... They are called .... The compounds in crude oil can be separated using ....

2 We drill crude oil from the ground or seabed. Why is this crude oil not very useful as a product itself?

3 a Write the formulae of the alkanes which have 6 to 10 carbon atoms. Then find out their names.

   b Draw the displayed formula of pentane (see Figure 2).

   c How many carbon atoms are there in an alkane which has 30 hydrogen atoms?

Key points

- Crude oil is a mixture of many different compounds.

- Many of the compounds in crude oil are hydrocarbons – they contain only hydrogen and carbon.

- Alkanes are saturated hydrocarbons. They contain as many hydrogen atoms as possible in their molecules.

C1 4.2

Fractional distillation

Learning objectives

- How do we separate crude oil into fractions?

- How are the properties of each fraction affected by the size of the molecules in it?

- What makes a fraction useful as a fuel?

Demonstration

Comparing fractions

Your teacher might compare the **viscosity** and flammability of some fractions (mixtures of hydrocarbons with similar boiling points) that we get from crude oil.

The compounds in crude oil

Hydrocarbon molecules can be very different. Some are quite small, with relatively few carbon atoms in short chains. These short-chain molecules are the hydrocarbons that tend to be most useful. They make good fuels as they burn well. They are very **flammable**. Other hydrocarbons have lots of carbon atoms, and may have branches or side-chains.

The boiling point of a hydrocarbon depends on the size of its molecules. We can use the differences in boiling points to separate the hydrocarbons in crude oil.

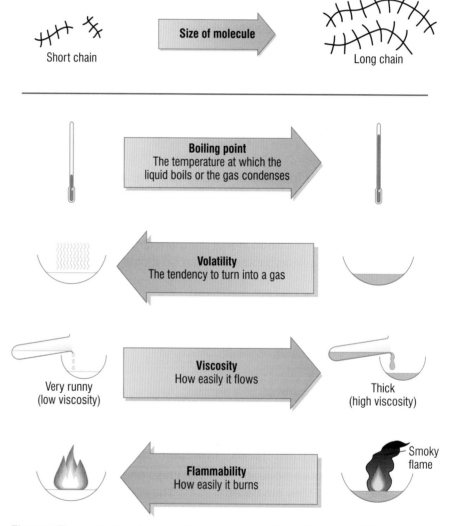

Figure 1 The properties of hydrocarbons depend on the chain-length of their molecules

a How does the length of the hydrocarbon chain affect:
  i the boiling point
  ii the viscosity (thickness) of a hydrocarbon?
b A hydrocarbon catches fire very easily. Is it likely to have molecules with long hydrocarbon chains or short ones?

Fractional distillation of crude oil

We separate out crude oil into hydrocarbons with similar boiling points, called fractions. We call this process **fractional distillation**. Each hydrocarbon fraction contains molecules with similar numbers of carbon atoms. Each of these fractions boils at a different temperature range. That is because of the different sizes of their molecules.

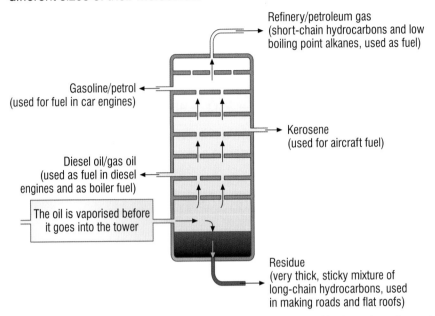

Figure 2 We use fractional distillation to separate the mixture of hydrocarbons in crude oil into fractions. Each fraction contains hydrocarbons with similar boiling points.

Crude oil is fed in near the bottom of a tall tower (a fractionating column) as hot vapour. The column is kept very hot at the bottom and much cooler at the top. The temperature decreases going up the column. The gases condense when they reach the temperature of their boiling points. So the different fractions are collected as liquids at different levels. Crude oil enters the fractionating column and fractions are collected in a continuous process.

Hydrocarbons with the smallest molecules have the lowest boiling points. They are collected at the cool top of the column. At the bottom of the column, the fractions have high boiling points. They cool to form very thick liquids or solids at room temperature.

Once collected, the fractions need more processing before they can be used.

??? Did you know ... ?

A quarter of the world's reserves of crude oil are found in Saudi Arabia.

There are many different types of crude oil. For example, crude oil from Venezuela contains many long-chain hydrocarbons. It is very dark and thick and we call it 'heavy' crude. Other countries, such as Nigeria and Saudi Arabia, produce crude oil which is much paler in colour and runnier. This is 'light' crude.

Figure 3 An oil refinery at night

Summary questions

1 Copy and complete using the words below:

   easily distillation fractions high mixture viscosity

   Crude oil is a .... of many different hydrocarbons. We can separate crude oil into different .... using fractional .... A hydrocarbon molecule with many carbon atoms will have a .... boiling point and .... Hydrocarbon molecules with few carbon atoms catch fire .... and burn with a cleaner flame.

2 a Explain the steps involved in the fractional distillation of crude oil.
   b Make a table to summarise how the properties of hydrocarbons depend on the size of their molecules.

Key points

● We separate crude oil into fractions using fractional distillation.

● The properties of each fraction depend on the size of their hydrocarbon molecules.

● Lighter fractions make better fuels as they ignite more easily and burn well, with cleaner (less smoky) flames.

175

C1 4.3 Burning fuels ⓚ

Learning objectives

- What are the products of combustion when we burn fuels in a good supply of air?
- What pollutants are produced when we burn fuels?

⊂⊃ **links**

For information on useful fractions from crude oil, look back at C1 4.2 Fractional distillation.

Figure 1 On a cold day we can often see the water produced when fossil fuels burn

The lighter fractions from crude oil are very useful as fuels. When hydrocarbons burn in plenty of air they release energy. The reaction produces two new substances – carbon dioxide and water.

For example, when propane burns we can write:

$$propane + oxygen \rightarrow carbon\ dioxide + water$$

or

$$C_3H_8 + 5O_2 \rightarrow 3CO_2 + 4H_2O$$

The carbon and hydrogen in the fuel are **oxidised** completely when they burn like this. 'Oxidised' means adding oxygen in a chemical reaction in which oxides are formed.

Practical

Products of combustion

We can test the products given off when a hydrocarbon burns as shown in Figure 2.

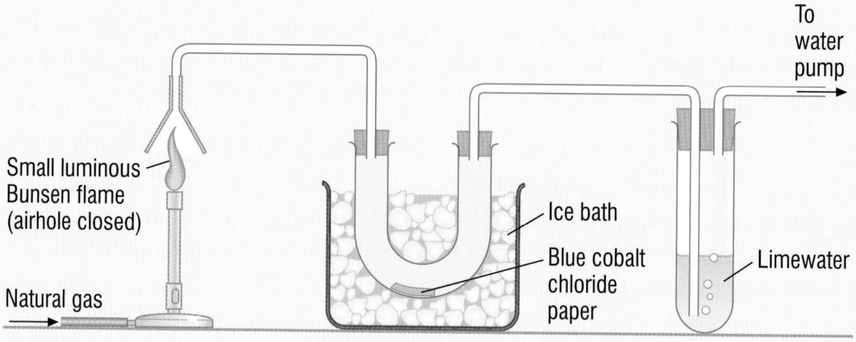

To water pump

Small luminous Bunsen flame (airhole closed)

Ice bath

Blue cobalt chloride paper

Limewater

Natural gas

Figure 2 Testing the products formed when a hydrocarbons burns

- What happens to the limewater? Which gas is given off?
- What happens in the U-tube? Which substance is present?

---

a What are the names of the two substances produced when hydrocarbons burn in plenty of air?

b Methane is the main gas in natural gas. Write a word equation for methane burning in plenty of air.

Pollution from fuels

All fossil fuels – oil, coal and natural gas – produce carbon dioxide and water when they burn in plenty of air. But as well as hydrocarbons, these fuels also contain other substances. Impurities containing sulfur found in fuels cause us major problems.

All fossil fuels contain at least some sulfur. This reacts with oxygen when we burn the fuel. It forms a gas called **sulfur dioxide**. This gas is poisonous. It is also acidic. This is bad for the environment, as it is a cause of acid rain. Sulfur dioxide can also cause engine corrosion.

c When fuels burn, what element present in the impurities in a fossil fuel may produce sulfur dioxide?

d Which pollution problem does sulfur dioxide gas contribute to?

When we burn fuels in a car engine, even more pollution can be produced.

- When there is not enough oxygen inside an engine, we get **incomplete combustion**. Instead of all the carbon in the fuel turning into carbon dioxide, we also get **carbon monoxide** gas (CO) formed.

  Carbon monoxide is a poisonous gas. Your red blood cells pick up this gas and carry it around in your blood instead of oxygen. So even quite small amounts of carbon monoxide gas are very bad for you.

- The high temperature inside an engine also allows the nitrogen and oxygen in the air to react together. This reaction makes **nitrogen oxides**. These are poisonous and can trigger some people's asthma. They also cause acid rain.

- Diesel engines burn hydrocarbons with much bigger molecules than petrol engines. When these big molecules react with oxygen in an engine they do not always burn completely. Tiny solid particles containing carbon and unburnt hydrocarbons are produced. These **particulates** get carried into the air. Scientists think that they may damage the cells in our lungs and even cause cancer.

Figure 3 A combination of many cars in a small area and the right weather conditions can cause smog to be formed. This is a mixture of SMoke and fOG.

Summary questions

1 Copy and complete using the words below:

monoxide   carbon   nitrogen   oxidised   particulates   sulfur   water

When hydrocarbons burn in a good supply of air, ... dioxide and .... are made, as the carbon and hydrogen in the fuel are .... As well as these compounds other substances such as .... dioxide may be made which causes acid rain. Other pollutants that may be formed include .... oxides, carbon .... and ....

2 Explain how **a** sulfur dioxide **b** nitrogen oxides and **c** particulates are produced when fuels burn in vehicles.

3 **a** Natural gas is mainly methane (CH_4). Write a balanced symbol equation for the complete combustion of methane.   [H]

b When natural gas burns in a faulty gas heater it can produce carbon monoxide (and water). Write a balanced symbol equation to show this reaction.   [H]

Key points

- When we burn hydrocarbon fuels in plenty of air the carbon and hydrogen in the fuel are completely oxidised. They produce carbon dioxide and water.

- Sulfur impurities in fuels burn to form sulfur dioxide which can cause acid rain.

- Changing the conditions in which we burn hydrocarbon fuels can change the products made.

- In insufficient oxygen, we get poisonous carbon monoxide gas formed. We can also get particulates of carbon (soot) and unburnt hydrocarbons, especially if the fuel is diesel.

- At the high temperatures in engines, nitrogen from the air reacts with oxygen to form nitrogen oxides. These cause breathing problems and can cause acid rain.

C1 4.4 Cleaner fuels

Learning objectives

- When we burn fuels, what are the consequences for our environment?
- What can we do to reduce the problems?

When we burn fuels, as well as producing carbon dioxide and water, we produce other substances. Many of these harm the environment, and can affect our health.

Pollution from our cars does not stay in one place but spreads through the atmosphere. For a long time the Earth's atmosphere seemed to cope with all this pollution. But the huge increase in our use of fossil fuels in the past 100 years means that pollution is a real concern now.

a Why is there more pollution in the air from fossil fuels now compared with 100 years ago?

What kinds of pollution?

When we burn any fuel containing carbon, it makes carbon dioxide. Carbon dioxide is the main greenhouse gas in the air. It absorbs energy released as radiation from the surface of the Earth. Most scientists think that this is causing **global warming**, which affects temperatures around the world. Look at the increase in our production of carbon dioxide and average global temperature data over recent times:

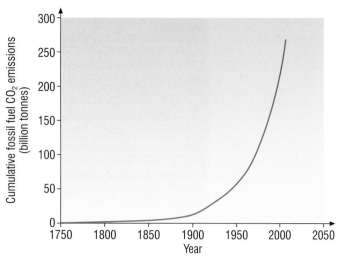

Figure 1 Cumulative carbon dioxide emissions from burning fossil fuels and the manufacture of cement

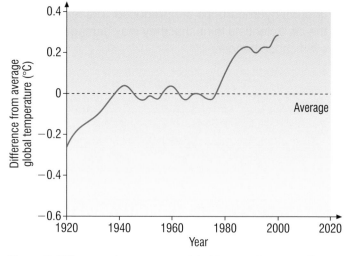

Figure 2 Differences from average global temperatures over time. People are worried about changing climates, and melting ice caps that could raise sea levels.

Burning fuels in engines also produces other substances. One group of pollutants is called the particulates. These are tiny solid particles made up of carbon (soot) and unburnt hydrocarbons. Scientists think that these may be especially bad for young children. Particulates may also be bad for the environment too. They travel into the upper atmosphere, reflecting sunlight back into space, causing **global dimming**.

Carbon monoxide is formed when there is not enough oxygen for complete combustion of a fuel. Then the carbon in it is partially oxidised to form carbon monoxide. Carbon monoxide is a serious pollutant because it affects the amount of oxygen that our blood is able to carry. This is particularly serious for people who have problems with their hearts.

Sulfur dioxide and nitrogen oxides from burning fuels damage us and our environment. In Britain, scientists think that the number of people who suffer from asthma has increased because of air pollution. Sulfur dioxide and nitrogen oxides also form acid rain. These gases dissolve in water droplets in the atmosphere and react with oxygen, forming sulfuric and nitric acids. The rain with a low pH can damage plant and animals.

b Name four harmful substances that may be produced when fuels burn.

Cleaning up our act

We can reduce the effects of burning fuels in several ways. For example, we can remove harmful substances from the gases that are produced when we burn fuels. For some time the exhaust systems of cars have been fitted with **catalytic converters**. A catalytic converter greatly reduces the carbon monoxide and nitrogen oxides produced by a car engine. They are expensive, as they contain precious metal catalysts, but once warmed up they are very effective.

The metal catalysts are arranged so that they have a very large surface area. This causes the carbon monoxide and nitrogen oxides in the exhaust gases to react together. They produce carbon dioxide and nitrogen:

carbon monoxide + nitrogen oxides → carbon dioxide + nitrogen

So although catalytic converters reduce the toxic gases given out, they do not help reduce levels of carbon dioxide in the air.

Filters can also remove most particulates from modern diesel engines. The filters need to burn off the trapped solid particles otherwise they get blocked.

In power stations, sulfur dioxide is removed from the waste or 'flue' gases by reacting it with calcium oxide or calcium hydroxide. This is called flue gas desulfurisation. The sulfur impurities can also be removed from a fuel *before* the fuel is burned. This happens in petrol and diesel for cars, as well as in the natural gas and oil used in power stations.

⚭ links

For more information on how we can also use alternative fuels to reduce pollution, see C1 4.5 Alternative fuels.

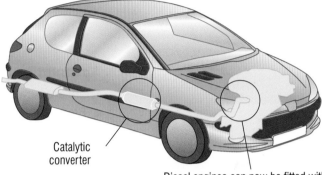

Catalytic converter

Diesel engines can now be fitted with filters to remove solid particulates

Figure 3 Modern cars are fitted with catalytic converters. Filters can also remove most of the particulates from diesel engine exhaust gases.

Key points

- Burning fuels releases substances that spread throughout the atmosphere.
- Sulfur dioxide and nitrogen oxides dissolve in droplets of water in the air and react with oxygen, and then fall as acid rain.
- Carbon dioxide produced from burning fuels is a greenhouse gas. It absorbs energy which is lost from the surface of the Earth by radiation.
- The pollution produced by burning fuels may be reduced by treating the pollutants from combustion. This can remove substances like nitrogen oxides, sulfur dioxide and carbon monoxide.
- Sulfur can also be removed from fuels before we burn them to prevent sulfur dioxide gas being formed.

Summary questions

1 **a** Why is carbon dioxide called a greenhouse gas?
  b How do you think particulates in the atmosphere might affect the Earth's temperature?
  c Which gases are mainly responsible for acid rain?

2 **a** Which pollutants from a car does a catalytic converter remove?
  b Why will catalytic converters not help to solve the problem of greenhouse gases in the atmosphere?

3 **a** Explain how acid rain is formed and how we are reducing the problem.
  b Compare the effects of global warming and global dimming.
  c Particulates in the atmosphere could eventually settle on the polar ice caps. What problem might this make worse?

C1 4.5 Alternative fuels

Biofuels

Biofuels are fuels that are made from plant or animal products. For example, **biodiesel** is made from oils extracted from plants. You can even use old cooking oil as a biofuel. Biogas is generated from animal waste. Biofuels will become more and more important as our supplies of crude oil run out.

> **a** What is biodiesel?

Advantages of biodiesel

There are advantages in using biodiesel as a fuel.

- Biodiesel is much less harmful to animals and plants than diesel we get from crude oil. If it is spilled, it breaks down about five times faster than 'normal' diesel.
- When we burn biodiesel in an engine it burns much more cleanly, reducing the particulates emitted. It also makes very little sulfur dioxide.
- As crude oil supplies run out, its price will increase and biodiesel will become cheaper to use than petrol and diesel.
- Another really big advantage over petrol and diesel is the fact that the crops used to make biodiesel absorb carbon dioxide gas as they grow. So biodiesel is in theory 'CO_2 neutral'. That means the amount of carbon dioxide given off when it burns is balanced by the amount absorbed as the plants it is made from grow. Therefore, biodiesel makes little contribution to the greenhouse gases in our atmosphere.

  However, we can't claim that biodiesel makes a zero contribution to carbon dioxide emissions. We should really take into account the CO_2 released when:
  - fertilising and harvesting the crops
  - extracting and processing the oil
  - transporting the plant material and biodiesel made.
- When we make biodiesel we also produce other useful products. For example, we get a solid waste material that we can feed to cattle as a high-energy food. We also get glycerine which we can use to make soap.

Disadvantages of biodiesel

There are however disadvantages in using biodiesel and other biofuels as a fuel.

- The use of large areas of farmland to produce fuel instead of food could pose problems. If we start to rely on oil-producing crops for our fuel, land once used for food crops will turn to growing biofuel crops.

Learning objectives

- What are biofuels?
- What are the advantages and disadvantages of using biodiesel?
- Why are scientists interested in developing hydrogen as a fuel?

Figure 1 This coach runs on biodiesel

Activity

Biodiesel – fuel from plants

In a group of three, each choose a different task:

A Write an article for a local newspaper describing the arguments for using biodiesel instead of other fuels made from crude oil.

B Write a letter to the newspaper pointing out why the article in **A** should not claim that biodiesel makes no overall contribution of global warming.

C Write an article for the newspaper focussing on the drawbacks of using biodiesel.

- Read each other's work and decide whether biodiesel will be a major fuel in 20 years time.

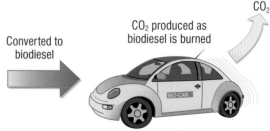

Plants absorb CO_2 as they grow → Converted to biodiesel → CO_2 produced as biodiesel is burned

Figure 2 Cars that run on biodiesel produce very little CO_2 overall, as CO_2 is absorbed by the plants used to make the fuel

This could result in famine in poorer countries if the price of staple food crops rises as demand overtakes supply. Forests, which absorb lots of carbon dioxide, might also be cleared to grow the biofuel crops if they get more popular.

- People are also worried about the destruction of habitats of endangered species. For example, orang-utans are under threat of extinction. Large areas of tropical forest where they live are being turned into palm plantations for palm oil used to make biodiesel.
- At low temperatures biodiesel will start to freeze before traditional diesel. It turns into a sludge. At high temperatures in an engine it can turn sticky as its molecules join together and can 'gum up' engines.

Using ethanol as a biofuel

Another biofuel is ethanol. We can make it by fermenting the sugar from sugar beet or sugar cane. In Brazil they can grow lots of sugar cane. They add the ethanol made to petrol, saving money as well as our dwindling supplies of crude oil. As with biodiesel, the ethanol gives off carbon dioxide (a greenhouse gas) when it burns, but the sugar cane absorbs CO_2 gas during photosynthesis.

b Why is burning ethanol a better choice of fuel than petrol if we want to reduce carbon dioxide emissions?

 How Science Works

Hydrogen – a fuel for the future

Scientists are very interested in developing hydrogen as a fuel. It burns well with a very clean flame as there is no carbon in the fuel:

$$\text{hydrogen} + \text{oxygen} \rightarrow \text{water}$$
$$2H_2 + O_2 \rightarrow 2H_2O$$

As you can see in the equation, water is the only product in the combustion of hydrogen. There are no pollutants made when hydrogen burns and no extra carbon dioxide is added to the air. Not only that, water is potentially a huge natural source of hydrogen. The hydrogen can be obtained from water by electrolysis. But the electricity must be supplied by a renewable energy source if we want to conserve fossil fuels and control carbon dioxide emissions.

However, there are problems to solve before hydrogen becomes a common fuel. When mixed with air and ignited it is explosive. So there are safety concerns in case of leaks, or accidents in vehicles powered by hydrogen. Vehicles normally run on liquid fuels but hydrogen is a gas. Therefore it takes up a much larger volume than liquid fuels. So storage is an issue. We can use high-pressure cylinders but these also have safety problems in crashes.

Figure 3 Ethanol can be made from sugar cane

⬭ links

For more information on ethanol, see C1 5.5 Ethanol.

Key points

- Biofuels are a renewable source of energy that could be used to replace some fossil fuels.

- Biodiesel can be made from vegetable oils.

- There are advantages, and some disadvantages, in using biodiesel.

- Ethanol is also a biofuel as it can be made from the sugar in plants.

- Hydrogen is a potential fuel for the future.

Summary questions

1 Copy and complete these sentences using the words below:
carbon dioxide   diesel   plants
Biodiesel is a fuel made from .... It produces less pollution than .... obtained from crude oil, and absorbs nearly as much .... when the plants that make it grow as it does when it burns.

2 Where does the energy in biodiesel come from?

3 a Explain why hydrogen is potentially a pollution-free fuel.
  b Why isn't hydrogen used as an everyday fuel at the moment?

Summary questions

1 This question is about the alkane family of compounds.

  a The alkanes are all 'saturated hydrocarbons'.
    i What is a hydrocarbon?
    ii What does saturated mean when describing an alkane?

  b i Give the name and formula of this alkane:

    ii What do the letters represent in this displayed formula?
    iii What do the lines between letters represent?

  c What is the general formula of the alkanes (where $n =$ the number of carbon atoms)?

  d Give the formula of the alkane with 20 carbon atoms.

2 One alkane, A, has a boiling point of 344 °C and another, B, has a boiling point of 126 °C.

  a Which one will be collected nearer the top of a fractionating column in an oil refinery? Explain your choice.

  b Which one will be the better fuel? Explain your choice.

  c Give another difference you expect between A and B.

3 a Name the two products formed when a hydrocarbon burns in enough oxygen to ensure that complete combustion takes place.

  b What problem is associated with the increased levels of carbon dioxide gas in the atmosphere?

  c i What gas is given off from fossil fuel power stations that can cause acid rain?
    ii Give **two** ways of stopping this gas getting into the atmosphere from power stations.
    iii Name the other cause of acid rain which comes from car engines?
    iv How do car makers stop the gases in part **iii** entering the air?

  d Why are diesel engines now fitted with a filter for their exhaust fumes?

4 a Which one of these fuels could be termed a 'biofuel'?
    Hydrogen Propane Ethanol Petrol Coal

  b Biodiesel is potentially 'CO_2 neutral'. What does this mean?

  c Scientists are concerned about the issue of global warming. Why is the use of hydrogen as a fuel one way to tackle the problem?

  d State **two** problems with the use of hydrogen as a fuel.

  e Write a word and a balanced symbol equation to show the combustion of hydrogen.   **[H]**

5 This apparatus can be used to compare the energy given out when different fuels are burned.

The burner is weighed before and after to determine the amount of fuel burned. The temperature of the water is taken before and after, to get the temperature rise. The investigation was repeated. From this the amount of energy released by burning a known amount of fuel can be calculated.

  a Design a table that could be used to collect the data as you carry out this experiment.

A processed table of results is given below.

Fuel	Mass burned (g)	Temperature rise (°C)	
Ethanol	4.9	48	47
Propanol	5.1	56	56
Butanol	5.2	68	70
Pentanol	5.1	75	76

  b List three variables that need to be controlled.

  c Describe how you would take the temperature of the water to get the most accurate measurement possible.

  d Do these results show precision? Explain your answer?

  e How might you present these results?

Practice questions

1 The table shows some information about the first four alkanes.

Name of alkane	Formula	Boiling point in °C
Methane	CH_4	−162
	C_2H_6	−88
Propane	C_3H_8	
Butane		0

a i Name the alkane missing from the table. (1)

ii What is the formula of butane? (1)

iii Estimate the boiling point of propane. (1)

b Which one of the following is the formula of the alkane with 6 carbon atoms?

C_6H_6 C_6H_{10} C_6H_{14} C_6H_{16} (1)

c Explain why alkanes are hydrocarbons. (1)

d A molecule of methane can be represented as:

$$
\begin{array}{c}
\text{H} \\
| \\
\text{H} - \text{C} - \text{H} \\
| \\
\text{H}
\end{array}
$$

Draw a molecule of propane in the same way. (2)

2 Some crude oil was distilled in a fractionating column. The table shows the boiling ranges of three of the fractions that were collected.

Fraction	Boiling range in °C
A	60–120
B	160–230
C	240–320

a Which of these fractions is the most flammable? (1)

b Which of these fractions is the most viscous? (1)

c Which of these fractions has the smallest hydrocarbon molecules? (1)

d Why do the fractions have boiling ranges and not boiling points? (1)

3 Some landfill sites produce a gas that can be collected and burned as a fuel. The gas is mainly methane.

a Choose the word from the list to complete the sentence.

condensed   distilled   oxidised

During the combustion of methane the elements in the fuel are .... (1)

b Write a word equation for the complete combustion of methane, CH_4. (2)

c Under what conditions could methane burn to produce carbon monoxide? (1)

d A sample of landfill gas was burned. The waste gases contained sulfur dioxide. Explain why. (1)

4 Crude oil is a resource from which fuels can be separated.

a The name of the main fuel fractions and one of the hydrocarbons in each fraction are shown in the table.

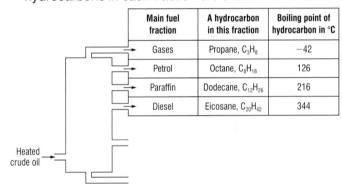

Main fuel fraction	A hydrocarbon in this fraction	Boiling point of hydrocarbon in °C
Gases	Propane, C_3H_8	−42
Petrol	Octane, C_8H_{18}	126
Paraffin	Dodecane, $C_{12}H_{26}$	216
Diesel	Eicosane, $C_{20}H_{42}$	344

Heated crude oil

i How does the number of carbon atoms in a hydrocarbon affect its boiling point? (1)

ii Suggest the lowest temperature to which crude oil needs to be heated to vaporise all the hydrocarbons in the table.
Temperature = ... °C? (1)

iii Dodecane boils at 216 °C. At what temperature will dodecane gas condense to liquid?
Temperature = ... °C? (1)

b *In this question you will be assessed on using good English, organising information clearly and using specialist terms where appropriate.*

Describe and explain how the fractions are separated in a fractionating column. (6)

AQA, 2009

C1 5.1 Cracking hydrocarbons

Learning objectives

- How do we make smaller, more useful molecules from larger, less useful molecules in crude oil?
- What are alkenes and how are they different from alkanes?

Figure 2 In an oil refinery, huge crackers like this are used to break down large hydrocarbon molecules into smaller ones

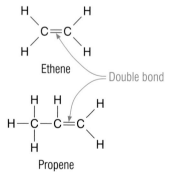

Figure 3 A molecule of ethene (C_2H_4) and a molecule of propene (C_3H_6). These are both alkenes – each molecule has a carbon–carbon double bond in it.

Some of the heavier fractions that we get by distilling crude oil are not very useful. The hydrocarbons in them are made up of large molecules. They are thick liquids or solids with high boiling points. They are difficult to vaporise and do not burn easily – so they are no good as fuels! Yet the main demand from crude oil is for fuels.

Luckily we can break down large hydrocarbon molecules in a process we call **cracking**.

The process takes place at an oil refinery in a steel vessel called a cracker.

In the cracker, a heavy fraction produced from crude oil is heated to vaporise the hydrocarbons. The vapour is then either passed over a hot catalyst or mixed with steam. It is heated to a high temperature. The hydrocarbons are cracked as thermal decomposition reactions take place. The large molecules split apart to form smaller, more useful ones.

> **a** Why is cracking so important?
> **b** How are large hydrocarbon molecules cracked?

Example of cracking

Decane is a medium-sized molecule with ten carbon atoms. When we heat it to 500 °C with a catalyst it breaks down. One of the molecules produced is pentane which is used in petrol.

Figure 1 Pentane (C_5H_{12}) can be used as a fuel. This is the displayed formula of pentane.

We also get propene and ethene which we can use to produce other chemicals.

$$C_{10}H_{22} \xrightarrow{800\,°C + catalyst} C_5H_{12} + C_3H_6 + C_2H_4$$
decane → pentane + propene + ethene

This reaction is an example of thermal decomposition.

Notice how this cracking reaction produces different types of molecules. One of the molecules is pentane. The first part of its name tells us that it has five carbon atoms (*pent-*). The last part of its name (*-ane*) shows that it is an alkane. Like all other alkanes, pentane is a saturated hydrocarbon. Its molecules have as much hydrogen as possible in them.

The other molecules in this reaction have names that end slightly differently. They end in *-ene*. We call this type of molecule an **alkene**. The different ending tells us that these molecules are **unsaturated**. They contain a **double bond** between two of their carbon atoms. Look at Figure 3. You can see that alkenes have one double bond and have the general formula C_nH_{2n}.

Practical

Cracking

Medicinal paraffin is a mixture of hydrocarbon molecules. You can crack it by heating it and passing the vapour over hot pieces of broken pot. The broken pot acts as a catalyst.

● Why must you remove the end of the delivery tube from the water before you stop heating?

If you carry out this practical, collect at least two test tubes of gas. Test one by putting a lighted splint into it. Test the other by shaking it with a few drops of bromine water.

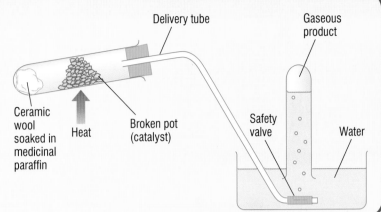

A simple experiment like the one above shows that alkenes burn. They also react with bromine water (which is orange). The products of this reaction are colourless. This means that we have a good test to see if a hydrocarbon is unsaturated:

Positive test:

unsaturated hydrocarbon + bromine water → products
           (orange-yellow)   (colourless)

Negative test:

saturated hydrocarbon + bromine water → no reaction
        (orange)   (orange)

Summary questions

1 Copy and complete using the words below:

alkenes   catalyst   cracking   double   heating   unsaturated

Large hydrocarbon molecules are broken down by .... them and passing them over a hot .... This is called .... Some of the molecules produced when we do this contain a .... bond. They are called .... hydrocarbons. They are examples of a group of hydrocarbons called the ....

2 Cracking a hydrocarbon makes two new hydrocarbons, A and B. When bromine water is added to A, nothing happens. Bromine water added to B turns from an orange solution to colourless.
  a Which hydrocarbon is unsaturated?
  b Which hydrocarbon is used as a fuel?
  c What type of reaction is cracking an example of?
  d Cracking can be carried out by passing large hydrocarbon molecules over a hot catalyst. State another way to crack a hydrocarbon in industry.

3 An alkene molecule with one double bond contains 7 carbon atoms. How many hydrogen atoms does it have? Write down its formula.

4 Decane (with 10 carbon atoms) is cracked into octane (with 8 carbon atoms) and ethene. Write a balanced equation for this reaction. **[H]**

?? Did you know ...?

Ethene gas makes fruits such as bananas ripen. Bananas are picked and stored as the unripe green fruit. When they are required for display in a shop ethene gas is passed over the stored bananas to start the ripening process.

Study tip

Remember:
alk**a**nes are s**a**turated
alk**e**nes have a double bond = (**e**quals)

Key points

● We can split large hydrocarbon molecules up into smaller molecules by:
  – mixing them with steam and heating them to a high temperature, or
  – by passing the vapours over a hot catalyst.

● Cracking produces saturated hydrocarbons which are used as fuels and unsaturated hydrocarbons (called alkenes).

● Alkenes react with orange bromine water, turning it colourless.

C1 5.2 Making polymers from alkenes

Learning objectives

● What are monomers and polymers?

● How do we make polymers from alkenes?

Figure 1 All of these products were manufactured using chemicals made from oil

The fractional distillation of crude oil and cracking produces a large range of hydrocarbons. These are very important to our way of life. Oil products are all around us. We simply cannot imagine life without them.

Hydrocarbons are our main fuels. We use them in our transport and at home to cook and for heating. We also use them to make electricity in oil-fired power stations.

Then there are the chemicals we make from crude oil. We use them to make things ranging from cosmetics to explosives. But one of the most important ways that we use chemicals from oil is to make plastics.

Plastics

Plastics are made up of huge molecules made from lots of small molecules joined together. We call the small molecules **monomers**. We call the huge molecules they make **polymers**. (*Mono* means 'one' and *poly* means 'many'). We can make different types of plastic which have very different properties by using different monomers.

> **a** List three ways that we use fuels.
> **b** What are the small molecules that make up a polymer called?

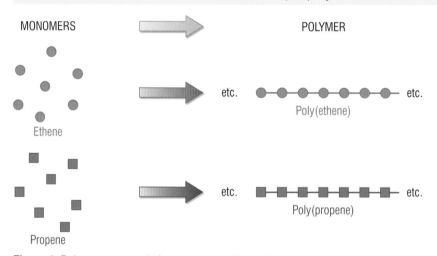

Figure 2 Polymers are made from many smaller molecules called monomers

Ethene (C_2H_4) is the smallest unsaturated hydrocarbon molecule. We can turn it into a polymer known as poly(ethene) or polythene. Poly(ethene) is a really useful plastic. It is easy to shape, strong and transparent (unless we add colouring material to it). 'Plastic' bags, plastic drink bottles, dustbins and clingfilm are all examples of poly(ethene).

Propene (C_3H_6) is another alkene. We can also make polymers with propene as the monomer. The polymer formed is called poly(propene). It forms a very strong, tough plastic. We can use it to make many things, including carpets, milk crates and ropes.

Figure 3 Polymers produced from oil are all around us and are part of our everyday lives

> **c** Is ethene an alkane or an alkene?
> **d** Which polymer can we make from propene monomers?

How do monomers join together?

When alkene molecules join together, the double bond between the carbon atoms in each molecule 'opens up'. It is replaced by single bonds as thousands of molecules join together. The reaction is called **polymerisation**.

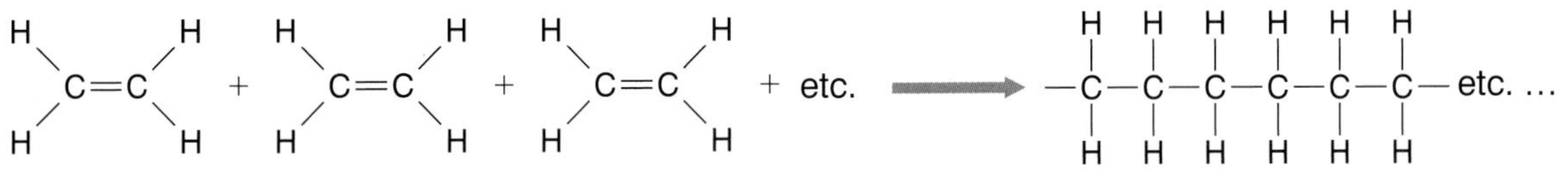

Ethene monomers   Poly(ethene)

We can also write this more simply as:

where n is a large number

Many single ethene monomers   Long chain of poly(ethene)

Activity

Modelling polymerisation

Use a molecular model kit to show how ethene molecules polymerise to form poly(ethene).

Make sure you can see how the equation shown above represents the polymerisation reaction you have modelled.

You should also be able to describe what happens to the bonds in the reaction.

Think up a model to demonstrate the polymerisation of ethene, using people in your class as monomers.

Evaluate the ideas of other groups.

Study tip

The double C=C bond in ethene (an alkene) makes it much more reactive than ethane (an alkane).

Summary questions

1 Copy and complete using the words below:

polymerisation   ethene   monomers   polymers

Plastics are made of large molecules called .... We make these by joining together lots of small, reactive molecules called .... One example of a polymer is poly(ethene), made from .... Poly(ethene) is formed as a result of a .... reaction.

2 Why is ethene the smallest possible unsaturated hydrocarbon molecule?

3 a Draw the displayed formula of a propene molecule, showing all its bonds.

 b Draw a diagram to show how propene molecules join together to form poly(propene).

 c Explain the polymerisation reaction in **b**.

Key points

● Plastics are made of polymers.

● Polymers are large molecules made when monomers (small, reactive molecules) join together. The reaction is called polymerisation.

C1 5.3

New and useful polymers

Learning objectives

- How are we using new polymers?
- What are smart polymers?

Chemists can design new polymers to make materials with special properties to do particular jobs. Medicine is one area where we are beginning to see big benefits from these 'polymers made to order'.

New polymer materials will eventually take over from fillings for teeth which contain mercury. Working with the toxic mercury every day is a potential hazard to dental workers. Other developments include:

- new softer linings for dentures (false teeth)
- new packaging material
- implants that can slowly release drugs into a patient.

a What do we mean by a 'designer polymer'?

Figure 1 A sticking plaster is often needed when we cut ourselves. Getting hurt isn't much fun – and sometimes taking the plaster off can be painful too.

Light-sensitive plasters

We all know how uncomfortable pulling a plaster off your skin can be. But for some of us taking off a plaster is really painful. Both very old and very young people have quite fragile skin. But now a group of chemists has made a plaster where the 'stickiness' can be switched off before the plaster is removed. The plaster uses a light-sensitive polymer.

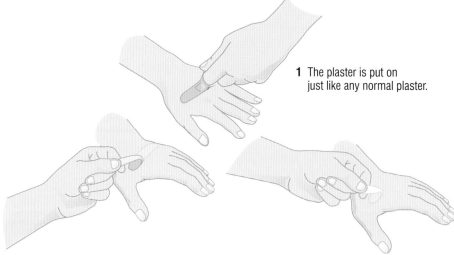

1 The plaster is put on just like any normal plaster.

2 To remove the plaster, the top layer is peeled away from the lower layer which stays stuck to the skin.

3 Once the lower layer is exposed to the light, the adhesive becomes less sticky, making it easy to peel off your skin.

Figure 2 This plaster uses a light-sensitive polymer

How Science Works

Evaluating plastics

Plan an investigation to compare and evaluate the suitability of different plastics for a particular use.

For example, you might look at treated and untreated fabrics for waterproofing and 'breatheability' (gas permeability) or different types of packaging.

Hydrogels

Hydrogels are polymer chains with a few cross-linking units between chains. This makes a matrix that can trap water. These hydrogels are used as wound dressings. They let the body heal in moist, sterile conditions. This makes them useful for treating burns.

The latest 'soft' contact lenses are also made from hydrogels. To change the properties of hydrogels, scientists can vary the amount of water in their matrix structure.

Shape memory polymers

New polymers can also come to our rescue when we are cut badly enough to need stitches. A new 'shape memory polymer' is being developed by doctors which will make stitches that keep the sides of a cut together. When a shape memory polymer is used to stitch a wound loosely, the temperature of the body makes the thread tighten and close the wound, applying just the right amount of force.

This is an example of a '**smart polymer**', i.e. one that changes in response to changes around it. In this case a change in temperature causes the polymer to change its shape. Later, after the wound is healed, the polymer is designed to dissolve and is harmlessly absorbed by the body. So there will be no need to go back to the doctor to have the stitches out.

Figure 3 A shape memory polymer uses the temperature of the body to make the thread tighten and close the wound

New uses for old polymers

The bottles that we buy fizzy drinks in are a good example of using a plastic because of its properties. These bottles are made out of a plastic called PET.

The polymer it is made from is ideal for making drinks bottles. It produces a plastic that is very strong and tough, and which can be made transparent. The bottles made from this plastic are much lighter than glass bottles. This means that they cost less to transport and are easier for us to carry around.

Do you recycle your plastic bottles? The PET from recycled bottles is used to make polyester fibres for clothing, such as fleece jackets, and the filling for duvet covers. School uniforms and football shirts are now also made from recycled drinks bottles.

b Why is PET used to make drinks bottles?

 Did you know ... ?

PET is an abbreviation for poly(ethene terephthalate). It takes 5 two-litre PET lemonade bottles to make one T-shirt.

○○ links

For more information on recycling, see C1 5.4 Plastic waste.

Key points

- New polymers are being developed all the time. They are designed to have properties that make them specially suited for certain uses.

- Smart polymers may have their properties changed by light, temperature or by other changes in their surroundings.

- We are now recycling more plastics and finding new uses for them.

Summary questions

1 Copy and complete using the words below:

cold hot PET properties shape strong transparent

We choose a polymer for a job because it has certain .... For example, we make drinks bottles out of a plastic called .... because it is .... and ....
Scientists can also design 'smart' polymers, for example .... memory polymers. These change their shape when they are .... or ....

2 a Give one advantage of using a polymer in sticking plasters that is switched off by light making the polymer less sticky.
  b Design a leaflet for a doctor to give to a patient, explaining how stitches made from smart polymers work.

C1 5.4 Plastic waste

- What are the problems caused by disposing of plastics?
- What does biodegradable mean?
- How can polymers be made biodegradable?

Figure 1 Finding space to dump and bury our waste is becoming a big problem

One of the problems with plastics is what to do with them when we've finished with them. Too much ends up as rubbish in our streets. Even the beaches in the remotest parts of the world can be polluted with plastic waste. Wildlife can get trapped in the waste or eat the plastics and die.

Not only that, just think of all the plastic packaging that goes in the bin after shopping. Most of it ends up as rubbish in landfill tips. Other rubbish in the tips rots away quite quickly. Microorganisms in the soil break it down. Many waste plastics last for hundreds of years before they are broken down completely. So they take up valuable space in our landfill sites. What was a useful property during the working life of the plastic (its lack of reactivity) becomes a disadvantage in a landfill site.

> **a** Why are waste plastics proving to be a problem for us?

Biodegradable plastics

Scientists are working to solve the problems of plastic waste. We are now making more plastics that do rot away in the soil when we dump them. These plastics are called **biodegradable**. They can be broken down by microorganisms.

Scientists have found different ways to speed up the decomposition. One way uses granules of cornstarch built into the plastic. The microorganisms in soil feed on the starch. This breaks the plastic up into small pieces more quickly.

Other types of plastic have been developed that are made from plant products. A plastic called PLA, poly(lactic acid), can be made from cornstarch. The plastic is totally biodegradable. It is used in food packaging. However, it cannot be put in a microwave which limits its use in ready-meal packaging.

We can also make plastic carrier bags using PLA. In carrier bags the PLA is mixed with a traditional plastic. This makes sure the bag is strong enough but will still biodegrade a lot more quickly.

Using plastics such as PLA also helps preserve our supplies of crude oil. Remember that crude oil is the raw material for many traditional plastics, such as poly(ethene).

Disadvantages of biodegradable plastics

However, the use of a food crop like corn to make plastics can raise the same issues as biofuels. Farmers who sell their crops to turn into fuel and plastics could cause higher food prices. The lack of basic food supplies could result in starvation in developing countries. Another problem is the destruction of tropical forests to create more farmland. This will destroy the habitats of wildlife and could affect global warming.

Other degradable plastics used for bags will break down in light. However, they will not decompose when buried in a landfill site. Probably the best solution is to reuse the same plastic carrier bags over and over again.

Figure 2 The breakdown of a biodegradable plastic. PLA can be designed to break down in a few months.

⬭ links

For information on the issues of using biofuels, look back at C1 4.5 Alternative fuels.

Figure 4 Recycling is becoming part of everyday life in the UK

Practical

Investigating cornstarch

Cornstarch can be fermented to make the starting material for PLA. However, cornstarch itself also has some interesting properties. You can make your own plastic material directly from cornstarch.

● How do varying the proportions of cornstarch and water affect the product?

Recycling plastics

Some plastics can be recycled. Once sorted into different types they can be melted down and made into new products. This can save energy and resources.

However, recycling plastics does tend to be more difficult than recycling paper, glass or metals. The plastic waste takes up a lot of space so is awkward to transport. Sorting out plastics into their different types adds another tricky step to the process. The energy savings are less than we get with other recycled materials. It would help recyclers if they could collect the plastics already sorted. You might have seen recycling symbols on some plastic products.

| PET (polyethene terephthalate) | HD PE (high density poly(ethene)) | PVC | LD PE | PP | PS | Others |

Figure 3 These symbols could help people sort out their plastic waste to help the recycling process

b How does recycling plastic waste help conserve our supplies of crude oil?

Activity

Sorting plastics

a Imagine you are the head of your council's waste collection department. You have to write a leaflet for householders persuading them to recycle more of their plastic waste. They will be provided with extra bins to sort the plastics out before they are collected, once every two weeks.

b Write a letter back to the council from an unhappy person who is not willing to do any more recycling than they do already.

c Take a class vote on which action, **a** or **b**, you would support.

Summary questions

1 What do we mean by a biodegradable plastic?

2 a Why are plastics whose raw materials are plants becoming more popular?

b PLA is a biodegradable plastic. What is its monomer?

3 Non-biodegradable plastics such as poly(ethene) can be made to decompose more quickly by mixing with additives. These enable the polymer chain to be broken down by reacting with oxygen. Why might this be a waste of money if the plastic is buried and compressed under other waste in a landfill site?

Key points

● Non-biodegradable plastics cause unsightly rubbish, can harm wildlife and take up space in landfill sites.

● Biodegradable plastics are decomposed by the action of microorganisms in soil. Making plastics with starch granules in their structure help the microorganisms break down a plastic.

● We can make biodegradable plastics from plant material such as cornstarch.

C1 5.5 Ethanol

Learning objectives

- What are the two methods used to make ethanol?

- What are the advantages and disadvantages of these two methods?

Ethanol is a member of the group of organic compounds called the alcohols. Its formula is C_2H_6O but it is more often written as C_2H_5OH. This shows the –OH group that all alcohols have in their molecules.

Making ethanol by fermentation

Ethanol is the alcohol found in alcoholic drinks. Ethanol for drinks is made by the **fermentation** of sugar from plants. Enzymes in yeast break down the sugar into ethanol and carbon dioxide gas:

$$\text{sugar} \xrightarrow{\text{yeast}} \text{ethanol} + \text{carbon dioxide}$$
$$\text{(glucose)}$$

$$C_6H_{12}O_6 \longrightarrow 2C_2H_5OH + 2CO_2$$

a Which gas is given off when sugar is fermented?
b Yeast is a living thing. It is a type of fungus. What type of molecules in yeast enable it to ferment sugar?

Figure 1 Some people brew their own alcoholic drinks. The fermentation stage is often carried out by leaving the fermenting mixture in a warm place. The enzymes in yeast work best in warm conditions.

?? Did you know ...?

The yeast in a fermenting mixture cannot survive in concentrations of ethanol beyond about 15%. Alcoholic spirits, such as whisky or vodka, need to be distilled to increase the ethanol content to about 40% of their volume. Ethanol in high concentrations is toxic, which is why ethanol in the lab should never be drunk!

Practical

Fermentation

In this experiment you can ferment sugar solution with yeast and test the gas given off.

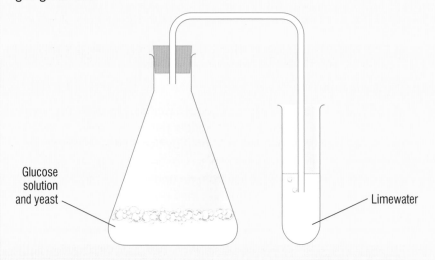

Glucose solution and yeast

Limewater

If you leave your apparatus till next lesson, your teacher can collect some fermented mixtures together and distil it to collect the ethanol formed. We use fractional distillation for the best separation as water and ethanol have similar boiling points. Ethanol boils at 78 °C. The ethanol collected will ignite and burn with a 'clean' blue flame.

∞ links

For information on using ethanol as a fuel, look back at C1 4.5 Alternative fuels.

Ethanol is also used as a solvent. Methylated spirit is mainly ethanol. Decorators can use it to clean brushes after using an oil-based paint. It is also used to make perfume. We have already seen how ethanol can be used as a fuel. It can be mixed with petrol or just used by itself to run cars.

Making ethanol from ethene (hydration)

Ethanol for industrial use as a fuel or solvent can be made from ethene gas instead of by fermentation. Remember that ethene is made when oil companies crack hydrocarbons to make fuels. Ethene is the main by-product made in cracking. Ethene gas can react with steam to make ethanol.

$$\text{ethene} + \text{steam} \xrightarrow{\text{catalyst}} \text{ethanol}$$
$$C_2H_4 + H_2O \longrightarrow C_2H_5OH$$

This reaction is called **hydration**.

> **c** Where do we get the ethene from to make industrial ethanol?

The reaction requires energy to heat the gases and to generate a high pressure. The reaction is reversible so ethanol can break down back into ethene and steam. So unreacted ethene and steam need to be recycled over the catalyst.

This process is continuous. It also produces no waste products. Both of these are advantages when making products in industry. When ethanol is made industrially by fermentation, the process is carried out in large vats which have to be left. This is called a batch process, which takes a lot longer than a continuous process. Carbon dioxide, a greenhouse gas, is also given off in fermentation.

However, using ethene to make ethanol relies on crude oil which is a **non-renewable** resource. Therefore making ethanol as a biofuel, by fermenting sugars from plant material (a renewable resource), will become ever more important. The sugars are from crops such as sugar cane or sugar beet. Any cereal crop can also be used as the raw material. These need their starch to be broken down to sugars before fermentation takes place. However as we have seen before there are issues that need to be addressed when using crops for large-scale industrial processes.

⊂⊃ links
For information on cracking, look back at C1 5.1 Cracking hydrocarbons.

Figure 2 Industrial fermentation is a slow batch process. The ethanol must be distilled off from the fermented mixture. This requires energy even though the fermentation process itself is energy efficient.

⊂⊃ links
For information on the issues of using crops for large scale industrial processes, look back at C1 4.5 Alternative fuels.

Summary questions

1 Copy and complete using the words below:

catalyst sugar yeast steam

Ethanol can be made by two processes, ethene reacting with ...., under pressure in the presence of a ...., or the fermentation of .... using enzymes in ....

2 Write a word equation to show the production of ethanol from:
  a ethene
  b glucose.

3 Why is a continuous process better than a batch process for making a product in industry?

4 How can people claim that the fermentation of plant materials does not contribute to the increase in carbon dioxide in the air?

Key points

- Ethanol can be made from ethene reacting with steam in the presence of a catalyst. This is called hydration.
- Ethanol is also made by fermenting sugar (glucose) using enzymes in yeast. Carbon dioxide is also made in this reaction.
- Using ethene to make ethanol needs non-renewable crude oil as its raw material whereas fermentation uses renewable plant material.

Summary questions Ⓚ

1 Write simple definitions for the following words:

  a hydrocarbon

  b cracking

  c distillation

  d saturated hydrocarbon

  e unsaturated hydrocarbon

  f monomer

  g polymer

  h biodegradable polymer

  i fermentation.

2 Propene is a hydrocarbon molecule containing three carbon atoms and six hydrogen atoms.

  a What is the chemical formula of propene?

  b Draw the display formula of propene, showing all its bonds.

  c Is propene a saturated molecule or an unsaturated molecule? Explain your answer.

  d You are given two unlabelled test tubes. One test tube contains propane gas, while the other test tube contains propene gas. Explain how you could test which tube contains which gas, stating clearly the results obtained in each case.

  e Propene molecules will react together to form long chains.

  i What do we call this type of reaction?

  ii Compare the properties of the reactants to those of the product.

  iii A molecule of ethene is a similar to a molecule of propene. Give an equation to show the reaction of ethene to make poly(ethene).

3 a Why does the disposal of much of our plastic waste cause problems?

  b How can chemists help to solve the issues in part **a** using a plant material such as starch from corn?

4 a Write a word equation and a balanced symbol equation for the reaction between ethene and steam. [H]

  b Write a word and balanced symbol equation for the fermentation of glucose. [H]

5 Draw a table showing the advantages and disadvantages of making ethanol from ethene or from sugar obtained from plant material.

6 Chemists have developed special waterproof materials made from polymers. The polymer materials have pores that are 2000 times smaller than a drop of water. However, the tiny pores are 700 times larger than a water molecule. Explain why these materials are described as 'breathable'.

7 Non-biodegradable plastic has been used for many years for growing melons. The plants are put into holes in the plastic and their shoots grow up above the plastic. The melons are protected from the soil by the plastic and grow with very few marks on them. Biodegradable plastic has been tested – to reduce the amount of non-recycled waste plastic.

In this investigation two large plots were used to grow melons. One using biodegradable plastic, the other using normal plastic. The results were as follows:

Plastic used	Total yield (kg/hectare)	Average mass of melons produced (kg)
Non-biodegradable	4829	2.4
Biodegradable	3560	2.2

  a This was a field investigation. Describe how the experimenter would have chosen the two plots.

  b What conclusion can you draw from this investigation?

  c How could the reliability of these results be tested?

  d How would you view these results if you were told that they were funded by the manufacturer of the traditional non-biodegradable plastic?

Practice questions (k)

1 Large alkanes from crude oil are broken down to give smaller molecules.

Large alkane (e.g. $C_{12}H_{26}$) → vaporised and passed over hot catalyst → smaller alkane (e.g. C_5H_{12}) + alkene (e.g. C_2H_4)

Choose the correct word from the list to complete each sentence.

a This process is called ....

cracking   distillation   fermentation   (1)

b The reaction is an example of thermal ....

decomposition   evaporation   polymerisation   (1)

c The smaller alkane can be used as a ....

plastic   monomer   fuel   (1)

d The alkene will turn bromine water ....

blue   colourless   orange   (1)

e The general formula for an alkene is ....

C_nH_{2n-2} C_nH_{2n} C_nH_{2n+2} (1)

2 Ethene is used to mke the plastic poly(ethene).

a Complete the equation to show the formation of poly(ethene).   (3)

$$n \quad \underset{\underset{H}{|}}{\overset{\overset{H}{|}}{C}} = \underset{\underset{H}{|}}{\overset{\overset{H}{|}}{C}} \longrightarrow$$

b In the equation, what does the letter *n* represent?   (1)

c What name is used for the small molecules that join to make a polymer?   (1)

d Name the polymer that is made from butene.   (1)

e Which one of the following could be used in a similar way to make a polymer?

C_3H_6 C_3H_8 C_4H_{10} (1)

3 Ethanol can be used as a fuel for cars. Pure ethanol (100%) can be used in specially adapted car engines. Petrol with up to 10% ethanol can be used in ordinary car engines. To mix with petrol the ethanol must not contain any water.

Ethanol can be made from plants or from crude oil.

        fermentation   distillation   dehydration
plants → sugars → 15% ethanol → 96% ethanol → 100%
                    in water   ethanol
        distillation   cracking   catalyst + steam
crude oil → fractions → ethene   →   100% ethanol

a Suggest **one** environmental advantage of making ethanol fuel from plants rather than from crude oil.   (1)

b Suggest **one** economic disadvantage of producing ethanol fuel from plants rather than from crude oil.   (1)

c Suggest **one** environmental disadvantage of producing ethanol fuel from plants.   (1)

d 10% ethanol in petrol can be used in ordinary car engines. Suggest **one** other advantage of using 10% ethanol in petrol as a fuel rather than pure ethanol.   (1)

4 Scientists develop new polymers and modify existing polymers.

a Polylactic acid (PLA) is a bioplastic that is biodegradable. It can be used to make sandwich containers, plastic cups and plastic cutlery. PLA is made from cornstarch. In the USA large amounts of maize are grown and used to make cornstarch, which has many uses. To make PLA the cornstarch is fermented with microbes to make lactic acid, which is then polymerised.

The structure of PLA is

$$\left(\!\! \begin{array}{c} \quad CH_3 \quad O \\ \quad | \quad\quad \| \\ O-CH-C \\ \end{array} \!\! \right)_{\!\! n}$$

i Give **one** way in which the structure of PLA is different from the structure of poly(ethene).   (1)

ii Give **one** way in which the structure of PLA is similar to the structure of poly(ethene).   (1)

iii Suggest what is meant by *bioplastic*.   (1)

iv Suggest **two** reasons why PLA was developed.   (2)

b *In this question you will be assessed on using good English, organising information clearly and using specialist terms where appropriate.*

Copper was considered to be the most suitable material to use for hot water pipes. PEX is now used as an alternative material for hot water pipes. PEX is made from poly(ethene).

Copper is extracted from its ore by a series of processes.

1 The low-grade ore is powdered and concentrated.
2 Smelting is carried out in an oxygen flash furnace. This furnace is heated to 1100 °C using a hydrocarbon fuel. The copper ore is blown into the furnace with air, producing impure, molten copper.
3 Oxygen is blown into the impure, molten copper to remove any sulfur. The copper is cast into rectangular slabs.
4 The final purification of copper is done by electrolysis.

Suggest the possible environmental advantages of using PEX instead of copper for hot water pipes.   (6)

AQA, 2009

C1 6.1 Extracting vegetable oil

Learning objectives

- How do we extract oils from plants?
- Why are vegetable oils important foods?
- What are unsaturated oils and how do we detect them?

Plants use the Sun's energy to produce glucose from carbon dioxide and water during photosynthesis:

$$\text{carbon dioxide} + \text{water} \xrightarrow[\text{energy (from sunlight)}]{\text{chlorophyll}} \text{glucose} + \text{oxygen}$$
$$6CO_2 + 6H_2O \longrightarrow C_6H_{12}O_6 + 6O_2$$

Plants then turn glucose into other chemicals they need using more chemical reactions. In some cases these other chemicals can also be very useful to us. For example, the **vegetable oils** from plants, such as oilseed rape, make biofuels and foodstuffs.

We find these oils in the seeds of the rape plant. Farmers collect the seeds from the plants using a combine harvester. The seeds are then taken to a factory where they are crushed and pressed to extract their oil. The impurities are removed from the oil. It is then processed to make it into useful products.

We extract other vegetable oils using steam. For example, we can extract lavender oil from lavender plants by distillation. The plants are put into water and boiled. The oil and water evaporate together and are collected by condensing them. The water and other impurities are removed to give pure lavender oil.

Figure 1 Oilseed rape is a common sight in our countryside. As its name tells us, it is a good source of vegetable oil.

Practical

Extracting plant oil by distillation (microscale)

Take care not to let the contents of the small vial boil over.

- What does the liquid collected look and smell like?

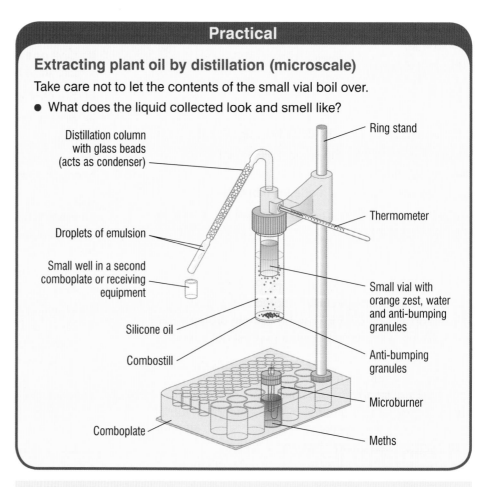

Figure 2 Norfolk lavender oil is extracted from lavender plants by distillation

a Write down two ways we can to extract vegetable oils from plants.

Vegetable oils as foods and fuels

Vegetable oils are very important foods. They provide important nutrients. For example olive oil is a source of vitamin E. They also contain a great deal of energy, as the table shows. This makes them useful foods and sources of biofuels, such as biodiesel.

There are lots of different vegetable oils. Each vegetable oil contains mixtures of compounds with slightly different molecules. However, all vegetable oils have molecules which contain chains of carbon atoms with hydrogen atoms:

In some vegetable oils the hydrocarbon chains contain carbon–carbon double bonds (C=C). We call these **unsaturated oils**. We can detect the double bonds in unsaturated oils with bromine water. You know the test for double bonds from your work on alkenes.

This provides us with an important way of detecting unsaturated oils:

unsaturated oil + bromine water (orange) → colourless solution

> **b** What will you see if you test a polyunsaturated margarine with bromine water?

links
For information on biofuels, look back at C1 4.5 Alternative fuels.

Energy in vegetable oil and other foods	
Food	**Energy in 100 g (kJ)**
vegetable oil	3900
sugar	1700
animal protein (meat)	1100

Figure 3 Vegetable oils have a high energy content

links
For information on the test for double bonds, look back at C1 5.1 Cracking hydrocarbons.

Practical

Testing for unsaturation

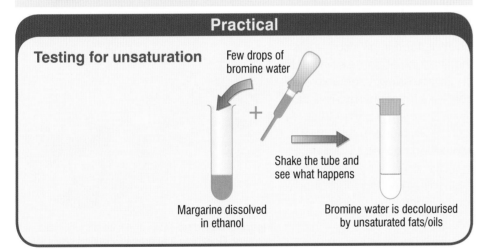

Few drops of bromine water

Shake the tube and see what happens

Margarine dissolved in ethanol

Bromine water is decolourised by unsaturated fats/oils

?? Did you know ...?

No more than 20% of the energy in your diet should come from fats.

Summary questions

1 Copy and complete using the words below:

   bromine decolorised distillation energy pressing unsaturated

   We can extract vegetable oils from some plants by .... or .... Vegetable oils are particularly important as foods because they contain a lot of .... Some vegetable oils contain carbon–carbon double bonds. We call these .... vegetable oils. They can be detected by reacting them with .... water, which will be ....

2 Why might a diet containing too much vegetable oil be unhealthy?

3 A sample of vegetable oils is tested with bromine water. The solution is decolorised. Which of the following statements is true?
   a The sample contains **only** unsaturated oils.
   b The sample contains **only** saturated oils.
   c The sample may contain a mixture of saturated and unsaturated oils. Explain your answer.

Key points

● Vegetable oils can be extracted from plants by pressing or by distillation.

● Vegetable oils provide nutrients and have a high energy content. They are important foods and can be used to make biofuels.

● Unsaturated oils contain carbon–carbon double bonds (C=C). We can detect them as they decolorise bromine water.

C1 6.2 Cooking with vegetable oils

Learning objectives

- What are the advantages and disadvantages of cooking with vegetable oils?

- What does it mean when we 'harden' vegetable oils? [H]

- How do we turn vegetable oils into spreads? [H]

When we cook food we heat it to a temperature where chemical reactions cause permanent changes to happen to the food. Cooking food in vegetable oil gives very different results to cooking food in water. This is because the boiling points of vegetable oils are much higher than the boiling point of water. Therefore, vegetable oils can be used at a much higher temperature than boiling water.

What's the difference?

So the chemical reactions that take place in the food are very different in oil and in water. When we cook using vegetable oil:

- the food cooks more quickly

- very often the outside of the food turns a different colour, and becomes crispier

- the inside of the food should be softer if you don't cook it for too long.

> a How does the boiling point of a vegetable oil compare to the boiling point of water?

Figure 1 An electric fryer like this one enables vegetable oil to be heated safely to a high temperature

Cooking food in oil also means that the food absorbs some of the oil. As you know, vegetable oils contain a lot of energy. This can make the energy content of fried food much higher than that of the same food cooked by boiling it in water. This is one reason why regularly eating too much fried food is unhealthy.

Figure 2 Boiled potatoes and fried potatoes are very different. One thing that probably makes chips so tasty is the contrast of crispy outside and soft inside, together with the different smell and taste produced by cooking at a higher temperature. The different colour may be important too as golden chips look more appetising than a pale boiled potato.

Practical

Investigating cooking

Compare the texture and appearance of potato pieces after equal cooking times in water and oil.

You might also compare the cooking times for boiling, frying and oven-baking chips.

If possible carry out some taste tests in hygienic conditions.

Study tip

No chemical bonds are broken when vegetable oils melt or boil – these are physical changes.

> b How is food cooked in oil different to food cooked in water?

Hardening unsaturated vegetable oils

Unsaturated vegetable oils are usually liquids at room temperature.

The boiling and melting points of these oils can be increased by adding hydrogen to the molecules. The reaction replaces some or all of the carbon–carbon double bonds with carbon–carbon single bonds.

With this higher melting point, the liquid oil becomes a solid at room temperature. We call changing a vegetable oil like this **hardening** it. We harden a vegetable oil by reacting it with hydrogen gas (H_2). To make the reaction happen, we must use a nickel catalyst, and carry it out at about 60 °C.

Figure 3 The hydrogen adds to the carbon–carbon double bonds in a vegetable oil when it is hardened and this can be used to make margarine

c What do we call it when we add hydrogen to a vegetable oil?

Oils that we have treated like this are sometimes called **hydrogenated oils**. They are solids at room temperature. This means that they can be made into spreads to be put on bread. We can also use them to make cakes, biscuits and pastry.

Figure 4 We can use hydrogenated vegetable oils in cooking to make a huge number of different, and delicious, foods

Summary questions

1 Copy and complete using the words below:

water energy higher tastes

The boiling points of vegetable oils are .... than the boiling point of water. This means that food cooked in oil .... different to food boiled in .... It also contains more ....

2 Copy and complete using the words below:

hydrogen hydrogenated hardening melting nickel

The .... points of oils may also be raised by adding .... to their molecules. We call this .... the oil. The reaction takes place at 60 °C in the presence of a .... catalyst. The reaction produces a .... oil.
[H]

3 **a** Why are hydrogenated vegetable oils more useful than oils that have not been hydrogenated?

b Explain how we harden vegetable oils. [H]

Study tip

When oils are hardened with hydrogen, a chemical change takes place, producing hydrogenated oils (which have a higher melting point than the original oil). These are used in margarines.

Key points

- Vegetable oils are useful in cooking because of their high boiling points. However, this increases the energy content of foods compared with cooking in boiling water.

- Vegetable oils are hardened by reacting them with hydrogen to increase their melting points. This makes them solids at room temperature which are suitable for spreading. [H]

- The hardening reaction takes place at 60 °C with a nickel catalyst. The hydrogen adds onto C=C bonds in the vegetable oil molecules. [H]

C1 6.3 Everyday emulsions

Learning objectives

- What are emulsions and how do we make them?
- Why are emulsions made from vegetable oils so important?
- What is an emulsifier?
- How do emulsifiers work? [H]

Emulsions in foods

The texture of food – what it feels like in your mouth – is a very important part of foods.

Some smooth foods are made from a mixture of oil and water. Everyone knows that oil and water don't mix. Just try it by pouring a little cooking oil into a glass of water. But we can get them to mix together by making the oil into very small droplets. These spread out throughout the water and produce a mixture called an **emulsion**.

A good example of this is milk. Milk is basically made up of small droplets of animal fat dispersed in water.

Figure 1 Mayonnaise is an emulsion. Smooth food has a good texture and looks as if it will taste nice – but it is not always easy to make, or to keep it smooth.

Figure 2 Milk is an emulsion made up of animal fat and water, together with some other substances

Emulsions often behave very differently to the things that we make them from. For example, mayonnaise is made from ingredients that include oil and water. Both of these are runny – but mayonnaise is not!

Another very important ingredient in mayonnaise is egg yolks. Apart from adding a nice yellow colour, egg yolks have a very important job to do in mayonnaise. They stop the oil and water from separating out into layers. Food scientists call this type of substance an **emulsifier**.

a What do we mean by 'an emulsifier'?

Emulsifiers make sure that the oil and water in an emulsion cannot separate out. This means that the emulsion stays thick and smooth. Any creamy sauce needs an emulsifier. Without it we would soon find blobs of oil or fat floating around in the sauce.

b How does an emulsifier help to make a good creamy sauce?

One very popular emulsion is ice cream. Everyday ice cream is usually made from vegetable oils, although luxury ice cream may also use animal fats.

Emulsifiers keep the oil and water mixed together in the ice cream while we freeze it. Without them, the water in the ice cream freezes separately, producing crystals of ice. That would make the ice cream crunchy rather than smooth. This happens if you allow ice cream to melt and then put it back in the freezer.

Figure 3 Ice cream contains emulsifiers

Other uses of emulsions

Emulsifiers are also important in the cosmetics industry. Face creams, body lotions, lipsticks and lip gloss are all emulsions.

Emulsion paint (often just called emulsion) is a water-based paint with oil droplets dispersed throughout. It is commonly used for painting indoor surfaces such as plastered walls.

Higher

How an emulsifier works

An emulsifier is a molecule with 'a tail' that is attracted to oil and 'a head' that is attracted to water. The 'tail' is a long hydrocarbon chain. This is called the **hydrophobic** part of the emulsifier molecule. The 'head' is a group of atoms that carry a charge. This is called the **hydrophilic** part of the molecule.

The 'tails' dissolve in oil making tiny droplets. The surface of each oil droplet is charged by the 'heads' sticking out into the water. As like charges repel, the oil droplets repel each other. This keeps them spread throughout the water, stopping the oil and water separating out into two layers.

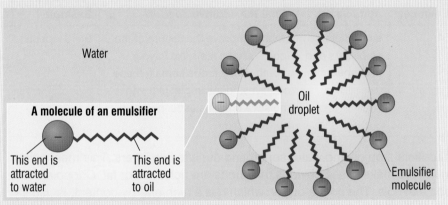

Figure 4 The structure of a typical emulsifier molecule with its water-loving (hydrophilic) head and its water-hating (hydrophobic) tail

Summary questions

1 Copy and complete using the words below:

emulsifier emulsion cosmetics ice mayonnaise mix separating small

Oil and water do not .... together. But if the oil droplets can be made very .... it is possible to produce a mixture of oil and water called an .... To keep the oil and water from .... we can use a chemical called an .... Important examples of food made like this include .... and .... cream. Emulsions are also important in paints and in ....

2 a Salad cream is an emulsion made from vegetable oil and water. In what ways is salad cream different from both oil and water?

b Why do we need to add an emulsifier to an emulsion like salad cream?

3 Explain how emulsifier molecules do their job. [H]

Practical

Making and testing emulsions

Detergents act as emulsifiers.

Add a little cooking oil to some water in a boiling tube. Stopper the tube and shake. Do the same in another boiling tube but also add a drop of washing-up liquid.

- Compare the mixtures when first shaken and when left standing a while.
- You can do some tests on other types of detergent to see which is the most effective emulsifier.

Key points

- Oils do not dissolve in water.
- Oils and water can be dispersed (spread out) in each other to produce emulsions which have special properties.
- Emulsions made from vegetable oils are used in many foods, such as salad dressings, ice creams, cosmetics and paints.
- Emulsifiers stop oil and water from separating out into layers.
- An emulsifier works because one part of its molecule dissolves in oil (hydrophobic part) and one part dissolves in water (hydrophilic part). [H]

C1 6.4

Food issues

- What are the benefits and drawbacks of using emulsifiers in our food?

- What are the good and bad points about vegetable oils in our food?

Figure 1 Modern foods contain a variety of additives to improve their taste, texture or appearance, and to give them a longer shelf-life

∞ links

For information on how an emulsifier works, look back at C1 6.3 Everyday emulsions.

Emulsifying additives

For hundreds of years we have added substances like salt or vinegar to food to help keep it longer. As our knowledge of chemistry has increased we have used other substances too, to make food look or taste better.

We call a substance that is added to food to preserve it or to improve its taste, texture or appearance a **food additive**. Additives that have been approved for use in Europe are given **E numbers**. These can be used to identify them.

> **a** What is a food additive?

Each group of additives is given a range of E numbers. These tell us what kind of additive it is. Emulsifiers are usually given E numbers in the range 400 to 500, along with stabilisers and thickeners.

E number	Additive	What the additive does	Example
E4 _ _	emulsifiers, stabilisers and thickeners	Help to improve the texture of the food – what it feels like in your mouth. Many foods contain these additives, for example, jam and the soya proteins used in veggie burgers.	E440 – pectin

Emulsifiers stop oil and water separating out into two layers. This means that emulsifiers make it less obvious that foods are rich in oil or fat. Chocolate is a good example. The cocoa butter, which has a high energy content, is usually mixed in well, often with the help of emulsifiers. However, have you ever left a bar of chocolate past its sell-by date? Then you can see a white haze on the surface of the chocolate. This is the fatty butter starting to separate out. Then most people will throw the bar away.

So emulsifiers make oil and fat more edible in foods. They can make a mixture that is creamier and thicker in texture than either oil or water. This makes it easier and more tempting for us to eat too much fatty food.

a

b

Figure 2 Which is more appetising – mayonnaise with emulsifier (**a**) or mayonnaise without emulsifier (**b**)?

Vegetable oils in our diet

Everyone knows the benefits of a healthy diet. But do you know the benefits of ensuring that you eat vegetable oils as part of your diet?

Scientists have found that eating vegetable oils instead of animal fats can do wonders for the health of your heart. The saturated fats you find in things like butter and cheese can make the blood vessels of your heart become clogged up.

However, the unsaturated fats in vegetable oils (like olive oil and corn oil) are very good for you. They are a source of nutrients such as vitamin E. They also help to keep your arteries clear and reduce the chance of you having heart disease. The levels of a special fat called cholesterol in your blood give doctors an idea about your risk of heart disease. People who eat vegetable oils rather than animal fats tend to have a much lower level of 'bad' cholesterol in their blood.

Figure 3 Butter contains saturated fats which raise health concerns

b Name a vitamin that we get from olive oil.

The fats used to cook chips and other fast foods often contain certain fats that are not good for us. Scientists are concerned that eating these fats might have caused an increase in heart disease.

Changes in food labelling are very important. But many products, including fast foods, often contain high levels of potentially harmful fats from the oil they were cooked in. Yet these are exempt from labelling regulations and may be advertised as 'cholesterol-free' and 'cooked in vegetable oil'.

Activity

Food for thought

1 Write an article for a family lifestyle magazine about 'Feeding your family'. Include in this article reasons for including vegetable oils in a balanced diet and their effect on people's health.
2 Design a poster with the title 'Vegetable oils – good or bad'?
3 Write the script for a two-minute slot on local radio about the benefits and drawbacks of using emulsifiers in foods.

Figure 4 Chips have a high energy content and may contain potentially harmful fats from cooking oil

Summary questions

1 Draw a table to summarise the advantages and disadvantages of vegetable oils in our diet.
2 **a** Give a list of five foods that can contain emulsifiers as additives.
  b Why could it be said that emulsifiers have played a role in increasing childhood obesity rates?

Key points

- Vegetable oils are high in energy and provide nutrients. They are unsaturated and believed to be better for your health than saturated animal fats and hydrogenated vegetable oils.

- Emulsifiers improve the texture of foods enabling water and oil to mix. This makes fatty foods more palatable and tempting to eat.

Summary questions

1 Write simple definitions for the following words:

  a vegetable oils

  b unsaturated oils

  c saturated oils

  d emulsion

  e emulsifier.

2 a A vegetable oil removes the colour from bromine water.

    It takes longer to decolourise the bromine water when the vegetable oil is partially hydrogenated.

    When the vegetable oil has been completely hardened it does not react with bromine water.

    Explain these observations.

  b Explain why plant oils need to be hardened and the effect this has on the melting point of the oil.   **[H]**

  c Give the conditions for the reaction between a plant oil and hydrogen.   **[H]**

3 Compare the cooking of a potato in boiling water and in vegetable oil.

4 a Some ice cream is left standing out on a table during a meal on a hot day. It is then put back in the freezer again. When it is taken out of the freezer a few days later, people complain that the ice cream tastes 'crunchy'. Why is this?

  b A recipe for making ice cream says: 'Stir the ice cream from time to time while it is freezing.' Why must you stir ice cream when freezing homemade ice cream?

  c Look at this list of ingredients for making ice cream:

    8 large egg yolks

    $\frac{3}{4}$ cup of sugar

    $2\frac{1}{2}$ cups of whipping cream

    $1\frac{1}{2}$ cups cold milk

    1 vanilla pod

    Which ingredient acts as an emulsifier in the mixture?

5 Draw a diagram of the structure of a typical emulsifier.   **[H]**

6 State a use of vegetable oils where their high energy content is:

  a an advantage.

  b a disadvantage.

7 A teacher decided that her class should do a survey of different cooking oils to find out the degree of unsaturated oils present in them. She chose five different oils and divided them among her students. This allowed each oil to be done twice, by two different groups. They were given strict instructions as to how to do the testing.

Bromine water was added to each oil from a burette. The volume added before the mixture in the conical flask was no longer colourless was noted.

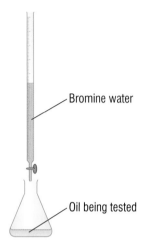

Bromine water

Oil being tested

The results are in this table.

Type of oil	Amount of bromine water added (cm³)	
	Group 1	**Group 2**
Ollio	24.2	23.9
Soleo	17.8	18.0
Spreo	7.9	8.1
Torneo	13.0	12.9
Margeo	17.9	17.4

  a Why was it important that the teacher gave strict instructions to all of the groups on how to carry out the tests?

  b List some control variables that should have been included in the instructions.

  c Are there any anomalous results? How did you decide?

  d What evidence is there in the results that indicate that they are reproducible?

  e How might the accuracy be checked?

  f How would you present these results on a graph? Explain your answer.

Practice questions k

1 Vegetable oils can be extracted from parts of plants that are rich in oils.

Choose the correct word from the list to complete each sentence.

a Sunflower oil is extracted from sunflower ....

leaves petals seeds (1)

b To extract olive oil the olives are crushed and ....

boiled evaporated pressed (1)

c The oil may contain small pieces of solid plant material that can be removed by ....

condensing distilling filtering (1)

d If the oil contains water it can be removed by leaving it to stand because oil and water ....

evaporate mix separate (1)

2 Lavender oil can be extracted from lavender plants by distillation.

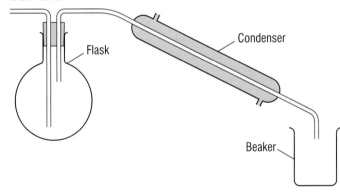

Put the following steps into the correct order, 1 to 6:

lavender plants are harvested → 1 → 2 → 3 → 4 → 5 → 6 → lavender oil is collected.

A Lavender oil and steam are condensed

B Lavender oil separates from water

C Steam is passed into the flask

D Lavender plants are put into the flask

E Lavender oil and water are collected

F Lavender oil and water evaporate (3)

3 Potatoes cooked in boiling water take about 20 minutes to cook. Potato chips can be cooked in less than 10 minutes by deep frying in hot oil. This is one reason why fast food outlets cook chips rather than potatoes.

a Explain why chips cook faster in hot oil than in boiling water. (2)

b Suggest another advantage for fast food outlets to cook chips. (1)

c Suggest a disadvantage for fast food outlets cooking chips. (1)

d Suggest an advantage for consumers who eat chips rather than boiled potatoes. (1)

e Suggest a disadvantage for consumers who eat chips rather than boiled potatoes. (1)

4 **a** A vegetable oil was shaken with water in flask 1 and with water and an emulsifier in flask 2. The diagrams show the results after leaving the mixtures to stand for 5 minutes.

Flask 1
Vegetable oil and water

Flask 2
Vegetable oil, water and an emulsifier

a **i** Give a reason for the result in Flask 1. (1)

ii Explain the result in Flask 2. (2)

b Give an example of a product that contains an emulsifier and give **two** ways in which its properties are better than those of the liquids from which it is made. (3)

c Explain how an emulsifier works. Your answer should include a diagram of a simple model of an emulsifier molecule. [H] (3)

AQA, 2007

C1 7.1

Structure of the Earth

Learning objectives

- What is the structure of the Earth?

- What are the relative sizes of each layer of the Earth's structure?

- In which parts of the Earth's structure are our minerals and other resources found?

How big do you think the Earth is? The deepest mines go down to about 3500 m, while geologists have drilled down to more than 12 000 m in Russia. Although these figures seem large, they are tiny compared with the diameter of the Earth. The Earth's diameter is about 12 800 km. That's more than one thousand times the deepest hole ever drilled!

What's inside?

The Earth is made up of layers that formed many millions of years ago, early in the history of our planet. Heavy materials sank towards the centre of the Earth while lighter material floated on top. This produced a structure consisting of a dense **core**, surrounded by the **mantle**. Outside the mantle there is a thin outer layer called the **crust**.

Above the Earth's crust we have a thin layer of gases called the **atmosphere**.

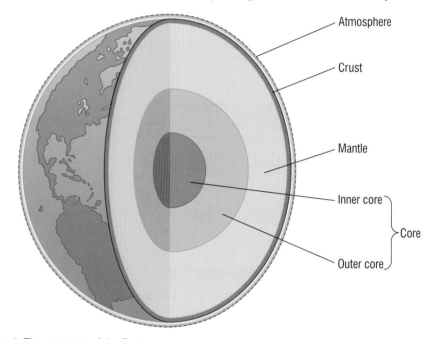

Figure 1 The structure of the Earth

The Earth's crust is a very thin layer compared to the diameter of the Earth. Its thickness can vary from as thin as 5 km under the oceans to 70 km under the continents.

Underneath the crust is the mantle. This layer is much, much thicker than the crust. It is nearly 3000 km thick. The mantle behaves like a solid, but it can flow in parts very slowly.

Finally, inside the mantle lies the Earth's core. This is about half the radius of the Earth. It is made of a mixture of the magnetic metals, nickel and iron. The core is actually two layers. The outer core is a liquid, while the inner core is solid.

> **a** What is the solid outer layer of the Earth called?
> **b** What is the next layer down of the Earth called?

Atmosphere	Crust	Mantle	Core
About 80% of the air in our atmosphere lies within 10 km of the surface. (Most of the rest is within 100 km but it is hard to judge exactly where our atmosphere ends and space begins).	The average thickness of the crust is about 6 km under the oceans; about 35 km under continental areas.	Starts underneath crust and continues to about 3000 km below Earth's surface. Behaves like a solid, but is able to flow very slowly.	Radius of about 3500 km. Made of nickel and iron. Outer core is liquid, inner core is solid.

Figure 2 All the resources that we depend on come from the thin crust of the Earth, its oceans and atmosphere

All the minerals and other resources that we depend on in our lives come from the thin crust of the Earth, its oceans and atmosphere. We get all the natural materials we use plus the raw materials for synthetic and processed materials from these sources. There is a limited supply of resources available to us so we should take care to conserve them for future generations.

 How Science Works

Developing scientific ideas from evidence

How do we know the structure inside the Earth if nobody has ever seen it?

Scientists use evidence from earthquakes. Following an earthquake, seismic waves travel through the Earth. The waves are affected by different layers in the Earth's structure. By observing how seismic waves travel, scientists have built up our picture of the inside of the Earth.

Also, by making careful measurements, physicists have been able to measure the mass of the Earth, and to calculate its density. The density of the Earth as a whole is much greater than the density of the rocks found in the crust. This suggests that the centre of the Earth must be made from a different material to the crust. This material must have a much greater density than the material in the crust.

 Did you know …?

The temperature at the centre of the Earth is between 5000 °C and 7000 °C.

Summary questions

1 Copy and complete using the words below:

core crust mantle slowly solid thin atmosphere

The structure of the Earth consists of three layers – the .... in the centre, then the .... and the outer layer of ...., with the .... above the surface. The outer layer of the Earth is very .... compared to the Earth's diameter. The layer below this is .... but can flow in parts very ....

2 Why do some people think that the mantle is best described as a 'very thick syrupy liquid'?

3 Why should we do our best to conserve the Earth's resources?

Key points

- The Earth consists of a series of layers. Starting from the surface we have the crust, the mantle then the core in the centre. A thin layer of gases called the atmosphere surrounds the Earth.

- The Earth's limited resources come from its crust, the oceans and the atmosphere.

C1 7.2

The restless Earth

Learning objectives

- What are tectonic plates?
- Why do the plates move?
- Why is it difficult for scientists to predict when earthquakes and volcanic eruptions will occur?

Figure 1 *Mesosaurus* was a reptile that existed million years ago. Its fossils have been found in Africa and in South America.

The continents are moving

The map of the world hasn't always looked the way it does today. Look at the map shown in Figure 2. Find the western coastline of Africa and the eastern coastline of South America. Can you see how the edges of the two continents look like they could slot together?

The fossils and rock structures that we find when we look in Africa and South America are also similar. Fossils show that the same reptiles and plants once lived on both continents. Also, the layers of rock in the two continents have been built up in the same sequence.

Scientists now believe that they can explain these facts. They think that the two continents were once joined together as one land mass.

> **a** What evidence is there that Africa and South America were once joined to each other?

Tectonic plates

Of course, the continents moved apart very, very slowly. In fact, they are still moving today, at the rate of a few centimetres each year. They move because the Earth's crust and uppermost part of the mantle is cracked into a number of huge pieces. We call these **tectonic plates**.

Deep within the Earth, radioactive atoms decay, producing vast amounts of energy. This heats up molten minerals in the mantle which expand. They become less dense and rise towards the surface. Cooler material sinks to take their place. Forces created by these **convection currents** move the tectonic plates slowly over the surface of the Earth.

Where the boundaries (edges) of the plates meet, huge stresses build up. These forces make the plates buckle and deform, and mountains may be formed. The plates may also suddenly slip past each other. These sudden movements cause earthquakes.

However, it is difficult for scientists to know exactly where and when the plates will suddenly slip like this.

> **b** What causes an earthquake to occur?

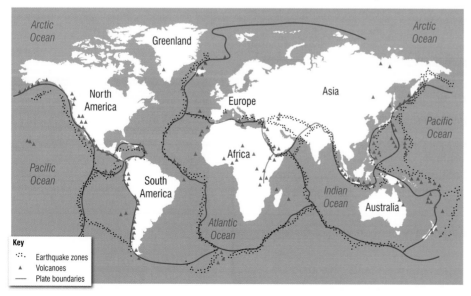

Figure 2 The distribution of volcanoes and earthquakes around the world largely follows the boundaries of the Earth's tectonic plates

How Science Works

Trying to predict the unpredictable!

Earthquakes and volcanic eruptions can be devastating. But making accurate predictions of when they will take place is difficult. Markers placed across a plate boundary or across the crater of a volcano can be monitored for movement. Scientists also monitor the angles of the slopes on volcanoes. The sides of some volcanoes start to bulge outwards before an eruption. Any abnormal readings can be used as a warning sign.

It has also been found that rocks heat up before earthquakes as a result of extreme compression. So satellites with infrared cameras can monitor the Earth's surface for unexpected rises in temperature. Our ability to predict these natural events, and evacuate people at risk, will improve as advances are made by scientists.

Figure 3 Earthquakes can be devastating to people living close by

 How Science Works

Wegener's revolutionary theory

In the past, scientists thought that features like mountain ranges were caused by the crust shrinking as the early molten Earth cooled down. They thought of it rather like the skin on the surface of a bowl of custard. It tends to shrink, then wrinkle, as the custard cools down.

The idea that huge land masses once existed before the continents we know today, was put forward in the late 19th century by the geologist Edward Suess. He thought that a huge southern continent had sunk. He suggested that this left behind a land bridge (since vanished) between Africa and South America.

The idea of continental drift was put forward first by Alfred Wegener in 1915. However, his fellow scientists found Wegener's ideas hard to accept. This was mainly because he could not explain *how* the continents had moved. So they stuck with their existing ideas.

His theory was finally shown to be right almost 50 years later. Scientists found that the sea floor is spreading apart in some places, where molten rock is spewing out between two continents. This led to a new theory, called plate tectonics.

Summary questions

1 Copy and complete using the words below:

convection earthquakes mantle tectonic volcanoes

The surface of the Earth is split up into a series of .... plates. These move across the Earth's surface due to .... currents in the .... Where the plates meet and rub against each other, we get .... and
....

2 a Explain how tectonic plates move.
  b Why are earthquakes and volcanic eruptions difficult to predict?

3 Imagine that you are a scientist who has just heard Wegener's ideas for the first time. Write a letter to another scientist explaining what Wegener has said and why you have chosen to reject his ideas.

Did you know ... ?

With the latest GPS (global positioning satellite) technology we can detect movement of tectonic plates down to 1 mm per year.

Study tip

The Earth's tectonic plates are made up of the crust and the upper part of the mantle (not just the crust).

Key points

- The Earth's crust and upper mantle is cracked into a number of massive pieces (tectonic plates) which are constantly moving slowly.

- The motion of the tectonic plates is caused by convection currents in the mantle, due to radioactive decay.

- Earthquakes and volcanoes happen where tectonic plates meet. It is difficult to know when the plates may slip past each other. This makes it difficult to predict accurately when and where earthquakes will happen.

C1 7.3 The Earth's atmosphere in the past

Learning objectives

● What was the Earth's atmosphere like in the past?

● How were the gases in the Earth's atmosphere produced?

● How was oxygen produced?

??? Did you know ...?

Comets could also have brought water to the Earth. As icy comets rained down on the surface of the Earth, they melted, adding to its water supplies. Even today many thousands of tonnes of water fall onto the surface of the Earth from space every year.

Scientists think that the Earth was formed about 4.5 billion years ago. To begin with it was a molten ball of rock and minerals. For its first billion years it was a very violent place. The Earth's surface was covered with volcanoes belching fire and gases into the atmosphere.

Figure 1 Volcanoes moved chemicals from inside the Earth to the surface and the newly forming atmosphere

The Earth's early atmosphere

There are several theories about the Earth's early atmosphere. One suggests that volcanoes released carbon dioxide, water vapour and nitrogen gas and these gases formed the early atmosphere.

The water vapour in the atmosphere condensed as the Earth gradually cooled down, and fell as rain. Water collected in hollows in the crust as the rock solidified and the first oceans were formed.

As the Earth began to stabilise, the atmosphere was probably mainly carbon dioxide. There could also have been some water vapour, and traces of methane and ammonia. There would have been very little or no oxygen at that time. Some scientists believe nitrogen was another gas present at this time.

This is very like the atmospheres which we know exist today on the planets Mars and Venus.

a What was the main gas in the Earth's early atmosphere?
b How much oxygen was there in the Earth's early atmosphere?

After the initial violent years of the history of the Earth, the atmosphere remained quite stable. That is until life first appeared on Earth.

Figure 2 The surface of one of Jupiter's moons, Io, with its active volcanoes releasing gases into its sparse atmosphere. This gives us a reasonable glimpse of what our own Earth was like billions of years ago.

Oxygen in the atmosphere

There are many theories as to how life was formed on Earth billions of years ago. Scientists think that life on Earth began about 3.4 billion years ago. That is when simple organisms similar to bacteria appeared. These could make food for themselves, using the breakdown of other chemicals as a source of energy.

Later, bacteria and other simple organisms, such as algae, evolved. They could use the energy from the Sun to make their own food by photosynthesis. This produced oxygen gas as a waste product.

By two billion years ago the levels of oxygen were rising steadily as algae and bacteria thrived in the seas. More and more plants evolved. All of them were photosynthesising, removing carbon dioxide and making oxygen.

$$\text{carbon dioxide} + \text{water} \xrightarrow{\text{(energy from sunlight)}} \text{sugar} + \text{oxygen}$$

As plants evolved, they successfully colonised most of the surface of the Earth. So the atmosphere became richer and richer in oxygen. This made it possible for animals to evolve. These animals could not make their own food and needed oxygen to respire.

On the other hand, many of the earliest living microorganisms could not tolerate a high oxygen concentration (because they had evolved without it). They largely died out, as there were fewer places where they could live.

Figure 3 Some of the first photosynthesising bacteria probably lived in colonies like these stromatolites. They grew in water and released oxygen into the early atmosphere.

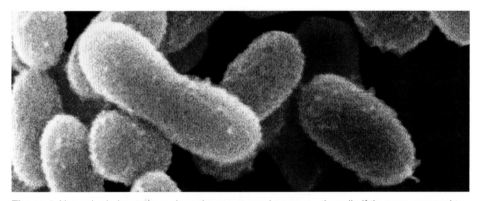

Figure 4 Not only do bacteria such as these not need oxygen – they die if they are exposed to it, but they can survive and breed in rotting tissue and other places where there is no oxygen

??? Did you know ...?

Scientists have reconstructed what they think the atmosphere must have been like millions of years ago based on evidence from gas bubbles trapped in ancient rocks. They also use data gathered from the atmospheres of other planets and their moons in the solar system.

Summary questions

1 Copy and complete using the words below:

dioxide   methane   oxygen   volcanoes   water

The Earth's early atmosphere probably consisted mainly of the gas carbon .... There could also have been .... vapour and nitrogen, plus small amounts of .... and ammonia. These gases were released by .... as they erupted. Plants removed carbon dioxide from the atmosphere and produced .... gas.

2 Describe how the Earth's early atmosphere was probably formed?

3 Why do scientists believe there was no life on Earth for 1.1 billion years?

4 Draw a chart that explains the early development of the Earth's atmosphere.

Key points

- The Earth's early atmosphere was formed by volcanic activity.

- It probably consisted mainly of carbon dioxide. There may also have been water vapour together with traces of methane and ammonia.

- As plants spread over the Earth, the levels of oxygen in the atmosphere increased.

C1 7.4   Life on Earth

Higher

Learning objectives

- Why are there many theories about how life began on Earth?

- Why does one theory involve hydrocarbons, ammonia and lightning? [H]

🔗 links

For information on theories about the Earth's early atmosphere, look back at C1 7.3 The Earth's atmosphere in the past.

Most theories of how our atmosphere developed include the arrival of living things on Earth. The oxygen in our atmosphere today is explained by photosynthesis in plants. The plants probably evolved from simple organisms like plankton and algae in the ancient oceans.

But where did the molecules that make up the cells of even the simplest living things come from? And how were they formed? Any theories to answer these questions are bound to be tentative. They will be based on assumptions. The best theories will be the ones that explain most of the widely-accepted evidence.

Miller–Urey experiment

We know the type of molecules that make up living things. To make these we need compounds called amino acids. These amino acids make proteins.

Most amino acids contain the elements carbon, hydrogen, nitrogen and oxygen. So one way forward is to try to re-create the conditions in the early atmosphere in an experiment. Could amino acids have been made in those conditions? That is the question the scientists Miller and Urey tried to answer in 1952. Figure 1 shows a diagram of their apparatus.

They used a mixture of water (H_2O), ammonia (NH_3), methane (CH_4) and hydrogen (H_2) to model the early atmosphere. Under normal conditions, these gases do not react together. However, Miller and Urey used a high voltage to produce a spark to provide the energy needed for a reaction. This simulated lightning in a storm. The experiment ran for a week then they analysed the mixture formed. It looked like a brown soup. In it they found 11 different amino acids.

a   Which elements make up most amino acids?

This experiment provided evidence that it was possible to make the molecules of life from gases that may have been in our early atmosphere. Miller and Urey published their findings in 1953. They froze some of the mixtures formed in their experiments and stored it. In 2008 other scientists analysed it using modern techniques. They found 22 amino acids, as well as other molecules important for life.

Theories of the composition of the early atmosphere have changed since the 1950s. For example many people think the atmosphere was mainly carbon dioxide and nitrogen before the first life on Earth. However when they carry out similar experiments to Miller and Urey, they still get similar biological molecules made.

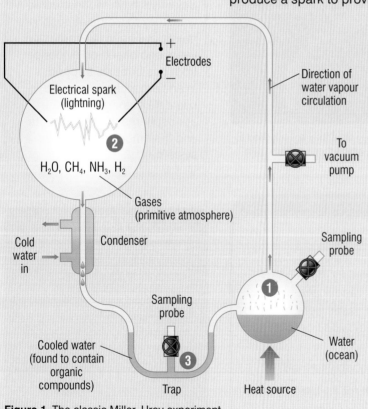

Figure 1 The classic Miller–Urey experiment

Higher

There are opponents of the theory that biological material can be made from non-biological material. They argue that the Miller–Urey experiment only works in the absence of oxygen. They believe that oxygen would have been present before the generally accepted time for its appearance. This would make any conclusions based on Miller–Urey or similar experimental results invalid.

Other theories

Another theory is based on analysis of meteors that crash to Earth from space. In 1969 a meteorite fell from the sky above Australia. Known as the Murchison meteorite, its mass was over 100 kg. However, more interesting were the range of organic molecules found in it.

The latest studies of fragments of the meteorite have identified about 70 different amino acids. This shows that the molecules capable of starting life on Earth might have arrived from outer space.

Figure 2 Part of the Murchison meteorite, which is rich in organic molecules – the molecules of life

b Why were scientists interested in the Murchison meteorite?

Another source of biological molecules could have been deep under the oceans. Near to volcanic vents on the seabed we get both the conditions and chemicals needed.

But just because the 'building blocks' of life might have been on Earth, it does not explain the really difficult step. How do they go on to form life?

The organic molecules, from whatever source, could have formed a 'primordial soup'. All the molecules needed to start life could have been in the seas. Then they would have had to react together to somehow make the first primitive cells. Protein molecules capable of replicating themselves might have been involved at this stage.

Others think that simple living organisms could have arrived on Earth in meteorites or comets. Their evolution had started elsewhere. This 'extraterrestrial seeding' from outer space supports the theory of life in other parts of the universe. Of course, nobody knows for sure but the search for evidence goes on.

Figure 3 Volcanic vents under the sea might have helped form a 'primordial soup' of organic molecules

Summary questions

1 Look at Figure 1.
   a Explain what Miller and Urey did in their experiment.
   b Which one of these statements best describes the outcome of their experiment:
      A It showed how life can be formed from simple molecules.
      B It showed how carbon dioxide and methane are essential parts of a living cell.
      C It showed that biological molecules can be made from substances that could have been in the early atmosphere.
      D It showed that the Earth's early atmosphere must have been made up of only carbon dioxide, ammonia, water vapour and methane.
   [H]

2 **a** What do we mean by a 'primordial soup'?
   b What role might a 'primordial soup' have played in developing life on Earth? [H]

Key points

● One theory states that the compounds needed for life on Earth came from reactions involving hydrocarbons, such as methane, and ammonia. The energy required for the reaction could have been provided by lightning. [H]

● All the theories about how life started on Earth are unproven. We can't be sure about the events that resulted in the first life-forms on Earth. [H]

C1 7.5 Gases in the atmosphere

Learning objectives

- What happened to most of the carbon dioxide in the early atmosphere?

- What are the main gases in the atmosphere today and what are their relative proportions?

- How can the gases in the air be separated? [H]

Figure 1 There is clear fossil evidence in carbonate rocks of the organisms which lived millions of years ago

We think that the early atmosphere of the Earth contained a great deal of carbon dioxide. Yet the Earth's atmosphere today only has around 0.04% of this gas. So where has it all gone? The answer is mostly into living organisms and into materials formed from living organisms.

Carbon 'locked into' rock

Carbon dioxide is taken up by plants during photosynthesis. The carbon can end up in new plant material. Then animals eat the plants and the carbon is transferred to the animal tissues, including bones, teeth and shells.

Over millions of years the dead bodies of huge numbers of these living organisms built up at the bottom of vast oceans. Eventually they formed sedimentary carbonate rocks like limestone (containing mainly calcium carbonate).

Some of these living things were crushed by movements of the Earth and heated within the crust. They formed the fossil fuels coal, crude oil and natural gas. In this way much of the carbon from carbon dioxide in the ancient atmosphere became locked up within the Earth's crust.

> **a** Where has most of the carbon dioxide in the Earth's early atmosphere gone?

Carbon dioxide also dissolved in the oceans. It reacted and made insoluble carbonate compounds. These fell to the seabed and helped to form more carbonate rocks.

Ammonia and methane

At the same time, the ammonia and methane, from the Earth's early atmosphere, reacted with the oxygen formed by the plants.

$$CH_4 + 2O_2 \rightarrow CO_2 + 2H_2O$$
$$4NH_3 + 3O_2 \rightarrow 2N_2 + 6H_2O$$

This got rid of methane and ammonia. The nitrogen (N_2) levels in the atmosphere built up as this is a very unreactive gas.

The atmosphere today

By 200 million years ago the proportions of gases in the Earth's atmosphere had stabilised. These were much the same as they are today. Look at the percentage of gases in the atmosphere today in the pie chart in Figure 2.

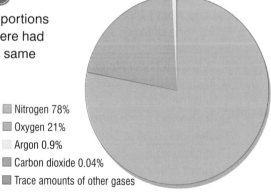

- Nitrogen 78%
- Oxygen 21%
- Argon 0.9%
- Carbon dioxide 0.04%
- Trace amounts of other gases

Figure 2 The relative proportions of nitrogen, oxygen and other gases in the Earth's atmosphere

Practical

Shelly carbonates

Carry out a test to see if crushed samples of shells contain carbonates. Think of the reaction that all carbonates undergo with dilute acid. How will you test any gas given off?

- Record your findings.

> **b** What gas did plants produce that changed the Earth's atmosphere?

Separating the gases in air

In industry the gases are separated by the fractional distillation of liquid air.

Fractional distillation is a process in which **liquids** with different boiling points are separated. So first we have to get air cold enough for it to condense into a liquid. It has to be cooled to a temperature below −200°C.

In industry they do this by compressing the air to about 150 times atmospheric pressure. This actually warms the air up. So it is cooled down to normal temperatures by passing the air over pipes carrying cold water.

But the main cooling takes place when the pressure is released. As this happens, the air is allowed to expand rapidly. This is similar to what happens in an aerosol can when pressure is released as the aerosol is sprayed. The temperature drops far enough for even the gases in air to condense to liquids. The carbon dioxide and water can be removed from the mixture as they are solids at this low temperature.

Here are the boiling points of the main substances left in the liquid air mixture: Nitrogen = −196°C, Argon = −186°C, Oxygen = −183°C.

The liquid is then allowed to warm up and at −196°C nitrogen boils off first. It is collected from the top of a tall fractionating column.

Liquid nitrogen is used to cool things down to very low temperatures. At these temperatures most things solidify. It is used to store sperm in hospitals to help in fertility treatment. Nitrogen gas is very unreactive so we use it in sealed food packaging to stop food going off. It is also used on oil tankers when the oil is pumped ashore to reduce the risk of explosion. In industry, nitrogen gas is used to make ammonia which we convert into fertilisers.

The oxygen separated off is used to help people breathe, often at the scene of an accident or in hospital. It is also used to help things react. Examples include high temperature welding and in the steel-making process.

Figure 4 Biological samples are preserved in liquid nitrogen until they are needed

⊙⊙ links

For information on the fractionating column used to separate crude oil into fractions, look back at C1 4.2 Fractional distillation.

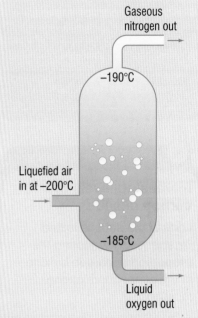

Figure 3 Fractional distillation of liquid air

Study tip

In a fractionating column the individual gases can be separated because of their different boiling points.

Summary questions

1 Copy and complete the table showing the proportion of gases in the Earth's atmosphere today.

nitrogen	oxygen	argon	carbon dioxide	other gases
%	%	%	%	%

2 a Which technique is used to separate the main gases in liquid air? [H]
  b How can water and carbon dioxide be removed from the air before the gases enter the fractionating column. [H]
  c Look at the boiling points of nitrogen, argon and oxygen above:
    i Which gas boils off after nitrogen?
    ii Why is it difficult to obtain 100% pure oxygen? [H]

Key points

- The main gases in the Earth's atmosphere are oxygen and nitrogen.

- About four-fifths (80%) of the atmosphere is nitrogen, and about one-fifth (20%) is oxygen.

- The main gases in the air can be separated by fractional distillation. These gases are used in industry as useful raw materials. [H]

C1 7.6 — Carbon dioxide in the atmosphere

Learning objectives

- How does carbon move in and out of the atmosphere?

- Why has the amount of carbon dioxide in the atmosphere increased recently?

How Science Works

Increasing levels of carbon dioxide

Look at the data collected by scientists monitoring the proportion of carbon dioxide in the atmosphere at one location:

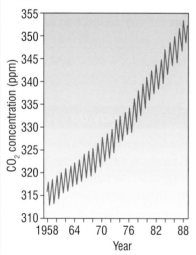

- Why is the line not a smooth curve?

- Explain the overall trend shown by the data.

Over the past 200 million years the levels of carbon dioxide in the atmosphere have not changed much. This is due to the natural cycle of carbon in which carbon moves between the oceans, rocks and the atmosphere.

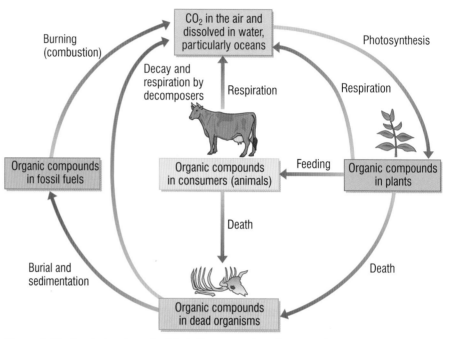

Figure 1 The level of carbon dioxide in the atmosphere has remained steady for the last 200 million years as a result of this natural cycle. However, over the past 200 years the carbon dioxide levels have risen as people started to burn more and more fossil fuels.

Left to itself, this cycle is self-regulating. The oceans act as massive reservoirs of carbon dioxide. They absorb excess CO_2 when it is produced and release it when it is in short supply. Plants also remove carbon dioxide from the atmosphere. We often call plants and oceans carbon dioxide 'sinks'.

a What has kept carbon levels roughly stable over the past 200 million years?

The changing balance

Over the recent past we have greatly increased the amount of carbon dioxide released into the atmosphere . We burn fossil fuels to make electricity, heat our homes and run our cars. This has enormously increased the amount of carbon dioxide we produce.

There is no doubt that the levels of carbon dioxide in the atmosphere are increasing.

We can record annual changes in the levels of carbon dioxide which are due to seasonal differences in the plants. The variations within each year show how important plants are for removing CO_2 from the atmosphere. But the overall trend over the recent past has been ever upwards.

The balance between the carbon dioxide produced and the carbon dioxide absorbed by 'CO_2 sinks' is very important.

links

For information about the effect humans have had on the levels of carbon dioxide in the atmosphere, look back at C1 4.4 Cleaner fuels.

Think about what happens when we burn fossil fuels. Carbon has been locked up for hundreds of millions of years in the fossil fuels. It is released as carbon dioxide into the atmosphere when used as fuel. For example:

$$propane + oxygen \rightarrow carbon\ dioxide + water$$
$$C_3H_8 + 5O_2 \rightarrow 3CO_2 + 4H_2O$$

As carbon dioxide levels in the atmosphere go up, the reactions of carbon dioxide in sea water also increase. The reactions make *insoluble* carbonates (mainly calcium carbonate). These are deposited as sediment on the bottom of the ocean. They also produce *soluble* hydrogencarbonates, mainly of calcium and magnesium. These compounds simply remain dissolved in the sea water.

In this way the seas and oceans act as a buffer, absorbing excess carbon dioxide but releasing it if necessary. However there are now signs that the seas cannot cope with all the additional carbon dioxide that we are currently producing. For example, coral reefs are dying in the more acidic conditions caused by excess dissolved carbon dioxide.

 How Science Works

Thinking of solutions but at what cost?

Most of the electricity that we use in the UK is made by burning fossil fuels. This releases carbon dioxide into the atmosphere. Scientists have come up with a number of solutions. One solution would be to pump carbon dioxide produced in fossil fuel power stations deep underground to be absorbed into porous rocks. This is called 'carbon capture and storage'. It is estimated that this would increase the cost of producing electricity by about 10%.

● Give an advantage and a disadvantage of reducing carbon dioxide emissions using 'carbon capture and storage'.

Figure 2 Most of the electricity that we use in the UK is made by burning fossil fuels

 Did you know ... ?

Some scientists predict that global warming may mean that the Earth's average temperature could rise by as much as 5.8 °C by the year 2100!

People are worried about changing climates (including increasingly common extreme weather events) and rising sea levels as a result of melting ice caps and expansion of the warmer oceans. Low-lying land then might disappear beneath the sea.

Summary questions

1 Match up the parts of sentences:

a	Carbon dioxide levels in the Earth's atmosphere ...	A	... carbon locked up long ago is released as carbon dioxide.
b	Plants and oceans are known as ...	B	... were kept steady by the natural recycling of carbon dioxide in the environment.
c	When we burn fossil fuels ...	C	... the reactions of carbon dioxide in sea water increase.
d	As carbon dioxide levels rise ...	D	... carbon dioxide sinks

2 Draw a labelled diagram to illustrate how boiling an electric kettle may increase the amount of carbon dioxide in the Earth's atmosphere.

3 Why has the amount of carbon dioxide in the Earth's atmosphere risen so much in the recent past?

Key points

● Carbon moves into and out of the atmosphere due to plants, animals, the oceans and rocks.

● The amount of carbon dioxide in the Earth's atmosphere has risen in the recent past largely due to the amount of fossil fuels we now burn.

Summary questions

1 Write simple definitions for the following words describing the structure of the Earth:

a mantle

b core

c atmosphere

d tectonic plate.

2 Wegener suggested that all the Earth's continents were once joined in a single land mass.

a Describe the evidence for this idea, and explain how the single land mass separated into the continents we see today.

b Why were other scientists slow to accept Wegener's ideas?

3 The pie charts show the atmosphere of a planet shortly after it was formed (A) and then millions of years later (B).

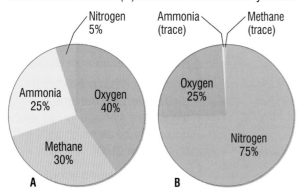

a How did the atmosphere of the planet change?

b What might have caused the change in part **a**?

c Copy and complete the word equations showing the chemical reactions that may have taken place in the atmosphere.

 i methane + ... → carbon dioxide + ...

 ii ammonia + ... → nitrogen + ...

d Why have levels of carbon dioxide in the Earth's atmosphere increased so dramatically over the past 200 years?

4 a Describe how the Miller–Urey experiment advanced our understanding of how life might have first formed on Earth. [H]

b Why didn't their experiment prove how life began on Earth? [H]

5 The Earth and its atmosphere are constantly changing. Design a poster to show this. It should be suitable for displaying in a classroom with children aged 10–11 years. Use diagrams and words to describe and explain ideas and to communicate them clearly to the children.

6 Core samples have been taken of the ice from Antarctica. The deeper the sample the longer it has been there. It is possible to date the ice and to take air samples from it. The air was trapped when the ice was formed. It is possible therefore to test samples of air that have been trapped in the ice for many thousands of years.

This table shows some of these results. The more recent results are from actual air samples taken from a Pacific island.

Year	CO_2 concentration (ppm)	Source
2005	379	Pacific island
1995	360	Pacific island
1985	345	Pacific island
1975	331	Pacific island
1965	320	Antarctica
1955	313	Antarctica
1945	310	Antarctica
1935	309	Antarctica
1925	305	Antarctica
1915	301	Antarctica
1905	297	Antarctica
1895	294	Antarctica
1890	294	Antarctica

a If you have access to a spreadsheet, enter the data and produce a line graph.

b Draw a line of best fit.

c What pattern can you detect?

d What conclusion can you make?

e Should the fact that the data came from two different sources affect your conclusion? Explain why.

Practice questions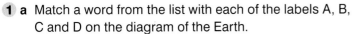

1 a Match a word from the list with each of the labels A, B, C and D on the diagram of the Earth.

atmosphere   core   crust   mantle

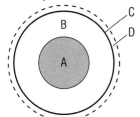

(4)

b From which parts of the Earth do we get all of our raw materials?

A atmosphere, core and crust

B atmosphere, crust and oceans

C atmosphere, core and mantle

D core, mantle and oceans   (1)

2 a About one hundred years ago there was a scientist called Alfred Wegener. He found evidence that the continents, such as South America and Africa, had once been joined and then drifted apart.

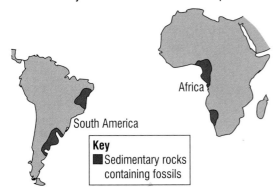

Africa

South America

Key
■ Sedimentary rocks containing fossils

Use the diagram to suggest **two** pieces of evidence that could be used to show that the continents had once been joined.   (2)

b About fifty years ago, new evidence convinced scientists that the Earth's crust is made up of tectonic plates that are moving very slowly.

Give **two** pieces of evidence that have helped to convince these scientists that the tectonic plates are moving.   (2)

c Describe as fully as you can what causes the Earth's tectonic plates to move.   (3)

AQA, 2009

3 a In the Earth's atmosphere the percentage of carbon dioxide has remained at about 0.03% for many thousands of years. The graph shows the percentage of carbon dioxide in the Earth's atmosphere over the last 50 years.

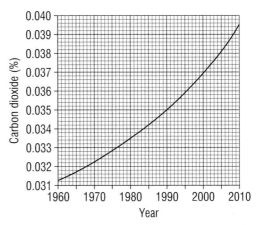

i What was the percentage of carbon dioxide in the Earth's atmosphere in 1965? .... %   (1)

ii What change has happened to the percentage of carbon dioxide in the Earth's atmosphere over the last 50 years?   (1)

iii Suggest **one** reason for this change.   (1)

iv Why does this change worry some people?   (1)

There are different theories about the Earth's early atmosphere.

b Some scientists believe the Earth's early atmosphere was mainly carbon dioxide and water vapour. What do the scientists believe produced these gases?   (1)

c In 1953 some scientists believed the Earth's early atmosphere was mainly water vapour, methane, ammonia and hydrogen. In the Miller–Urey experiment, electricity was passed through a mixture of these gases and produced amino acids, the building blocks for proteins and life. Give two reasons why the experiment does not prove that life began in this way.   **[H]** (2)

d Most scientists agree that there was very little oxygen in the Earth's early atmosphere. Explain how the oxygen that is now in the atmosphere was produced.
          [H] (3)

4 The elements oxygen, nitrogen and argon can be separated from the air. Carbon dioxide and water vapour are removed from air, which is then cooled to −200°C. The liquid obtained is a mixture of oxygen, nitrogen and argon. The table shows the boiling points of these elements.

Element	Boiling point in °C
argon	−186
nitrogen	−196
oxygen	−183

Explain how these elements can be separated by fractional distillation of the liquid.   **[H]** (3)

1 a The diagram shows the parts of a hydrogen atom.
Use words from the list to label the diagram.
electron   group   nucleus   symbol (2)

b Hydrogen can be used as a *clean fuel* for cars.
  i When hydrogen burns in air, it reacts with another element.
Complete the word equation for this reaction. (1)
hydrogen + ... → water
  ii Suggest **one** reason why hydrogen is called a *clean fuel*. (1)

AQA, 2008

2 Use a periodic table to help you to answer this question.
Oxygen is in Group 6 of the periodic table.

a i How many protons are in an atom of oxygen? (1)
  ii How many electrons are in an atom of oxygen? (1)

b Chlorine is in Group 7 of the periodic table.
Complete the electronic structure of chlorine: 2, ... (1)

c Fluorine is also in Group 7.
Explain why in terms of electronic structure. (1)

d Neon and argon are in Group 0 of the periodic table. They are very unreactive elements. What does this tell you about their electronic structures? (2)

3 When calcium carbonate is heated it decomposes. The equation for this reaction is:
$CaCO_3 \rightarrow CaO + CO_2$

a Use numbers from the list to complete the sentences.
  2   3   4   5   6
  i The number of products in the equation is .... (1)
  ii The formula $CaCO_3$ shows that calcium carbonate was made from .... different elements. (1)
  iii The equation is balanced because there are .... atoms on both sides. (1)

b Other metal carbonates decompose in a similar way.
  i Name the solid produced when zinc carbonate decomposes. (1)
  ii Name the gas produced when copper carbonate decomposes. (1)

4 Farmers can use calcium hydroxide to neutralise soils that are too acidic. Limestone is mainly calcium carbonate, $CaCO_3$.
Limestone is used to make calcium hydroxide, $Ca(OH)_2$.

a What are the two reactions used to make calcium hydroxide from limestone? (2)
b Explain why calcium hydroxide neutralises soils that are too acidic. (2)
c Farmers can also use powdered limestone to neutralise soils that are too acidic. Explain why. (2)
d Suggest one reason why it may be safer for farmers to use powdered limestone instead of calcium hydroxide. (1)
e Suggest one reason why powdered limestone costs less than calcium hydroxide. (1)
f Write balanced equations for the reactions in 4(a) **[H]** (2)
g The formula of calcium chloride is $CaCl_2$. Write a balanced equation for the reaction of calcium hydroxide with hydrochloric acid, HCl. **[H]** (2)
h Write a balanced equation for the reaction of calcium carbonate with hydrochloric acid. **[H]** (2)

> **Study tip**
>
> When you are asked to complete a word equation for a reaction, read the information in the question carefully and you should find the names of the reactants and products.

> **Study tip**
>
> The AQA data sheet that you will have in the exam has a periodic table.

> **Study tip**
>
> Remember that each symbol represents one atom of an element and that small (subscript) numbers in a formula multiply only the atom they follow.

> **Study tip**
>
> Attempt all parts of a question. If you come to a part you cannot answer do not be put off reading the next parts.

5 Titanium is as strong as steel but is much more expensive. It is used to make jet engines for aircraft and to make replacement hip joints for people.

a Give two properties that make titanium better than steel for making jet engines and replacement hip joints. (2)

b *In this question you will be assessed on using good English, organising information clearly and using specialist terms where appropriate.*

Titanium is made in batches of about 10 tonnes that takes up to 15 days. The main steps to make titanium are:

- Titanium oxide is reacted with chlorine to produce titanium chloride.
- Titanium chloride is reacted with magnesium at 900 °C in a sealed reactor for three days to give a mixture of titanium and magnesium chloride.
- The reactor is cooled for 7 days, and then the mixture is removed.
- The magnesium chloride is removed from the mixture by distillation at very low pressure.
- The titanium is melted in an electric furnace and poured into moulds.

Steel is produced at about 8000 tonnes per day. The main steps to make steel are:

- Iron oxide is reacted with carbon (coke) in a blast furnace that runs continuously.
- The molten impure iron flows to the bottom of the furnace and is removed every four hours.
- Oxygen is blown into the molten iron for about 20 minutes to produce steel.
- The steel is poured into moulds.

Explain why titanium costs more than steel. (6)

AQA, 2008

6 Olives are the fruits of the olive tree. Olive oil is extracted from olives.

a Use a word from the list to complete the sentence.

condensed   evaporated   pressed

In the first step to extract the oil the olives are crushed and .... (1)

b This gives a mixture of liquids and solids that is left to settle.

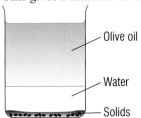

Olive oil

Water

Solids

Why does the olive oil separate from the water? (1)

c The olive oil is removed from the water and filtered to remove any small pieces of solids.

Suggest **two** reasons why separating olive oil by this method is better than separating it by distilling. (2)

d Olive oil can be used as a fuel. Explain why. (2)

e Food can be cooked in olive oil. Give one advantage and one disadvantage of cooking food in olive oil. (2)

f Olive oil can be used with vinegar to make salad dressings. Name the type of substance that is added to salad dressings to stop them from separating. (1)

Study tip

You may be given information about familiar or unfamiliar applications of chemistry. The information you are given should help you to answer the questions. Q5(b) requires you to organise information clearly. Think about the points in the information and decide which ones make titanium more expensive than steel. Underline or circle the points you are going to use on the question paper. Add brief notes, perhaps numbers for the order that you will use. Think about how you are going to write your answer. Rehearse it in your head before you write your answer.

Study tip

Always be aware of the number of marks for a question. If it is two marks, you need to make two points in your answer. Sometimes this is obvious, as in Q6(c), but in Q6(d) you need to make sure you have not given just a single simple statement.

P1 1.1 | Infrared radiation

Learning objectives

- What is infrared radiation?
- Do all objects give off infrared radiation?
- How does infrared radiation depend on the temperature of an object?

Figure 1 Keeping watch in darkness

 links

For more information on infrared heaters, see P1 1.9 Heating and insulating buildings.

Did you know ... ?

A **passive infrared (PIR) detector** in a burglar alarm circuit will 'trigger' the alarm if someone moves in front of the detector. The detector contains sensors that detect infrared radiation from different directions.

Seeing in the dark

We can use special cameras to 'see' animals and people in the dark. These cameras detect **infrared radiation**. Every object gives out (**emits**) infrared radiation.

The hotter an object is, the more infrared radiation it emits in a given time.

Look at the photo in Figure 1. The rhinos are hotter than the ground.

- **a** Why is the ground darker than the rhinos?
- **b** Which part of each rhino is coldest?

Practical

Detecting infrared radiation

You can use a thermometer with a blackened bulb to detect infrared radiation. Figure 2 shows how to do this.

- The glass prism splits a narrow beam of white light into the colours of the spectrum.
- The thermometer reading rises when it is placed just beyond the red part of the spectrum. Some of the infrared radiation in the beam goes there. Our eyes cannot detect it but the thermometer can.
- Infrared radiation is beyond the red part of the visible spectrum.

What would happen to the thermometer reading if the thermometer were moved away from the screen?

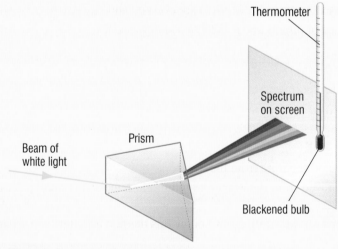

Figure 2 Detecting infrared radiation

The electromagnetic spectrum

Radio waves, **microwaves**, infrared radiation and **visible light** are parts of the electromagnetic spectrum. So too are ultraviolet rays and X-rays. Electromagnetic waves are electric and magnetic waves that travel through space.

 links

For more information on electromagnetic waves, see P1 6.1 The electromagnetic spectrum.

Energy from the Sun

The Sun emits all types of electromagnetic radiation. Fortunately for us, the Earth's atmosphere blocks most of the radiation that would harm us. But it doesn't block infrared radiation from the Sun.

Figure 3 shows a solar furnace. This uses a giant reflector that focuses sunlight.

The temperature at the focus can reach thousands of degrees. That's almost as hot as the surface of the Sun, which is 5500 °C.

The greenhouse effect

The Earth's atmosphere acts like a greenhouse made of glass. In a greenhouse:

● short wavelength infrared radiation (and light) from the Sun can pass through the glass to warm the objects inside the greenhouse

● infrared radiation from these warm objects is trapped inside by the glass because the objects emit infrared radiation of longer wavelengths that can't pass through the glass.

So the greenhouse stays warm.

Gases in the atmosphere, such as water vapour, methane and carbon dioxide, trap infrared radiation from the Earth. This makes the Earth warmer than it would be if it had no atmosphere.

But the Earth is becoming too warm. If the polar ice caps melt, it will cause sea levels to rise. Reducing our use of fossil fuels will help to reduce the production of 'greenhouse gases'.

Figure 3 A solar furnace in the Eastern Pyrenees, France

How Science Works

A huddle test

Design an investigation to model the effect of penguins huddling together. You could use beakers of hot water to represent the penguins.

Figure 4 Penguins keeping warm

Summary questions

1 Copy and complete **a** and **b** using the words below. Each word can be used more than once.

temperature   radiation   waves

a Infrared .... is energy transfer by electromagnetic ....

b The higher the .... of an object is, the more .... it emits each second.

2 **a** Copy and complete the table to show if the object emits infrared radiation or light or both.

Object	Infrared	Light
A hot iron		
A light bulb		
A TV screen		
The Sun		

b How can you tell if an electric iron is hot without touching it?

3 Explain why penguins huddle together to keep warm.

Key points

● Infrared radiation is energy transfer by electromagnetic waves.

● All objects emit infrared radiation.

● The hotter an object is, the more infrared radiation it emits in a given time.

P1 1.2 Surfaces and radiation

Learning objectives

Learning objectives

- Which surfaces are the best emitters of infrared radiation?

- Which surfaces are the best absorbers of infrared radiation?

- Which surfaces are the best reflectors of infrared radiation?

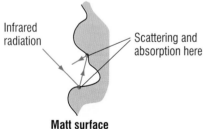

Figure 1 A thermal blanket in use

Smooth surface

Matt surface

Figure 3 Absorbing infrared radiation

∞ links

Infrared heaters use light, shiny surfaces to reflect infrared radiation. See P1 3.4 Cost effectiveness matters.

Which surfaces are the best emitters of radiation?

Rescue teams use light shiny thermal blankets to keep accident survivors warm (see Figure 1). A light, shiny outer surface emits much less radiation than a dark, matt surface.

Practical

Testing radiation from different surfaces

To compare the radiation from two different surfaces, you can measure how fast two cans of hot water cool. The surface of one can is light and shiny and the other has a dark matt surface (see Figure 2). At the start, the volume and temperature of the water in each can must be the same.

Thermometer to measure water temperature at intervals as it cools

- Why should the volume and temperature of the water be the same at the start?

- Which can will cool faster?

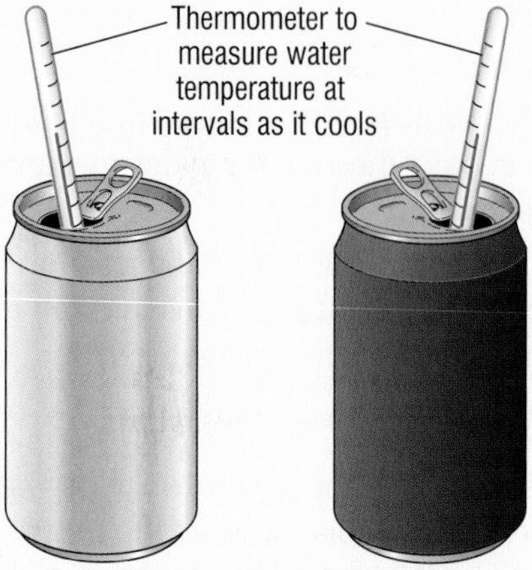

Figure 2 Testing different surfaces

Your tests should show that:

- **Dark, matt surfaces are better at emitting radiation than light, shiny surfaces.**

Which surfaces are the best absorbers and reflectors of radiation?

When you use a photocopier, infrared radiation from a lamp dries the ink on the paper. Otherwise, the copies would be smudged. Black ink absorbs more infrared radiation than white paper.

A light, shiny surface absorbs less radiation than a dark, matt surface. A matt surface has lots of cavities, as shown in Figure 3.

- The radiation reflected from the matt surface hits the surface again.

- The radiation reflected from the shiny surface travels away from the surface.

So the shiny surface absorbs less and reflects more radiation than a matt surface.

In general:

- **Light, shiny surfaces absorb less radiation than dark, matt surfaces.**

- **Light, shiny surfaces reflect more radiation than dark, matt surfaces.**

a Why does ice on a road melt faster in sunshine if sand is sprinkled on it?

b Why are solar heating panels painted matt black?

Practical

Absorption tests

Figure 4 shows how we can compare absorption by two different surfaces.

- The front surfaces of the two metal plates are at the same distance from the heater.
- The back of each plate has a coin stuck on with wax. The coin drops off the plate when the wax melts.
- The coin at the back of the matt black surface drops off first. The matt black surface absorbs more radiation than the light shiny surface.

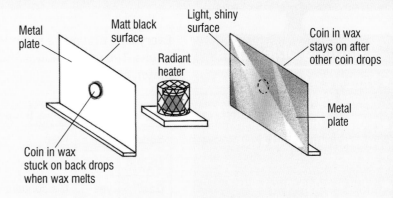

Figure 4 Testing different absorbers of infrared radiation

Summary questions

1 Copy and complete **a** and **b** using the words below:

absorber   emitter   reflector

   a A dark, matt surface is a better .... and a better .... of infrared radiation than a light, shiny surface.

   b A light, shiny surface is a better .... of infrared radiation than a dark, matt surface.

2 A black car and a metallic silver car are parked next to each other on a sunny day. Why does the temperature inside the black car rise more quickly than the temperature inside the silver car?

3 A metal cube filled with hot water was used to compare the infrared radiation emitted from its four vertical faces, A, B, C and D.

   An infrared sensor was placed opposite each face at the same distance, as shown in Figure 5. The sensors were connected to a computer. The results of the test are shown in the graph below.

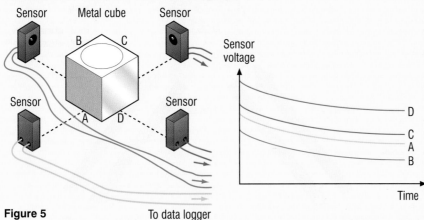

Figure 5   To data logger

   a Why was it important for the distance from each sensor to the face to be the same?

   b One face was light and shiny, one was light and matt, one was dark and shiny, and one was dark and matt.
   Which face A, B, C or D emits the **i** least radiation, **ii** most radiation?

   c What are the advantages of using data logging equipment to collect the data in this investigation?

 Did you know ...?

Scientists are developing blacker and blacker materials. These new materials have very tiny pits in the surface to absorb almost all the light that hits them. They can be used to coat the insides of telescopes so that there are no reflections.

Key points

- Dark, matt surfaces emit more infrared radiation than light, shiny surfaces.

- Dark, matt surfaces absorb more infrared radiation than light, shiny surfaces.

- Light, shiny surfaces reflect more infrared radiation than dark matt surfaces.

P1 1.3 | States of matter

Learning objectives

- How are solids, liquids and gases different?
- How are the particles in a solid, liquid and a gas arranged?
- Why is a gas much less dense than a solid or a liquid?

Everything around us is made of matter in one of three states – solid, liquid or gas. The table below summarises the main differences between the three **states of matter**.

	Flow	Shape	Volume	Density
Solid	no	fixed	fixed	much higher than a gas
Liquid	yes	fits container shape	fixed	much higher than a gas
Gas	yes	fills container	can be changed	low compared with a solid or liquid

a We can't see it and yet we can fill objects like balloons with it. What is it?

b When an ice cube melts, what happens to its shape?

Change of state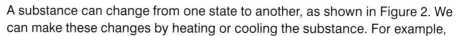

A substance can change from one state to another, as shown in Figure 2. We can make these changes by heating or cooling the substance. For example,

- when water in a kettle boils, the water turns to steam. Steam, also called water vapour, is water in its gaseous state
- when solid carbon dioxide or 'dry ice' warms up, the solid turns into gas directly
- when steam touches a cold surface, the steam condenses and turns to water.

Figure 1 Spot the three states of matter

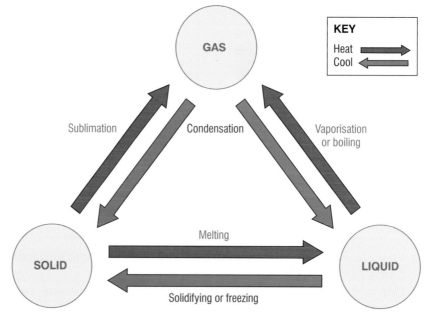

Figure 2 Change of state

c What change of state occurs when hailstones form?

Practical

Changing state

1 Heat some water in a beaker using a Bunsen burner, as shown in Figure 3. Notice that:
   ● steam or 'vapour' leaves the water surface before the water boils
   ● when the water boils, bubbles of vapour form inside the water and rise to the surface to release steam.
2 Switch the Bunsen burner off and hold a cold beaker or cold metal object above the boiling water. Observe condensation of steam from the boiling water on the cold object. Take care with boiling water.

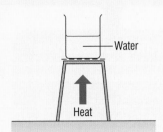

Figure 3 Changing state

The kinetic theory of matter

Solids, liquids and gases consist of particles. Figure 4 shows the arrangement of the particles in a solid, a liquid and a gas. When the temperature of the substance is increased, the particles move faster.

● The particles in a solid are held next to each other in fixed positions. They vibrate about their fixed positions so the solid keeps its own shape.
● The particles in a liquid are in contact with each other. They move about at random. So a liquid doesn't have its own shape and it can flow.
● The particles in a gas move about at random much faster. They are, on average, much further apart from each other than in a liquid. So the density of a gas is much less than that of a solid or liquid.
● The particles in solids, liquids and gases have different amounts of energy. In general, the particles in a gas have more energy than those in a liquid, which have more energy than those in a solid.

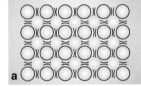 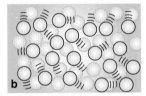 

Figure 4 The arrangement of particles in **a** a solid, **b** a liquid and **c** a gas

??? Did you know ... ?

Random means unpredictable. Lottery numbers are chosen at random.

Summary questions

1 Copy and complete **a** to **d** using the words below. Each word can be used more than once.

   gas   liquid   solid

   a A .... has a fixed shape and volume.
   b A .... has a fixed volume but no shape.
   c A .... and a .... can flow.
   d A .... does not have a fixed volume.

2 State the scientific word for each of the following changes.
   a A mist appears on the inside of a window in a bus full of people.
   b Steam is produced from the surface of the water in a pan when the water is heated before it boils.
   c Ice cubes taken from a freezer thaw out.
   d Water put into a freezer gradually turns to ice.

3 Describe the changes that take place in the movement and arrangement of the particles in an ice cube when the ice melts.

Key points

● Flow, shape, volume and density are the properties used to describe each state of matter.

● The particles in a solid are held next to each other in fixed positions.

● The particles in a liquid move about at random and are in contact with each other.

● The particles in a gas move about randomly and are much further apart than particles in a solid or liquid.

P1 1.4 Conduction

Learning objectives

- What materials make the best conductors?
- What materials make the best insulators?
- Why are metals good conductors?
- Why are non-metals poor conductors?

Figure 1 At a barbecue – the steel cooking utensils have wooden or plastic handles

When you have a barbecue, you need to know which materials are good **conductors** and which are good **insulators**. If you can't remember, you are likely to burn your fingers!

Testing rods of different materials as conductors

The rods need to be the same width and length for a fair test. Each rod is coated with a thin layer of wax near one end. The uncoated ends are then heated together.

Look at Figure 2. The wax melts fastest on the rod that conducts best.

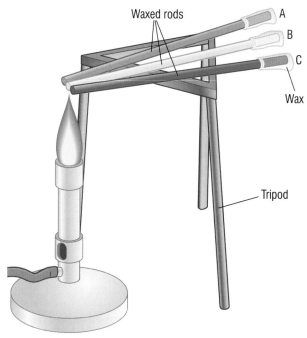

Figure 2 Comparing conductors

- Metals conduct energy better than non-metals.
- Copper is a better conductor than steel.
- Wood conducts better than glass.

> a Why do steel pans have handles made of plastic or wood?
> b Name the independent and the dependent variables investigated in Figure 2.

⊂⊃ **links**

For more information on independent and dependent variables, look back at H3 Starting an investigation.

⚙️ Practical

Testing sheets of materials as insulators

Use different materials to insulate identical cans (or beakers) of hot water. The volume of water and its temperature at the start should be the same.

Use a thermometer to measure the water temperature after a fixed time. The results should tell you which insulator was best.

The table below gives the results of comparing two different materials using the method explained in the practical.

Material	Starting temperature (°C)	Temperature after 300 s (°C)
paper	40	32
felt	40	36

c Which material, felt or paper, was the better insulator?
d Which variable shown in the table was controlled to make this a fair test?

Figure 3 Insulating a loft. The air trapped between fibres make fibreglass a good insulator.

Conduction in metals

Metals contain lots of **free electrons**. These electrons move about at random inside the metal and hold the positive metal ions together. They collide with each other and with the positive ions. (Ions are charged particles.)

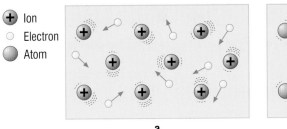

⊕ Ion
○ Electron
⬤ Atom

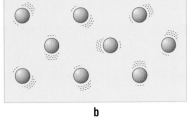

a   b

Figure 4 Energy transfer in **a** a metal, **b** a non-metal

When a metal rod is heated at one end, the free electrons at the hot end gain kinetic energy and move faster.

- These electrons **diffuse** (i.e. spread out) and collide with other free electrons and ions in the cooler parts of the metal.
- As a result, they transfer kinetic energy to these electrons and ions.

So energy is transferred from the hot end of the rod to the colder end.

In a non-metallic solid, all the electrons are held in the atoms. Energy transfer only takes place because the atoms vibrate and shake each other. This is much less effective than energy transfer by free electrons. This is why metals are much better conductors than non-metals.

?? Did you know ...?

Materials like wool and fibreglass are good insulators. This is because they contain air trapped between the fibres. Trapped air is a good insulator. We use materials like fibreglass for loft insulation and for lagging water pipes.

∞ **links**

For more information about ions, see C1 1.4 Forming bonds.

Summary questions

1 Copy and complete **a** to **c** using the words below:

fibreglass plastic steel wood

  a A material called .... is used to insulate a house loft.
  b The handle of a frying pan is made of .... or ....
  c A radiator in a central heating system is made from ....

2 **a** Choose a material you would use to line a pair of winter boots. Explain your choice of material.
  b How could you carry out a test on three different lining materials?

3 Explain why metals are good conductors of energy.

Key points

- Metals are the best conductors of energy.
- Materials such as wool and fibreglass are the best insulators.
- Conduction of energy in a metal is due mainly to free electrons transferring energy inside the metal.
- Non-metals are poor conductors because they do not contain free electrons.

P1 1.5

Convection

Learning objectives

- What is convection?
- Where can convection take place?
- Why does convection occur?

??? Did you know ...?

The Gulf Stream is a current of warm water that flows across the Atlantic Ocean from the Gulf of Mexico to the British Isles. If it ever turned away from us, our winters would be much colder!

Figure 1 A natural glider – birds use convection currents to soar high above the ground

Gliders and birds use convection to stay in the air. **Convection currents** can keep them high above the ground for hours.

Convection happens whenever we heat **fluids**. A fluid is a gas or a liquid. Look at the diagram in Figure 2. It shows a simple demonstration of convection.

- The hot gases from the burning candle go straight up the chimney above the candle.
- Cold air is drawn down the other chimney to replace the air leaving the box.

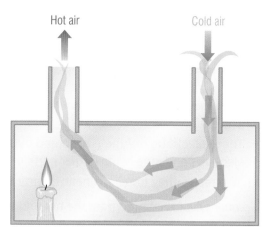

Figure 2 Convection

Using convection

Hot water at home

Many homes have a hot water tank. Hot water from the boiler rises and flows into the tank where it rises to the top. Figure 3 shows the system. When you use a hot water tap at home, you draw off hot water from the top of the tank.

a What would happen if we connected the hot taps to the bottom of the tank?

Sea breezes

Sea breezes keep you cool at the seaside. On a sunny day, the ground heats up faster than the sea. So the air above the ground warms up and rises. Cooler air from the sea flows in as a 'sea breeze' to take the place of the rising warm air (see Figure 4).

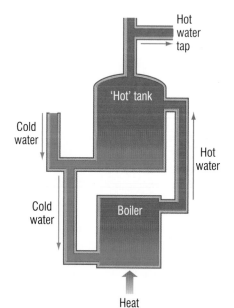

Figure 3 Hot water at home

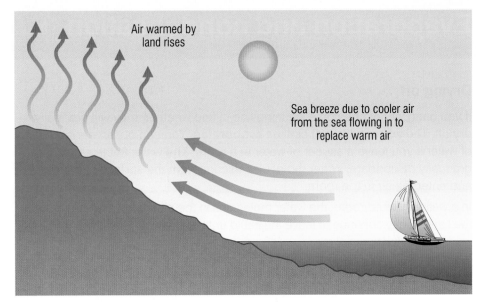

Figure 4 Sea breezes

How convection works

Convection takes place:

● only in fluids (liquids and gases)

● due to circulation (convection) currents within the fluid.

The circulation currents are caused because fluids rise where they are heated (as heating makes them less dense). Then they fall where they cool down (as cooling makes them more dense). Convection currents transfer energy from the hotter parts to the cooler parts.

So why do fluids rise when heated?

Most fluids expand when heated. This is because the particles move about more, taking up more space. Therefore the **density** decreases because the same mass of fluid now occupies a bigger volume. So heating part of a fluid makes that part less dense and therefore it rises.

Study tip

When you explain convection, remember it is the hot fluid that rises, NOT 'heat'.

Summary questions

1 Copy and complete **a** and **b** using the words below:

 cools falls mixes rises

 a When a fluid is heated, it .... and .... with the rest of the fluid.

 b The fluid circulates and .... then it ....

2 Figure 5 shows a convector heater. It has an electric heating element inside and a metal grille on top.

 a What does the heater do to the air inside the heater?

 b Why is there a metal grille on top of the heater?

 c Where does air flow into the heater?

Hot air

Figure 5 A convector heater

3 Describe how you could demonstrate convection currents in water using a strongly coloured crystal or a suitable dye. Explain in detail what you would see.

Key points

● Convection is the circulation of a fluid (liquid or gas) caused by heating it.

● Convection takes place only in liquids and gases.

● Heating a liquid or a gas makes it less dense so it rises and causes circulation.

P1 1.6   Evaporation and condensation ⓚ

Learning objectives

- What is evaporation?
- What is condensation?
- How does evaporation cause cooling?
- What factors affect the rate of evaporation from a liquid?
- What factors affect the rate of condensation on a surface?

Drying off

If you hang wet clothes on a washing line in fine weather, they will gradually dry off. The water in the wet clothes **evaporates**. You can observe evaporation of water if you leave a saucer of water in a room. The water in the saucer gradually disappears. Water molecules escape from the surface of the water and enter the air in the room.

In a well-ventilated room, the water molecules in the air are not likely to re-enter the liquid. They continue to leave the liquid until all the water has evaporated.

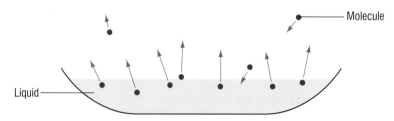

Figure 1 Water molecules escaping from a liquid

Condensation

In a steamy bathroom, a mirror is often covered by a film of water. There are lots of water molecules in the air. Some of them hit the mirror, cool down and stay there. We say water vapour in the air **condenses** on the mirror.

  a Why does opening a window in a steamy room clear the condensation?

Cooling by evaporation

If you have an injection, the doctor or nurse might 'numb' your skin by dabbing it with a liquid that easily evaporates. As the liquid evaporates, your skin becomes too cold to feel any pain.

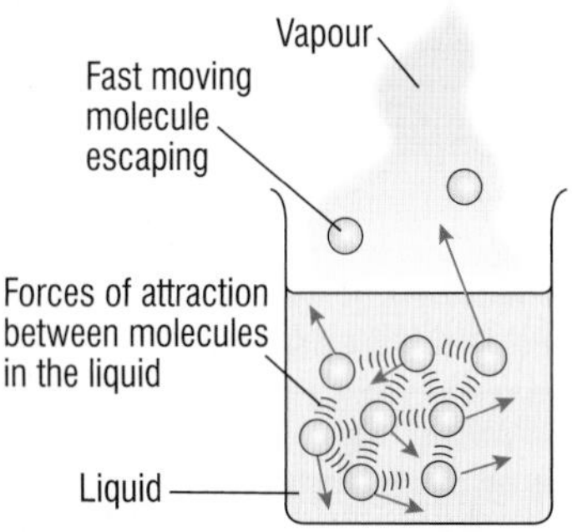

Figure 2 Condensation

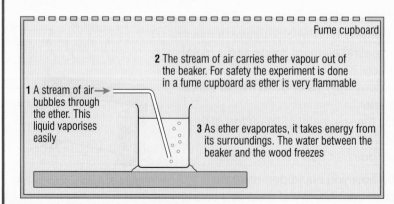

Figure 3 Explanation of cooling by evaporation

Demonstration

Cooling by evaporation

Watch your teacher carry out this experiment in a fume cupboard.

- Why is ether used in this experiment?

Fume cupboard

1 A stream of air bubbles through the ether. This liquid vaporises easily

2 The stream of air carries ether vapour out of the beaker. For safety the experiment is done in a fume cupboard as ether is very flammable

3 As ether evaporates, it takes energy from its surroundings. The water between the beaker and the wood freezes

Figure 4 A demonstration of cooling by evaporation

Figure 3 explains why evaporation causes this cooling effect.

- Weak attractive forces exist between the molecules in the liquid.
- The faster molecules, which have more kinetic energy, break away from the attraction of the other molecules and escape from the liquid.
- After they leave, the liquid is cooler because the average kinetic energy of the remaining molecules in the liquid has decreased.

Factors affecting the rate of evaporation

Clothes dry faster on a washing line:

- if each item of wet clothing is spread out when it is hung on the line. This increases the area of the wet clothing that is in contact with dry air.
- if the washing line is in sunlight. Wet clothes dry faster the warmer they are.
- if there is a breeze to take away the molecules that escape from the water in the wet clothes.

The example above shows that the rate of evaporation from a liquid is increased by:

- increasing the surface area of the liquid
- increasing the temperature of the liquid
- creating a draught of air across the liquid's surface.

Factors affecting the rate of condensation

In a steamy kitchen, water can often be seen trickling down a window pane. The glass pane is a cold surface so water vapour condenses on it. The air in the room is moist or 'humid'. The bigger the area of the window pane, or the colder it is, the greater the rate of condensation. This example shows that the rate of condensation of a vapour on a surface is increased by:

- increasing the surface area
- reducing the surface temperature.

b Why does washing on a line take longer to dry on a damp day?

Summary questions

1 Copy and complete **a** to **c** using the words below. Each word can be used more than once.

condenses   cools   evaporates

a A liquid .... when its molecules escape into the surrounding air.
b When water .... on glass, water molecules in the air form a liquid on the glass.
c When a liquid ...., it loses its faster-moving molecules and it ....

2 Why do the windows on a bus become misty when there are lots of people on the bus?

3 Explain the following statements.
a Wet clothes on a washing line dry faster on a hot day than on a cold day.
b A person wearing wet clothes on a cold windy day is likely to feel much colder than someone wearing dry clothes.

Did you know ...?

Air conditioning

An **air conditioning unit** in a room transfers energy from inside the room to the outside. The unit contains a 'coolant' liquid that easily evaporates. The coolant is pumped round a sealed circuit of pipes that go through the unit and the outside.

- The liquid coolant evaporates in the pipes in the room and cools the room.
- The evaporated coolant condenses in the pipes outside and transfers energy to the surroundings.

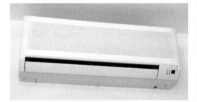

Figure 5 An air-conditioning unit

Key points

- Evaporation is when a liquid turns into a gas.
- Condensation is when a gas turns into a liquid.
- Cooling by evaporation of a liquid is due to the faster-moving molecules escaping from the liquid.
- Evaporation can be increased by increasing the surface area of the liquid, by increasing the liquid's temperature, or by creating a draught of air across the liquid's surface.
- Condensation on a surface can be increased by increasing the area of the surface or reducing the temperature of the surface.

P1 1.7 — Energy transfer by design

Learning objectives

- What design factors affect the rate at which a hot object transfers energy?
- What can we do to control the rate of energy transfer to or from an object?

Figure 1 A car radiator

Figure 2 Motorcycle engine fins

??? Did you know ...?

Some electronic components get warm when they are working, but if they become too hot they stop working. Such components are often fixed to a metal plate to keep them cool. The metal plate increases the effective surface area of the component. We call the metal plate a **heat sink**.

Figure 3 A heat sink in a computer

How Science Works

Cooling by design

Lots of things can go wrong if we don't control energy transfer. For example, a car engine that overheats can go up in flames.

- The cooling system of a car engine transfers energy from the engine to a radiator. The radiator is shaped so it has a large surface area. This increases the rate of energy transfer through convection in the air and through radiation.
- A motorcycle engine is shaped with **fins** on its outside surface. The fins increase the surface area of the engine in contact with air so the engine transfers energy to its surroundings faster than if it had no fins.
- Most cars also have a cooling fan that switches on when the engine is too hot. This increases the flow of air over the surface of the radiator.

a Why do car radiators have a large surface area?
b What happens to the rate of energy transfer when the cooling fan switches on?

The vacuum flask

If you are outdoors in cold weather, a hot drink from a vacuum flask keeps you warm. In the summer the same vacuum flask keeps your drinks cold.

In Figure 4, the liquid you drink is in the double-walled glass container.

- The vacuum between the two walls of the container cuts out energy transfer by conduction and convection between the walls.
- Glass is a poor conductor so there is little energy transfer by conduction through the glass.
- The glass surfaces are silvery to reduce radiation from the outer wall.
- The spring supporting the double-walled container is made of plastic which is a good insulator.
- The plastic cap stops cooling by evaporation as it stops vapour loss from the flask. In addition, energy transfer by conduction is cut down because the cap is made from plastic.

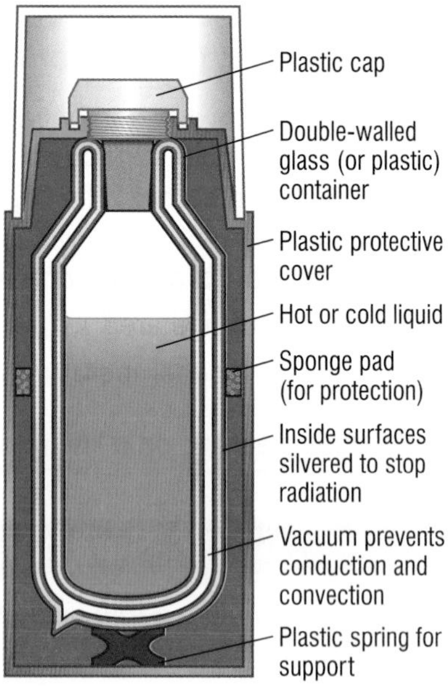

Labels on Figure 4:
- Plastic cap
- Double-walled glass (or plastic) container
- Plastic protective cover
- Hot or cold liquid
- Sponge pad (for protection)
- Inside surfaces silvered to stop radiation
- Vacuum prevents conduction and convection
- Plastic spring for support

Figure 4 A vacuum flask

So why does the liquid in the flask eventually cool down?

The above features cut down but do not totally stop the transfer of energy from the liquid. Energy transfer occurs at a very low rate due to radiation from the silvery glass surface and conduction through the cap, spring and glass walls. The liquid transfers energy slowly to its surroundings so it eventually cools.

c List the other parts of the flask that are good insulators. What would happen if they weren't good insulators?

Factors affecting the rate of energy transfer

The bigger the **temperature difference** between an object and its surroundings, the faster the rate at which energy is transferred. In addition, the above examples show that the rate at which an object transfers energy depends on its design. The design factors that matter are:

- the materials the object is in contact with
- the object's shape
- the object's surface area.

In addition, the object's mass and the material it is made from are important. That is because they affect how quickly its temperature changes (and therefore the rate of transfer of energy to or from it) when it loses or gains energy.

How Science Works

Foxy survivors

A desert fox has much larger ears than an arctic fox. Blood flowing through the ears transfers energy from inside the body to the surface of the ears. Big ears have a much larger surface area than little ears so they transfer energy to the surroundings more quickly than little ears.

- A desert fox has big ears so it keeps cool by transferring energy quickly to its surroundings.
- An arctic fox has little ears so it transfers energy more slowly to its surroundings. This helps keep it warm.

links

For more information on factors affecting energy transfer, see P1 1.8 Specific heat capacity.

Practical

Investigating the rate of energy transfer

You can plan an investigation using different beakers and hot water to find out what affects the rate of cooling.

- Write a question that you could investigate.
- Identify the independent, dependent and control variables in your investigation.

Figure 5 Fox ears **a** A desert fox **b** An arctic fox

Key points

- The rate of energy transferred to or from an object depends on:
  - the shape, size and type of material of the object
  - the materials the object is in contact with
  - the temperature difference between the object and its surroundings.

Summary questions

1 Hot water is pumped through a radiator like the one in Figure 6.
Copy and complete **a** to **c** using the words below:

conduction   radiation   convection

Figure 6 A central heating radiator

  a Energy transfer through the walls of the radiator is due to ....
  b Hot air in contact with the radiator causes energy transfer to the room by ....
  c Energy transfer to the room takes place directly due to ....

2 An electronic component in a computer is attached to a heat sink.
  a i Explain why the heat sink is necessary.
    ii Why is a metal plate used as the heat sink?
  b Plan a test to show that double glazing is more effective at preventing energy transfer than single glazing.

3 Describe, in detail, how the design of a vacuum flask reduces the rate of energy transfer.

P1 1.8

Specific heat capacity (k)

Learning objectives

- How does the mass of a substance affect how quickly its temperature changes when it is heated?

- What else affects how quickly the temperature of a substance changes when it is heated?

- How do storage heaters work?

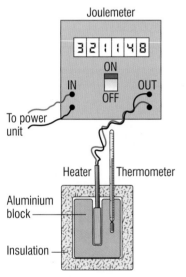

Joulemeter

3 2 1 1 4 8

ON
OFF

IN   OUT

To power unit

Heater   Thermometer

Aluminium block

Insulation

Figure 1 Heating an aluminium block

A car in strong sunlight can become very hot. A concrete block of equal **mass** would not become as hot. Metal heats up more easily than concrete. Investigations show that when a substance is heated, its temperature rise depends on:

- the amount of energy supplied to it
- the mass of the substance
- what the substance is.

Practical

Investigating heating

Figure 1 shows how we can use a low voltage electric heater to heat an aluminium block.

Energy is measured in units called joules (J).

Use the energy meter (or joulemeter) to measure the energy supplied to the block. Use the thermometer to measure its temperature rise.

Replace the block with an equal mass of water in a suitable container. Measure the temperature rise of the water when the same amount of energy is supplied to it by the heater.

Your results should show that aluminium heats up more than water.

The following results were obtained using two different amounts of water. They show that:

- 1600 J was used to heat 0.1 kg of water by 4 °C
- 3200 J was used to heat 0.2 kg of water by 4 °C.

Using these results we can say that:

- 16 000 J of energy would have been needed to heat 1.0 kg of water by 4 °C
- 4000 J of energy is needed to heat 1.0 kg of water by 1 °C.

More accurate measurements would give 4200 J per kg per °C for water. This is its **specific heat capacity**.

The specific heat capacity of a substance is the energy needed or energy transferred to 1 kg of the substance to raise its temperature by 1 °C.

The unit of specific heat capacity is the joule per kilogram per °C.

For a known change of temperature of a known mass of a substance:

$$E = m \times c \times \theta$$

Where:

E is the energy transferred in joules, J; m is the mass in kilograms, kg; c is the specific heat capacity, J/kg°C; θ is the temperature change in degrees Celsius, °C

To find the specific heat capacity you need to rearrange the above equation:

$$c = \frac{E}{m \times \theta}$$

a How much energy is needed to heat 5.0 kg of water from 20 °C to 60 °C?

??? Did you know ...?

Coastal towns are usually cooler in summer and warmer in winter than towns far inland. This is because water has a very high specific heat capacity. Energy from the Sun (or lack of energy) affects the temperature of the sea much less than the land.

Practical

Measuring the specific heat capacity of a metal

Use the arrangement shown in Figure 1 to heat a metal block of known mass. Here are some measurements using an aluminium block of mass 1.0 kg.

Starting temperature = 14 °C
Final temperature = 22 °C
Energy supplied = 7200 J

To find the specific heat capacity of aluminium, the measurements above give:
E = energy transferred = energy supplied = 7200 J
θ = temperature change = 22 °C − 14 °C = 8 °C

Inserting these values into the rearranged equation gives:

$$c = \frac{E}{m \times \theta} = \frac{7200\,J}{1.0\,kg \times 8\,°C} = 900\,J/kg\,°C$$

The table below shows the values for some other substances.

Substance	water	oil	aluminium	iron	copper	lead	concrete
Specific heat capacity (joules per kg per °C)	4200	2100	900	390	490	130	850

Storage heaters

A storage heater uses electricity at night (off-peak) to heat special bricks or concrete blocks in the heater. Energy transfer from the bricks keeps the room warm. The bricks have a high specific heat capacity so they store lots of energy. They warm up slowly when the heater element is on and cool down slowly when it is off.

Electricity consumed at off-peak times is sometimes charged for at a cheaper rate, so storage heaters are designed to be cost effective.

Figure 2 A storage heater

b How would the temperature of the room change if the bricks cooled quickly?

Key points

- The greater the mass of an object, the more slowly its temperature increases when it is heated.

- The rate of temperature change of a substance when it is heated depends on:
  – the energy supplied to it
  – its mass
  – its specific heat capacity.

- Storage heaters use off-peak electricity to store energy in special bricks.

Summary questions

1 A small bucket of water and a large bucket of water are left in strong sunlight. Which one warms up faster? Give a reason for your answer.

2 Use the information in the table above to answer this question.
 a Explain why a mass of lead heats up more quickly than an equal mass of aluminium.
 b Calculate the energy needed
  i to raise the temperature of 0.20 kg of aluminium from 15 °C to 40 °C
  ii to raise the temperature of 0.40 kg of water from 15 °C to 40 °C.

3 State two ways in which a storage heater differs from a radiant heater.

P1 1.9 Heating and insulating buildings

Learning objectives

- How can we reduce the rate of energy transfer from our homes?
- What are U-values?
- Is solar heating free?

??? Did you know …?

A duvet is a bed cover filled with 'down' or soft feathers or some other suitable thermal insulator such as wool. Because the filling material traps air, a duvet on a bed cuts down the transfer of energy from the sleeper. The 'tog' rating of a duvet tells us how effective it is as an insulator. The higher its tog rating is, the more effective it is as an insulator.

How Science Works

Reducing the rate of energy transfers at home 🅚

Home heating bills can be expensive. Figure 1 shows how we can reduce the rate of energy transfer at home and reduce our home heating bills.

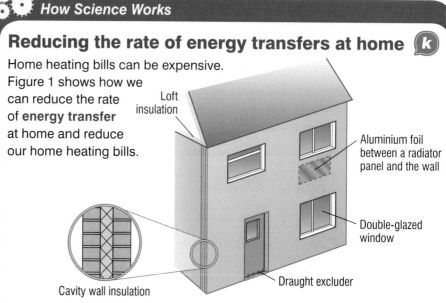

Figure 1 Saving money

- **Loft insulation** such as fibreglass reduces the rate of energy transfer through the roof. Fibreglass is a good insulator. The air between the fibres also helps to reduce the rate of energy transfer by conduction.
- **Cavity wall insulation** reduces energy loss through the outer walls of the house. The 'cavity' of an outer wall is the space between the two layers of brick that make up the wall. The insulation is pumped into the cavity. It is a better insulator than the air it replaces. It traps the air in small pockets, reducing convection currents.
- **Aluminium foil** between a radiator panel and the wall reflects radiation away from the wall.
- **Double-glazed windows** have two glass panes with dry air or a vacuum between the panes. Dry air is a good insulator so it reduces the rate of energy transfer by conduction. A vacuum cuts out energy transfer by convection as well.

a Why is cavity wall insulation better than air in the cavity between the walls of a house?

U-values

We can compare different insulating materials if we know their U-values. This is the energy per second that passes through one square metre of material when the temperature difference across it is 1 °C.

The lower the U-value, the more effective the material is as an insulator.

For example, replacing a single-glazed window with a double-glazed window that has a U-value four times smaller would make the energy loss through the window four times smaller.

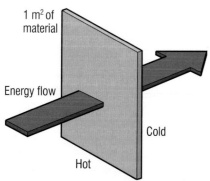

U-value of the material = energy/s passing per m² for 1°C temperature difference

Figure 2 U-values

b The U-value of 'MoneySaver' loft insulation is twice that of 'Staywarm'. Which type is more effective as an insulator?

Solar heating panels

Heating water at home using electricity or gas can be expensive. A **solar heating panel** uses solar energy to heat water. The panel is usually fitted on a roof that faces south, making the most of the Sun's energy. Figure 3 shows the design of one type of solar heating panel.

The panel is a flat box containing liquid-filled copper pipes on a matt black metal plate. The pipes are connected to a heat exchanger in a water storage tank in the house.

A transparent cover on the top of the panel allows solar radiation through to heat the metal plate. Insulating material under the plate stops energy being transferred through the back of the panel.

On a sunny day, the metal plate and the copper pipes in the box become hot. Liquid pumped through the pipes is heated when it passes through the panel. The liquid may be water or a solution containing antifreeze. The hot liquid passes through the heat exchanger and transfers energy to the water in the storage tank.

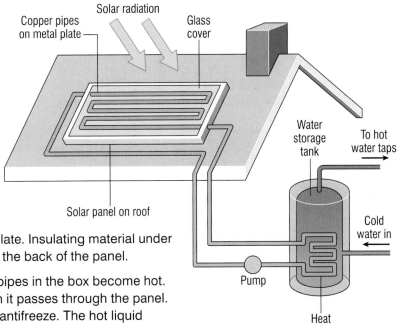

Figure 3 A solar heating panel

How Science Works

Payback time

Solar heating panels save money because no fuel is needed to heat the water. But they are expensive to buy and to install.

Suppose you pay £2000 to buy and install a solar panel and you save £100 each year on your fuel bills. After 20 years you would have saved £2000. In other words, the **payback time** for the solar panel is 20 years. This is the time taken to recover the up-front costs from the savings on fuel bills.

links

For more information on payback times, see P1 3.4 Cost effectiveness matters.

Summary questions

1 Copy and complete **a** to **c** using the words below. Each word can be used more than once.

   conduction   convection   radiation

   a Cavity wall insulation reduces the rate of energy transfer due to ....

   b Aluminium foil behind a radiator reduces the rate of energy transfer due to ....

   c Closing the curtains in winter reduces the rate of energy transfer due to .... and ....

2 Some double-glazed windows have a plastic frame and a vacuum between the panes.

   a Why is a plastic frame better than a metal frame?

   b Why is a vacuum between the panes better than air?

3 A manufacturer of loft insulation claimed that each roll of loft insulation would save £10 per year on fuel bills. A householder bought 6 rolls of the loft insulation at £15 per roll and paid £90 to have the insulation fitted in her loft.

   a How much did it cost to buy and install the loft insulation?

   b What would be the saving each year on fuel bills?

   c Calculate the payback time.

Key points

- Energy transfer from our homes can be reduced by fitting:
  – loft insulation
  – cavity wall insulation
  – double glazing
  – draught proofing
  – aluminium foil behind radiators.

- U-values tell us how much energy per second passes through different materials.

- Solar heating panels do not use fuel to heat water but they are expensive to buy and install.

Summary questions

1 **a** Why does a matt surface in sunshine get hotter than a shiny surface?

b What type of surface is better for a flat roof – a matt dark surface or a smooth shiny surface? Explain your answer.

c A solar heating panel is used to heat water. Why is the top surface of the metal plate inside the panel painted matt black?

d Why is a car radiator painted matt black?

2 Copy and complete **a** and **b** using the words below:

collide   electrons   atoms   vibrate

a Energy transfer in a metal is due to particles called .... moving about freely inside the metal. They transfer energy when they .... with each other.

b Energy transfer in a non-metallic solid is due to particles called .... inside the non-metal. They transfer energy because they ....

3 A heat sink is a metal plate or clip fixed to an electronic component to stop it overheating.

Figure 1 A heat sink

a When the component becomes hot, how does energy transfer from where it is in contact with the plate to the rest of the plate?

b Why does the plate have a large surface area?

4 Copy and complete **a** to **d** using the words below. Each word can be used more than once.

conduction   convection   radiation

a .... cannot happen in a solid or through a vacuum.

b Energy transfer from the Sun is due to ....

c When a metal rod is heated at one end, energy transfer due to .... takes place in the rod.

d .... is energy transfer by electromagnetic waves.

5 **a** In winter, why do gloves keep your hands warm outdoors?

b Why do your ears get cold outdoors in winter if they are not covered?

6 Energy transfer takes place in each of the following examples. In each case, state where the energy transfer occurs and if the energy transfer is due to conduction, convection or radiation.

a The metal case of an electric motor becomes warm due to friction when the motor is in use.

b A central heating radiator warms up first at the top when hot water is pumped through it.

c A slice of bread is toasted under a red-hot electric grill.

7 A glass tube containing water with a small ice cube floating at the top was heated at its lower end. The time taken for the ice cube to melt was measured. The test was repeated with a similar ice cube weighted down at the bottom of the tube of water. The water in this tube was heated near the top of the tube. The time taken for the ice cube to melt was much longer than in the first test.

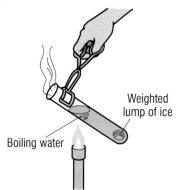

Figure 2 Energy transfer in water

a Energy transfer in the tube is due to conduction or convection or both.

  i Why was convection the main cause of energy transfer to the ice cube in the first test?

  ii Why was conduction the only cause of energy transfer in the second test?

b Which of the following conclusions about these tests is true?

  1 Energy transfer due to conduction does not take place in water.

  2 Energy transfer in water is mainly due to convection.

  3 Energy transfer in water is mainly due to conduction.

Practice questions *k*

1 Convection takes place in fluids.

Use words from the list to complete each sentence. Each word can be used once, more than once or not at all.

contracts   expands   rises   sinks   transfers

When a fluid is heated it ...., becomes less dense, and .... The warm fluid is replaced by cooler, denser, fluid. The resulting convection current .... energy throughout the fluid. (3)

2 There are three states of matter: solid, liquid and gas.

Complete each sentence.

a A solid has
a fixed shape and a fixed volume.
a fixed shape but not a fixed volume.
a fixed volume but not a fixed shape.
neither a fixed shape nor a fixed volume. (1)

b A liquid has
a fixed shape and a fixed volume.
a fixed shape but not a fixed volume.
a fixed volume but not a fixed shape.
neither a fixed shape nor a fixed volume. (1)

c A gas has
a fixed shape and a fixed volume.
a fixed shape but not a fixed volume.
a fixed volume but not a fixed shape.
neither a fixed shape nor a fixed volume. (1)

d Fluids are
solids or liquids.
solids or gases.
liquids or gases. (1)

e The particles in a solid
move about at random in contact with each other.
move about at random away from each other.
vibrate about fixed positions. (1)

3 In an experiment a block of copper is heated from 25 °C to 45 °C.

a Give the name of the process by which energy is transferred through the copper block. (1)

b The mass of the block is 1.3 kg.
Calculate the energy needed to increase the temperature of the copper from 25 °C to 45 °C.
Specific heat capacity of copper = 380 J/kg °C.
Show clearly how you work out your answer. (3)

4 The diagram shows some water being heated with a solar cooker.

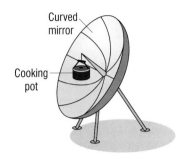

The curved mirror reflects the sunlight that falls on it. The sunlight can be focused on to the cooking pot. The energy from the sunlight is absorbed by the pot, heating up the water inside.

a Suggest **one** reason why a matt black pot has been used. (2)

b When the water has been heated, equal amounts of the water are poured into two metal pans. The pans are identical except one has a matt black surface and the other has a shiny metal surface.

Which pan will keep the water warm for the longer time? Explain your answer. (2)

5 The continuous movement of water from the oceans to the air and land and back to the oceans is called the water cycle.

a The Sun heats the surface of the oceans, which causes water to evaporate.
How does the rate of evaporation depend on
i the wind speed (1)
ii the temperature (1)
iii the humidity? (1)

b Explain how evaporation causes a cooling effect. (3)

6 Double-glazed windows are used to reduce the rate of energy transfer from buildings. The diagrams show cross-sections of single-glazed and double-glazed windows.

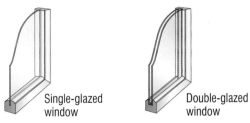

Single-glazed window   Double-glazed window

Give two reasons why a double-glazed window reduces conduction more effectively than a single-glazed window. (2)

7 *In this question you will be assessed on using good English, organising information clearly and using specialist terms where appropriate.*

Compare the similarities and differences between the process of conduction in metals and non-metals. (6)

P1 2.1 Forms of energy

Learning objectives

- What forms of energy are there?
- How can we describe energy changes?
- What energy changes take place when an object falls to the ground?

On the move

Cars, buses, planes and ships all use energy from fuels. They carry their own fuel. Electric trains use energy from fuel in power stations. Electricity transfers energy from the power station to the train.

Figure 1 The French TGV (Train à grande vitesse) electric train can reach speeds of more than 500 km/hour

We describe energy stored or transferred in different ways as **forms of energy**.

Here are some examples of forms of energy:

- **Chemical energy** is energy stored in fuel (including food). This energy is released when chemical reactions take place.
- **Kinetic energy** is the energy of a moving object.
- **Gravitational potential energy** is the energy of an object due to its position.
- **Elastic potential energy** is the energy stored in a springy object when we stretch or squash it.
- **Electrical energy** is energy transferred by an electric current.

a What form of energy is supplied to the train in Figure 1?

Energy may be transferred from one form to another.

In the torch in Figure 2, the torch's battery pushes a current through the bulb. This makes the torch bulb emit light and it also gets hot. We can show the energy transfers using a flow diagram:

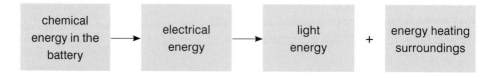

| chemical energy in the battery | → | electrical energy | → | light energy | + | energy heating surroundings |

b What happens to the energy of the torch bulb?

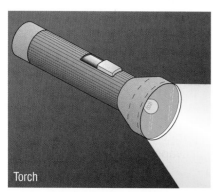
Torch

Skier

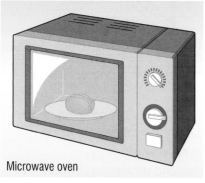
Microwave oven

Figure 2 Energy transfers

Practical

Energy transfers

When an object starts to fall freely, it gains kinetic energy because it speeds up as it falls. So its gravitational potential energy is transferred to kinetic energy as it falls.

Look at Figure 3. It shows a box that hits the floor with a thud. All of its kinetic energy is transferred by heating and to **sound** energy at the point of impact. The proportion of kinetic energy transferred to sound is much smaller than that transferred by heating.

● Draw an energy flow diagram to show the changes in Figure 3.

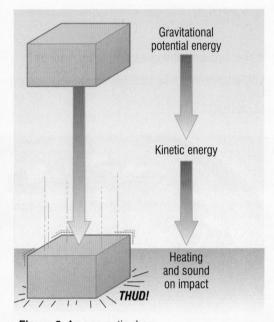

Gravitational potential energy

Kinetic energy

Heating and sound on impact

THUD!

Figure 3 An energetic drop

Tall buildings need firm foundations. Engineers make the foundations using a pile driver to hammer steel girders end-on into the ground. The pile driver lifts a heavy steel block above the top end of the girder. Then it lets the block crash down onto the girder. The engineers keep doing this until the bottom end of the girder reaches solid rock.

Figure 4 A pile driver in action

Summary questions

1 Copy and complete **a** and **b** using the words below:

electrical   kinetic   gravitational potential

   a When a ball falls in air, it loses .... energy and gains .... energy.
   b When an electric heater is switched on, it transfers .... energy by heating.

2 **a** List two different objects you could use to light a room if you have a power cut. For each object, describe the energy transfers that happen when it lights up the room.
   b Which of the two objects in **a** is:
      i easier to obtain energy from?
      ii easier to use?

3 Read the 'Did you know?' box on this page about the pile driver.
   a What form of energy does the steel block have after it has been raised?
   b Draw an energy flow diagram for the steel block from the moment it is released to when it stops moving.

Key points

● Energy exists in different forms.

● Energy can change from one form into another form.

● When an object falls and gains speed, its gravitational potential energy decreases and its kinetic energy increases.

P1 2.2 Conservation of energy

Learning objectives

- What do we mean by 'conservation of energy'?

- Why is conservation of energy a very important idea?

At the funfair

Funfairs are very exciting places because lots of energy transfers happen quickly. A roller coaster gains gravitational potential energy when it climbs. This energy is then transferred as the roller coaster races downwards.

As it descends:

its gravitational potential energy → kinetic energy + sound + energy transfer by heating due to air resistance and friction

The energy transferred by heating is 'wasted' energy, which you will learn more about in P1 2.3.

> **a** When a roller coaster gets to the bottom of a descent, what energy transfers happen if:
> **i** we apply the brakes to stop it
> **ii** it goes up and over a second 'hill'?

Study tip

Never use the term 'movement energy' in the exam; you will only gain marks for using 'kinetic energy'.

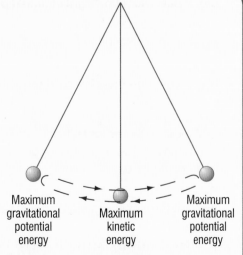

Figure 1 On a roller coaster – having fun with energy transfers!

Practical

Investigating energy changes

Pendulum swinging

When energy changes happen, does the total amount of energy stay the same? We can investigate this question with a simple pendulum.

Figure 2 shows a pendulum bob swinging from side to side.

As it moves towards the middle, its gravitational potential energy is transferred to kinetic energy.

As it moves away from the middle, its kinetic energy transfers back to gravitational potential energy. If the air resistance on the bob is very small, you should find that the bob reaches the same height on each side.

- What does this tell you about the energy of the bob when it goes from one side at maximum height to the other side at maximum height?

- Why is it difficult to mark the exact height the pendulum bob rises to? How could you make your judgement more accurate?

Maximum gravitational potential energy

Maximum kinetic energy

Maximum gravitational potential energy

Figure 2 A pendulum in motion

Conservation of energy

Scientists have done lots of tests to find out if the total energy after a transfer is the same as the energy before the transfer. All the tests so far show it is the same.

This important result is known as the **conservation of energy**.

It tells us that **energy cannot be created or destroyed**.

Bungee jumping

What energy transfers happen to a bungee jumper after jumping off the platform?

- When the rope is slack, some of the gravitational potential energy of the bungee jumper is transferred to kinetic energy as the jumper falls.
- Once the slack in the rope has been used up, the rope slows the bungee jumper's fall. Most of the gravitational potential energy and kinetic energy of the jumper is transferred into elastic strain energy.
- After reaching the bottom, the rope pulls the jumper back up. As the jumper rises, most of the elastic strain energy of the rope is transferred back to gravitational potential energy and kinetic energy of the jumper.

The bungee jumper doesn't return to the same height as at the start. This is because some of the initial gravitational potential energy has been transferred to the surroundings by heating as the rope stretched then shortened again.

Figure 3 Bungee jumping

b What happens to the gravitational potential energy lost by the bungee jumper?

c Draw a flow diagram to show the energy changes.

Practical

Bungee jumping

You can try out the ideas about bungee jumping using the experiment shown in Figure 4.

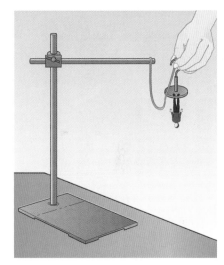

Figure 4 Testing a bungee jump

Summary questions

1 Copy and complete using the words below:

electrical   gravitational potential   kinetic

A person going up in a lift gains .... energy. The lift is driven by electric motors. Some of the .... energy supplied to the motors is wasted instead of being transferred to .... energy.

2 **a** A ball dropped onto a trampoline returns to almost the same height after it bounces. Describe the energy transfer of the ball from the point of release to the top of its bounce.

b What can you say about the energy of the ball at the point of release compared with at the top of its bounce?

c You could use the test in **a** above to see which of three trampolines was the bounciest.

 i Name the independent variable in this test.

 ii Is this variable categoric or continuous?

3 One exciting fairground ride acts like a giant catapult. The capsule, in which you are strapped, is fired high into the sky by the rubber bands of the catapult. Explain the energy transfers taking place in the ride.

∞ links

For more information on variables, look back at H2 Fundamental ideas about how science works.

Key points

- Energy cannot be created or destroyed.

- Conservation of energy applies to all energy changes.

P1 2.3

Useful energy

Learning objectives

- What is 'useful' energy?
- What do we mean by 'wasted' energy?
- What eventually happens to wasted energy?
- Does energy become less useful after we use it?

Figure 1 Using energy

??? Did you know ...?

Lots of energy is transferred in a car crash. The faster the car travels the more kinetic energy it has and the more it has to transfer before stopping. In a crash, kinetic energy is quickly transferred to elastic strain energy, distorting the car's shape, and energy is transferred by heating the metal. There is usually quite a lot of sound energy too!

Energy for a purpose

Where would we be without **machines**? We use washing machines at home. We use machines in factories to make the goods we buy. We use them in the gym to keep fit and we use them to get us from place to place.

> **a** What eventually happens to all the energy you use in a gym?

A machine transfers energy for a purpose. Friction between the moving parts of a machine causes the parts to warm up. So not all of the energy supplied to a machine is usefully transferred. Some energy is wasted.

- **Useful energy** is energy transferred to where it is wanted, in the form it is wanted.
- **Wasted energy** is energy that is not usefully transferred.

> **b** What eventually happens to the kinetic energy of a machine when it stops?

Practical

Investigating friction

Friction in machines always causes energy to be wasted. Figure 2 shows two examples of friction in action. Try one of them out.

In **a**, friction acts between the drill bit and the wood. The bit becomes hot as it bores into the wood. Some of the electrical energy supplied to the bit heats up the drill bit (and the wood).

In **b**, when the brakes are applied, friction acts between the brake blocks and the wheel. This slows the bicycle and the cyclist down. Some of the kinetic energy of the bicycle and the cyclist is transferred to energy heating the brake blocks (and the bicycle wheel).

You can practise your skills in 'How Science Works' by investigating friction on different surfaces.

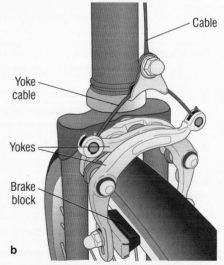

Cable

Yoke cable

Yokes

Brake block

b

a

Figure 2 Friction in action
a Using a drill **b** Braking on a bicycle

Disc brakes at work

The next time you are in a car slowing down at traffic lights, think about what is making the car stop. Figure 3 shows how the disc brakes of a car work. When the brakes are applied, the pads are pushed on to the disc in each wheel. Friction between the pads and each disc slows the wheel down. Some of the kinetic energy of the car is transferred to energy heating the disc pads and the discs. In Formula One racing cars you can sometimes see the discs glow red hot.

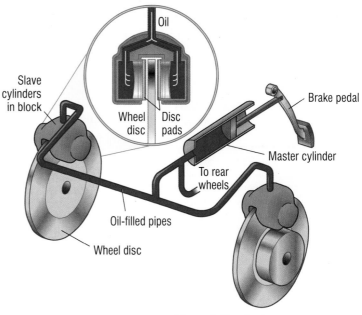

Figure 3 Disc brakes

Spreading out

- **Wasted energy is dissipated (spreads out) to the surroundings.**
  For example, the gears of a car get hot due to friction when the car is running. So energy transfers from the gear box to the surrounding air.

- **Useful energy eventually transfers to the surroundings too.**
  For example, the useful energy supplied to the wheels of a car is transferred to energy heating the tyres. This energy is then transferred to the road and the surrounding air.

- **Energy becomes less useful, the more it spreads out.**
  For example, the hot water from the cooling system of a CHP (combined heat and power) power station gets used to heat nearby buildings. The energy supplied to heat the buildings will eventually be transferred to the surroundings.

  c The hot water from many power stations flows into rivers or lakes. Why is this wasteful?

Summary questions

1 Copy and complete the table below.

Energy transfer by	Useful energy output	Wasted energy output
a An electric heater		
b A television		
c An electric kettle		
d Headphones		

2 What would happen, in terms of energy transfer, to
  a a gear box that was insulated so it could not transfer energy heating the surroundings?
  b a jogger wearing running shoes, which are well-insulated?
  c a blunt electric drill if you use it to drill into hard wood?

3 **a** Describe the energy transfers of a pendulum as it swings from one side to the middle then to the opposite side.
  b Explain why a swinging pendulum eventually stops.

Key points

- Useful energy is energy in the place we want it and in the form we need it.

- Wasted energy is energy that is not useful energy.

- Useful energy and wasted energy both end up being transferred to the surroundings, which become warmer.

- As energy spreads out, it gets more and more difficult to use for further energy transfers.

P1 2.4 Energy and efficiency

Learning objectives

- What do we mean by efficiency?

- How efficient can a machine be?

- How can we make machines more efficient?

When you lift an object, the useful energy from your muscles goes to the object as gravitational potential energy. This depends on its weight and how high it is raised.

- Weight is measured in **newtons (N)**. The weight of a 1 kilogram object on the Earth's surface is about 10 N.
- Energy is measured in **joules (J)**. The energy needed to lift a weight of 1 N by a height of 1 metre is equal to 1 joule.

Your muscles get warm when you use them so they do waste some energy.

> **a** Think about lowering a weight. What happens to its gravitational potential energy?

Sankey diagrams

Figure 1 represents the energy transfer through a device. It shows how we can represent any energy transfer where energy is wasted. This type of diagram is called a **Sankey diagram**.

Because energy cannot be created or destroyed:

Input energy (energy supplied) = useful energy delivered + energy wasted

For any device that transfers energy:

$$\text{Efficiency} = \frac{\text{useful energy transferred by the device}}{\text{total energy supplied to the device}} \text{ (×100\%)}$$

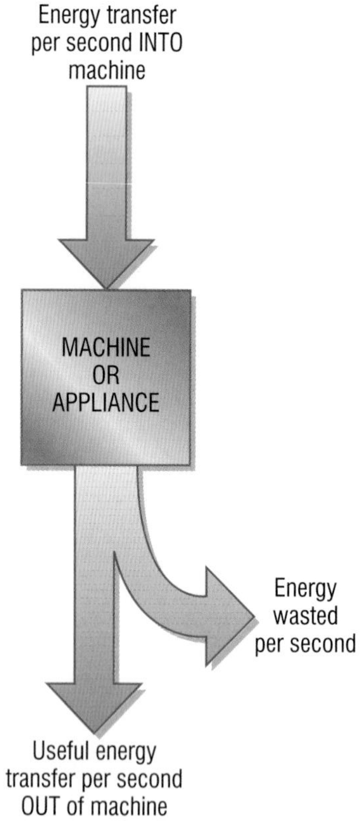

Maths skills

Efficiency can be written as a number (which is never more than 1) or as a percentage.

For example, a light bulb with an efficiency of 0.15 would radiate 15 J of energy as light for every 100 J of electrical energy we supply to it.

- Its efficiency (as a number) $= \dfrac{15}{100} = 0.15$
- Its percentage efficiency $= 0.15 \times 100\% = 15\%$

> **b** In the example above, how much energy is wasted for every 100 J of electrical energy supplied?
> **c** What happens to the wasted energy?

Energy transfer per second INTO machine

MACHINE OR APPLIANCE

Energy wasted per second

Useful energy transfer per second OUT of machine

Figure 1 A Sankey diagram

Maths skills

Worked example

An electric motor is used to raise an object. The object gains 60 J of gravitational potential energy when the motor is supplied with 200 J of electrical energy. Calculate the percentage efficiency of the motor.

Solution

Total energy supplied to the device = 200 J

Useful energy transferred by the device = 60 J

Percentage efficiency of the motor

$= \dfrac{\text{useful energy transferred by the motor}}{\text{total energy supplied to the motor}} \times 100\%$

$= \dfrac{60 \text{ J}}{200 \text{ J}} \times 100\% = 0.30 \times 100\% = 30\%$

Efficiency limits

No machine can be more than 100% efficient because we can never get more energy from a machine than we put into it.

Practical

Investigating efficiency

Figure 2 shows how you can use an electric winch to raise a weight. You can use the joulemeter to measure the electrical energy supplied.

- If you double the weight for the same increase in height, do you need to supply twice as much electrical energy to do this task?

The gravitational potential energy gained by the weight = weight in newtons × height increase in metres.

- Use this equation and the joulemeter measurements to work out the percentage efficiency of the winch.

Safety: Protect the floor and your feet. Stop the winch before the masses wrap round the pulley.

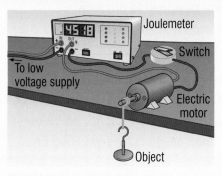

Figure 2 An electric winch

Improving efficiency

	Why machines waste energy	How to reduce the problem
1	Friction between the moving parts causes heating.	Lubricate the moving parts to reduce friction.
2	The resistance of a wire causes the wire to get hot when a current passes through it.	In circuits, use wires with as little electrical resistance as possible.
3	Air resistance causes energy transfer to the surroundings.	Streamline the shapes of moving objects to reduce air resistance.
4	Sound created by machinery causes energy transfer to the surroundings.	Cut out noise (e.g. tighten loose parts to reduce vibration).

d Which of the above solutions would hardly reduce the energy supplied?

Study tip

- The greater the percentage of the energy that is usefully transferred in a device, the more efficient the device is.
- Efficiency and percentage efficiency are numbers without units. The maximum efficiency is 1 or 100%, so if a calculation produces a number greater than this it must be wrong.

Summary questions

1 Copy and complete **a** to **c** using the words below. Each term can be used more than once.

supplied to   wasted by

   a The useful energy from a machine is always less than the total energy .... it.

   b Friction between the moving parts of a machine causes energy to be .... the machine.

   c Because energy is conserved, the energy .... a machine is the sum of the useful energy from the machine and the energy .... the machine.

2 An electric motor is used to raise a weight. When you supply 60 J of electrical energy to the motor, the weight gains 24 J of gravitational potential energy. Work out:

   a the energy wasted by the motor

   b the efficiency of the motor.

3 A machine is 25% efficient. If the total energy supplied to the machine is 3200 J, how much useful energy can be transferred?

Key points

- The efficiency of a device = useful energy transferred by the device ÷ total energy supplied to the device (× 100%).
- No machine can be more than 100% efficient.
- Measures to make machines more efficient include reducing friction, air resistance, electrical resistance and noise due to vibrations.

Summary questions (k)

1 The devices listed below transfer energy in different ways.

   1 Car engine   2 Electric bell   3 Electric light bulb

The next list gives the useful form of energy the devices are designed to produce.

Match words A, B and C with the devices numbered 1 to 3.

   A Light   B Kinetic energy   C Sound

2 Copy and complete using the words below:

useful wasted light electrical

When a light bulb is switched on, ... energy is transferred into .... energy and energy that heats the surroundings. The energy that radiates from the light bulb is .... energy. The rest of the energy supplied to the light bulb is .... energy.

3 You can use an electric motor to raise a load. In a test, you supply the motor with 10 000 J of electrical energy and the load gains 1500 J of gravitational potential energy.

  a Calculate its efficiency.

  b How much energy is wasted?

  c Copy and complete the Sankey diagram below for the motor.

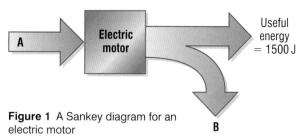

Figure 1 A Sankey diagram for an electric motor

4 A ball gains 4.0 J of gravitational potential energy when it is raised to a height of 2.0 m above the ground. When it is released, it falls to the ground and rebounds to a height of 1.5 m.

  a How much kinetic energy did it have just before it hit the ground? Assume air resistance is negligible.

  b How much gravitational potential energy did it transfer when it fell to the ground?

  c The ball gained 3.0 J of gravitational potential energy when it moved from the ground to the top of the rebound. How much energy did it transfer in the impact at the ground?

  d What happened to the energy it transferred on impact?

5 A low energy light bulb has an efficiency of 80%. Using an energy meter, a student found the light bulb used 1500 J of electrical energy in 100 seconds.

  a How much useful energy did the light bulb transfer in this time?

  b How much energy was wasted by the light bulb?

  c Draw a Sankey diagram for the light bulb.

6 A bungee jumper jumps from a platform and transfers 12 000 J of gravitational potential energy before the rope attached to her becomes taut and starts to stretch. She then transfers a further 24 000 J of gravitational potential energy before she stops falling and begins to rise.

  a Describe the energy changes:
    i after she jumps before the rope starts to stretch
    ii after the rope starts to stretch until she stops falling.

  b What is the maximum kinetic energy she has during her descent?

7 On a building site, an electric winch and a pulley were used to lift bricks from the ground.

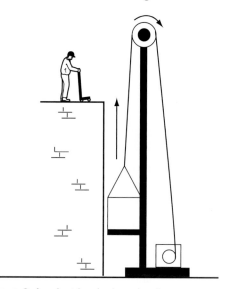

Figure 2 An electric winch and pulley

The winch transferred 12 000 J of electrical energy to raise a load through a height of 3.0 m. The load gained 1500 J of gravitational potential energy when it was raised.

  a **i** How much useful energy was transferred by the motor ?
    ii Calculate the energy wasted.
    iii Calculate the percentage efficiency of the system.

  b How could the efficiency of the winch be improved?

Practice questions Ⓚ

1 A television transfers electrical energy.

Use words from the list to complete each sentence. Each word can be used once, more than once or not at all.

electrical   light   sound   warmer

A television is designed to transfer .... energy into light and .... energy. Some energy is transferred to the surroundings, which become .... (3)

2 A hairdryer contains an electrical heater and a fan driven by an electric motor. The hairdryer transfers electrical energy into other forms.

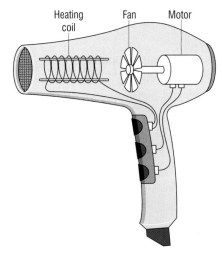

Heating coil   Fan   Motor

a Apart from energy by heating, name **two** of the other forms of energy. (2)

b Not all of the energy supplied to the fan is usefully transferred. Name **one** form of energy that is wasted by the fan. (1)

c Which of the following statements about the energy wasted by the fan is true?
   A It eventually becomes very concentrated.
   B It eventually makes the surroundings warmer.
   C It is eventually completely destroyed.
   D It is eventually transferred into electrical energy. (1)

d The fan in another hairdryer transfers useful energy at the same rate but wastes more of the energy supplied to it. What does this tell you about the efficiency of this hairdryer? (1)

3 In a hot water system water is heated by burning gas in a boiler. The hot water is then stored in a tank. For every 111 J of energy released from the gas, 100 J of energy is absorbed by the water in the boiler.

a Calculate the percentage efficiency of the boiler.

Write down the equation you use. Show clearly how you work out your answer. (4)

b The energy released from the gas but **not** absorbed by the boiler is 'wasted'. Explain why this energy is of little use for further energy transfers. (1)

c The tank in the hot water system is surrounded by a layer of insulation. Explain the effect of the insulation on the efficiency of the hot water system. (3)

4 A chairlift carries skiers to the top of a mountain. The chairlift is powered by an electric motor.

a What type of energy have the skiers gained when they reach the top of the mountain? (1)

b The energy required to lift two skiers to the top of the mountain is 240 000 J.
   The electric motor has an efficiency of 40%.

   Calculate the energy wasted in the motor.

   Write down the equation you use. Show clearly how you work out your answer and give the unit.   [H] (4)

c Explain why some energy is wasted in the motor. (2)

5 A light bulb transfers electrical energy into useful light energy and wasted energy to the surroundings. For every 100 J of energy supplied to the bulb, 5 J of energy is transferred into light.

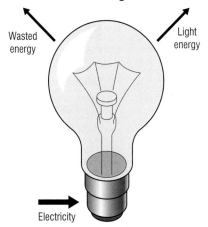

Wasted energy   Light energy

Electricity

Draw and label a Sankey diagram for the light bulb. (3)

6 *In this question you will be assessed on using good English, organising information clearly and using specialist terms where appropriate.*

Explain why an electric heater is the only appliance that can possibly be 100% efficient. (6)

P1 3.1

Electrical appliances

Learning objectives

- Why are electrical appliances so useful?

- What do we use most everyday electrical appliances for?

- How do we choose an electrical appliance for a particular job?

Everyday electrical appliances 🔊

We use **electrical appliances** every day. They transfer electrical energy into useful energy at the flick of a switch. Some of the electrical energy we supply to them is wasted.

Figure 1 Electrical appliances – how many can you see in this photo?

Practical

Energy transfers

Carry out a survey of electrical appliances you find at school or at home.

Record the useful and wasted energy transfers of each appliance.

Table 1

Appliance	Useful energy	Energy wasted
Light bulb	Light from the glowing filament.	Energy transfer from the filament heating surroundings.
Electric heater	Energy heating the surroundings.	Light from the glowing element.
Electric toaster	Energy heating bread.	Energy heating the toaster case and the air around it.
Electric kettle	Energy heating water.	Energy heating the kettle itself.
Hairdryer	Kinetic energy of the air driven by the fan. Energy heating air flowing past the heater filament.	Sound of fan motor (energy heating the motor heats the air going past it, so is not wasted). Energy heating the hairdryer itself.
Electric motor	Kinetic energy of object driven by the motor. Potential energy of objects lifted by the motor.	Energy heating the motor and sound energy of the motor.
Computer disc drive	Energy stored in magnetic dots on the disc.	Energy heating the motor that drives the disc.

??? Did you know ... ?

Unlike high voltage electrical injuries, people do not get many burns when they are struck by lightning. Damage is usually to the nervous system. The brain is frequently damaged as the skull is the most likely place to be struck. Lightning that strikes near the head can enter the body through the eyes, ears and mouth and flow internally through the body.

a What energy transfers happen in an electric toothbrush?

How Science Works

Clockwork radio

People without electricity supplies can now listen to radio programmes – thanks to the British inventor Trevor Baylis. In the early 1990s, he invented and patented the clockwork radio. When you turn a handle on the radio, you wind up a clockwork spring in the radio. When the spring unwinds, it turns a small electric generator in the radio. It doesn't need batteries or mains electricity. So people in remote areas where there is no mains electricity can listen to their radios without having to walk miles for a replacement battery. But they do have to wind up the spring every time it runs out of energy.

Figure 2 Clockwork radios are now mass-produced and sold all over the world

Choosing an electrical appliance

We use electrical appliances for many purposes. Each appliance is designed for a particular purpose and it should waste as little energy as possible. Suppose you were a rock musician at a concert. You would need appliances that transfer sound energy into electrical energy and then back into sound energy. But you wouldn't want them to produce lots of energy heating the appliance itself and its surroundings. See if you can spot some of these appliances in Figure 3.

b What electrical appliance transfers:
  i sound energy into electrical energy?
  ii electrical energy into sound energy?
c What other electrical appliance would you need at a concert?

Figure 3 On stage

Summary questions

1 Copy and complete using the words below:

electrical   light   heating

When a battery is connected to a light bulb, ... energy is transferred from the battery to the light bulb. The filament of the light bulb becomes hot and so energy transfers to its surroundings by .... and as .... energy.

2 Match each electrical appliance in the list below with the energy transfer A or B it is designed to bring about.
  1 Electric drill
  2 Food mixer
  3 Electric bell

  Energy transfer A Electrical energy → sound energy
                     B Electrical energy → kinetic energy

3 **a** Why does a clockwork radio need to be wound up before it can be used?
  b What energy transfers take place in a clockwork radio when it is wound up then switched on?
  c Give an advantage and a disadvantage of a clockwork radio compared with a battery-operated radio.

Key points

- Electrical appliances can transfer electrical energy into useful energy at the flick of a switch.

- Uses of everyday electrical appliances include heating, lighting, making objects move (using an electric motor) and creating sound and visual images.

- An electrical appliance is designed for a particular purpose and should waste as little energy as possible.

P1 3.2 Electrical power

Learning objectives

- What do we mean by power?
- How can we calculate the power of an appliance?
- How can we calculate the efficiency of an appliance in terms of power?

Figure 1 A lift motor

When you use a lift to go up, a powerful electric motor pulls you and the lift upwards. The lift motor transfers energy from electrical energy to gravitational potential energy when the lift goes up at a steady speed. We also get electrical energy transferred to wasted energy heating the motor and the surroundings, and sound energy.

- The energy we supply per second to the motor is the **power** supplied to it.
- The more powerful the lift motor is, the faster it moves a particular load.

In general, we can say that:

the more powerful an appliance, the faster the rate at which it transfers energy.

We measure the power of an appliance in watts (W) or kilowatts (kW).

1 **watt** is a rate of transfer of energy of 1 joule per second (J/s).

1 **kilowatt** is equal to 1000 watts (i.e. 1000 joules per second or 1 kJ/s).

You can calculate power using:

$$P = \frac{E}{t}$$

Where:

P is the power in watts, W

E is the energy transferred to the appliance in joules, J

t is the time taken for the energy to be transferred in seconds, s.

 Maths skills

Worked example

A motor transfers 10 000 J of energy in 25 s. What is its power?

Solution

$$P = \frac{E}{t}$$

$$P = \frac{10\ 000\ J}{25\ s} = 400\,W$$

a What is the power of a lift motor that transfers 50 000 J of energy from the electricity supply in 10 s?

Power ratings

Here are some typical values of power ratings for different energy transfers:

Appliance	Power rating
A torch	1 W
An electric light bulb	100 W
An electric cooker	10 000 W = 10 kW (where 1 kW = 1000 watts)
A railway engine	1 000 000 W = 1 megawatt (MW) = 1 million watts
A Saturn V rocket	100 MW
A very large power station	10 000 MW
World demand for power	10 000 000 MW
A star like the Sun	100 000 000 000 000 000 000 000 MW

Figure 2 Rocket power

b How many 100 W electric light bulbs would use the same amount of power as a 10 kW electric cooker?

Muscle power

How powerful is a weightlifter?

A 30 kg dumbbell has a weight of 300 N. Raising it by 1 m would give it 300 J of gravitational potential energy. A weightlifter could lift it in about 0.5 seconds. The rate of energy transfer would be 600 J/s (= 300 J ÷ 0.5 s). So the weightlifter's power output would be about 600 W in total!

c An inventor has designed an exercise machine that can also generate 100 W of electrical power. Do you think people would buy this machine in case of a power cut?

Efficiency and power

For any appliance

- its useful power out (or output power) is the useful energy **per second** transferred by it.
- its total power in (or input power) is the energy **per second** supplied to it.

In P1 2.4 Energy and efficiency, we saw that the efficiency of an appliance

$$= \frac{\text{useful energy transferred by the device}}{\text{total energy supplied to it}} (\times\ 100\%)$$

Because power = energy **per second** transferred or supplied, we can write the efficiency equation as:

$$\text{Efficiency} = \frac{\text{useful power out}}{\text{total power in}} (\times\ 100\%)$$

For example, suppose the useful power out of an electric motor is 20 W and the total power in is 80 W, the percentage efficiency of the motor is:

$$\frac{\text{useful power out}}{\text{total power in}} \times 100\% = \frac{20\ W}{80\ W} \times 100\% = 25\%$$

Figure 3 Muscle power

Summary questions

1 **a** Which is more powerful?
   i A torch bulb or a mains filament bulb.
   ii A 3 kW electric kettle or a 10 000 W electric cooker.
   b There are about 20 million occupied homes in England. If a 3 kW electric kettle was switched on in 1 in 10 homes at the same time, how much power would need to be supplied?

2 The total power supplied to a lift motor is 5000 W. In a test, it transfers 12 000 J of electrical energy to gravitational potential energy in 20 seconds.
   a How much electrical energy is supplied to the motor in 20 s?
   b What is its efficiency in the test?

3 A machine has an input power rating of 100 kW. If the useful energy transferred by the machine in 50 seconds is 1500 kJ, calculate
   a its output power in kilowatts
   b its percentage efficiency.

Key points

- Power is rate of transfer of energy.

- $P = \frac{E}{t}$

- Efficiency = $\frac{\text{useful power out}}{\text{total power in}}$ ($\times$ 100%)

P1 3.3

Using electrical energy

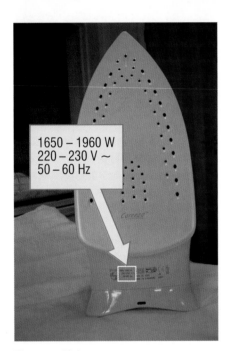

1650 – 1960 W
220 – 230 V ~
50 – 60 Hz

Figure 1 Mains power

When you use an electric heater, how much electrical energy is transferred from the mains? You can work this out if you know its power and how long you use it for.

For any appliance, the energy supplied to it depends on:

- how long it is used for

- the power supplied to it.

A 1 kilowatt heater uses the same amount of electrical energy in 1 hour as a 2 kilowatt heater would use in half an hour. For ease, we say that:

the energy supplied to a 1 kW appliance in 1 hour is 1 **kilowatt-hour (kWh)**.

We use the kilowatt-hour as the unit of energy supplied by mains electricity. You can use this equation to work out the energy, in kilowatt-hours, transferred by a mains appliance in a certain time:

$$E = P \times t$$

Where:

E is the energy transferred in kilowatt hours, kWh

P is the power in kilowatts, kW

t is the time taken for the energy to be transferred in hours, h.

Maths skills

Worked example

You have used this equation before in P1 3.2 to calculate the power of an appliance. It is the same equation, just rearranged and with different units.

$$E = P \times t$$

Divide both sides by t $\dfrac{E}{t} = P$

This is the same as $P = \dfrac{E}{t}$

For example:

- a 1 kW heater switched on for 1 hour uses 1 kWh of electrical energy (= 1 kW × 1 hour)

- a 1 kW heater switched on for 10 hours uses 10 kWh of electrical energy (= 1 kW × 10 hours)

- a 0.5 kW or 500 W heater switched on for 6 hours uses 3 kWh of electrical energy (= 0.5 kW × 6 hours).

If we want to calculate the energy transferred in joules, we can use the equation:

$$E = P \times t$$

Where:

E is the energy transferred in joules, J

P is the power in watts, W

t is the time taken for the energy to be transferred in seconds, s.

Did you know ...?

One kilowatt-hour is the amount of electrical energy supplied to a 1 kilowatt appliance in 1 hour.

So **1 kilowatt-hour**

= 1000 joules per second × 60 × 60 seconds

= 3 600 000 J

= **3.6 million joules**.

a   How many kWh of energy are used by a 100 W lamp in 24 hours?

b   How many joules of energy are used by a 5 W torch lamp in 3000 seconds (= 50 minutes)?

Paying for electrical energy

The **electricity meter** in your home measures how much electrical energy your family uses. It records the total energy supplied, no matter how many appliances you all use. It gives us a reading of the number of kilowatt-hours (kWh) of energy supplied by the mains.

In most houses, somebody reads the meter every three months. Look at the electricity bill in Figure 2.

NELEB

L. Jones
26 Homewood Road
Otwood M51 9YZ

Meter readings		units	pence per unit	amount	VAT %
present	previous				
31534	30092	1442	10.89	157.03	Zero
				17.30	
Standing charge					
TOTAL NOW DUE				174.33	
PERIOD ENDED				31.03.10	

Figure 2 Checking your bill

Figure 3 An electricity meter

The difference between the two readings is the number of kilowatt-hours supplied since the last bill.

> **c** Check for yourself that 1442 kWh of electrical energy is supplied in the bill shown.

We use the kilowatt-hour to work out the cost of electricity. For example, a cost of 12p per kWh means that each kilowatt-hour of electrical energy costs 12p. Therefore:

total cost = number of kWh used × cost per kWh

> **d** Work out the cost of 1442 kWh at 12p per kWh.

Study tip

Remember that a kilowatt-hour (kWh) is a unit of energy.

Summary questions

1 Copy and complete **a** to **c** using the words below. Each word can be used more than once.

hours   kilowatt   kilowatt-hours

a The .... is a unit of power.
b Electricity meters record the mains electrical energy transferred in units of ....
c Two .... is the energy transferred by a 1 .... appliance in 2 ....

2 **a** Work out the number of kWh transferred in each case below.
   i A 3 kilowatt electric kettle is used 6 times for 5 minutes each time.
   ii A 1000 watt microwave oven is used for 30 minutes.
   iii A 100 watt electric light is used for 8 hours.
   b Calculate the total cost of the electricity used in part **a** if the cost of electricity is 12p per kWh.

3 An electric heater is left on for 3 hours. During this time it uses 12 kWh of electrical energy.
   a What is the power of the heater?
   b How many joules are supplied?

Key points

- The kilowatt-hour is the energy supplied to a 1 kW appliance in 1 hour.
- $E = P \times t$
- Total cost = number of kWh used × cost per kWh

P1 3.4   Cost effectiveness matters

Learning objectives

- What do we mean by cost effectiveness?

- How can we compare the cost effectiveness of different energy-saving measures?

Costs

When we compare the effectiveness of different energy-saving appliances that do the same job, we need to make sure we get value for money. In other words, we need to make sure the appliance we choose is **cost effective**.

To compare the cost effectiveness of different cost-cutting measures, we need to consider:

- the capital costs such as buying and installing equipment
- the running costs, including fuel and maintenance
- environmental costs, for example
  - removal or disposal of old equipment (e.g. refrigerators, used batteries)
  - tax charges such as carbon taxes of fossil fuels
- other costs such as interest on loans.

Payback time again!

A householder wants to cut her fuel bills by reducing energy losses from her home. This would save fuel and reduce fuel bills. She is comparing loft insulation with cavity wall insulation in terms of payback time.

- The loft insulation costs £200 (including gloves and a safety mask) and she would fit the insulation herself. This could save £100 per year on the fuel bill. So the payback time would be 2 years.

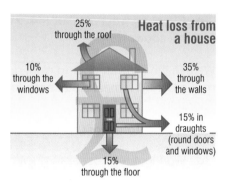

Figure 1 Heat loss from a house

- The cavity wall insulation for a house costs £500 and an additional £100 to fit the insulation. This could save £200 per year on the fuel bill. It would pay for itself after 3 years.

⚙ links

For more information on payback time, look back at P1 1.9 Heating and insulating buildings.

??? Did you know ... ?

Infrared cameras can be used to identify heat losses from a house in winter. The camera image shows hot spots as a different colour.

Figure 2 Heat losses at home

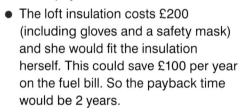

a For each type of insulation, how much would the householder have saved after 5 years?

b A double-glazed window costs £200. It saves £10 per year on the fuel bill. How long is the payback time?

Activity

Buying a heater

An artist wants to buy an electric heater to provide instant heating in his workshop when he starts work on a cold morning. He can't decide between a fan heater, a radiant heater and a tubular heater.

Table 1 shows how each type of heater works and its main drawback.

Assuming the heaters cost the same to buy, write a short report advising the artist which type of heater would be most suitable for him.

Table 1

Heater type	Input power	How the heater works	Drawbacks
Fan heater	2.0 kW	blows warm air from the hot element round the room	energy needed to run the fan motor
Radiant heater	1.0 kW	uses a reflector to direct radiation from the glowing element	the radiation only heats the air and objects in front of the heater
Tubular heater	0.5 kW	the heater element is inside a metal tube which heats the room	provides background heat gradually

Lighting costs

Low-energy bulbs use much less electrical energy than filament bulbs. This is why the UK government has banned the sale of filament bulbs. Table 2 gives some data about different types of mains bulbs.

Table 2

Type	Power in watts	Efficiency	Lifetime in hours	Cost of bulb	Typical use	Drawbacks
Filament bulb	100 W	20%	1000	50p	room lighting	inefficient, gets hot
Halogen bulb	100 W	25%	2500	£2.00	spotlight	inefficient, gets hot
Low-energy compact fluorescent bulb	25 W	80%	15000	£2.50	room lighting	takes a few minutes for full brightness, disposal must be in a sealed bag due to mercury (which is toxic) in it
Low-energy light-emitting diode	2 W	90%	30000	£7.00	spotlight	expensive to buy, brightness of one halogen bulb needs several LEDs

c Which bulb has the greatest output in terms of useful energy?
d Which type of spotlight wastes the least energy by heating?

Summary questions

1 State with a reason which type of heater from Table 1 you would choose to keep a bedroom warm at night in winter.

2 Use the information in Table 2 to answer the following questions.
  a State one advantage and one disadvantage of a CFL bulb compared with a filament bulb.
  b State one advantage and one disadvantage of an LED compared with a halogen bulb.

Key points

- Cost effectiveness means getting the best value for money.

- To compare the cost effectiveness of different appliances, we need to take account of costs to buy it, running costs and other costs such as environmental costs.

Summary questions 🄚

1 a Name an appliance that transfers electrical energy into:
    i light and sound energy
    ii kinetic energy.

b Complete the sentences below.
    i In an electric bell, electrical energy is transferred into useful energy in the form of .... energy, and .... energy.
    ii In a dentist's drill, electrical energy is transferred into useful energy in the form of .... energy and sometimes as .... energy.

2 a Which two words in the list below are units that can be used to measure energy?

    joule   kilowatt   kilowatt-hour   watt

b Rank the electrical appliances below in terms of energy used from highest to lowest.
    A a 0.5 kW heater used for 4 hours
    B a 100 W lamp left on for 24 hours
    C a 3 kW electric kettle used 6 times for 10 minutes each time
    D a 750 W microwave oven used for 10 minutes.

3 a The readings of an electricity meter at the start and the end of a month are shown below.

0	9	3	7	2

0	9	6	1	5

    i Which is the reading at the end of the month?
    ii How many kilowatt-hours of electricity were used during the month?
    iii How much would this electricity cost at 12p per kWh?

b A pay meter in a holiday home supplies electricity at a cost of 12p per kWh.
    i How many kWh would be supplied for £1.20?
    ii How long could a 2 kW heater be used for after £1.20 is put in the meter slot?   **[H]**

4 An escalator in a shopping centre is powered by a 50 kW electric motor. The escalator is in use for a total time of 10 hours every day.

a How much electrical energy in kWh is supplied to the motor each day?

b The electricity supplied to the motor costs 12p per kWh. What is the daily cost of the electricity supplied to the motor?

c How much would be saved each day if the motor was replaced by a more efficient 40 kW motor?

5 The data below show the electrical appliances used in a house in one evening.
    A a 1.0 kW heater for 4 hours
    B a 0.5 kW television for 2 hours
    C a 3 kW electric kettle three times for 10 minutes each time.

a Which appliance uses most energy?

b How many kWh of electrical energy is used by each appliance?

c Each kWh costs 12p. How much did it cost to use the three appliances?

6 The battery of a laptop computer is capable of supplying 60 watts to the computer circuits for 2 hours before it needs to be recharged.

a Calculate the electrical energy the battery can supply in two hours in:
    i kilowatt-hours
    ii joules.

b Describe the energy transfers that take place when the computer is being used.

c A mains charging unit can be connected to the computer when in use to keep its battery fully charged. Would the computer use less energy with the charging unit connected than without it connected?

7 A student has an HD television at home that uses 120 watts of electrical power when it is switched on. He monitors its usage for a week and finds it is switched on for 30 hours.

Figure 1 An HD TV in use

a How many kilowatt-hours of electrical energy are supplied to it in this time?

b Calculate the cost of this electrical energy at 12p per kilowatt-hour.

Practice questions

1 The pictures show six different household appliances.

Fan heater Vacuum cleaner Washing machine

Iron Kettle Blender

Name the **four** appliances in which electrical energy is usefully transferred into kinetic energy. (4)

2 An electric motor is used to lift a load. The useful power output of the motor is 30 W. The total input power to the motor is 75 W.

Calculate the efficiency of the motor.

Write down the equation you use. Show clearly how you work out your answer. (3)

3 Which **two** of the following units are units of energy?

a J

b J/s

c kWh

d W (1)

4 The diagram shows the readings on a household electricity meter at the beginning and end of one week.

| 5 | 2 | 3 | 4 | 0 | | 5 | 2 | 5 | 5 | 5 |

Beginning of the week **End of the week**

a How many kWh of electricity were used during the week? (1)

b On one day 35 kWh of electricity were used. The total cost of this electricity was £5.25.

Calculate how much the electricity cost per kWh.

Write down the equation you use. Show clearly how you work out your answer and give the unit. [H] (3)

c During the week a 2.4 kW kettle was used for 2 hours.

Calculate how much energy was transferred by the kettle.

Write down the equation you use. Show clearly how you work out your answer and give the unit. (3)

5 A student uses some hair straighteners.

a The hair straighteners have a power of 90 W.

What is meant by *a power of 90 W*? (2)

b Calculate how many kilowatt-hours of electricity are used when the straighteners are used for 15 minutes.

Write down the equation you use. Show clearly how you work out your answer and give the unit. (3)

c The electricity supplier is charging 14p per kWh.

Calculate how much it will cost to use the straighteners for 15 minutes a day for one year.

Write down the equation you use. Show clearly how you work out your answer and give the unit. (2)

6 Filament bulbs are being replaced by compact fluorescent bulbs.

A compact fluorescent bulb costs £12, a filament bulb costs 50p.

A 25 W compact fluorescent bulb gives out as much light as a 100 W filament bulb.

A filament bulb lasts for about 1000 hours; a compact fluorescent bulb lasts for about 8000 hours, although this time is significantly shorter if the bulb is turned on and off very frequently.

A compact fluorescent bulb contains a small amount of poisonous mercury vapour.

a Explain how a 25 W compact fluorescent bulb provides the same amount of light as a 100 W filament bulb but use less electricity. (2)

b *In this question you will be assessed on using good English, organising information clearly and using specialist terms where appropriate.*

Compare the advantages and disadvantages of buying compact fluorescent bulbs rather than filament bulbs. (6)

P1 4.1 Fuel for electricity

Learning objectives

- How is electricity generated in a power station?
- Which fossil fuels do we burn in power stations?
- How do we use nuclear fuels in power stations?
- What other fuels can be used to generate electricity?

Figure 2 Inside a gas-fired power station

Practical

Turbines

See how we can use water to drive round the blades of a turbine.

- Why is steam better than water?

Figure 3 Using biofuel to generate electricity

Inside a power station

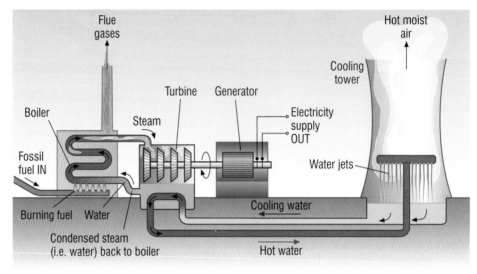

Figure 1 Inside a fossil fuel power station

Almost all the electricity you use is generated in power stations.

- In **coal-** or **oil-fired power stations**, and in most **gas-fired power stations**, the burning fuel heats water in a boiler. This produces steam. The steam drives a **turbine** that turns an electricity **generator**. Coal, oil and gas are fossil fuels, which are fuels obtained from long-dead biological material.

  a What happens to the steam after it has been used?
  b What happens to the energy of the steam after it has been used?

- In some gas-fired power stations, we burn natural gas directly in a gas turbine engine. This heats the air drawn into the engine. It produces a powerful jet of hot gases and air that drives the turbine. A gas-fired turbine can be switched on very quickly.

Biofuels

We can get methane gas from cows or animal manure and from sewage works, decaying rubbish and other sources. It can be used in small-scale gas-fired power stations. Methane is an example of a **biofuel**.

A biofuel is any fuel obtained from living or recently living organisms such as animal waste or woodchip. Other biofuels include ethanol (from fermented sugar cane), straw, nutshells and woodchip.

A biofuel is:

- **renewable** because its biological source continues to exist and never dies out as a species
- **carbon-neutral** because, in theory, the carbon it takes in from the atmosphere as carbon dioxide can 'balance' the amount released when it is burned.

Nuclear power k

Figure 4 shows you that every atom contains a positively charged nucleus surrounded by electrons. The **atomic nucleus** is composed of two types of particles: neutrons and protons. Atoms of the same element can have different numbers of neutrons in the nucleus.

How is electricity obtained from a nuclear power station?

The fuel in a nuclear power station is uranium (or plutonium). The uranium fuel is in sealed cans in the core of the reactor. The nucleus of a uranium atom is unstable and can split in two. Energy is released when this happens. We call this process **nuclear fission**. Because there are lots of uranium atoms in the core, it becomes very hot.

The energy of the core is transferred by a fluid (called the 'coolant') that is pumped through the core.

● The coolant is very hot when it leaves the core. It flows through a pipe to a 'heat exchanger', then back to the reactor core.
● The energy of the coolant is used to turn water into steam in the heat exchanger. The steam drives turbines that turn electricity generators.

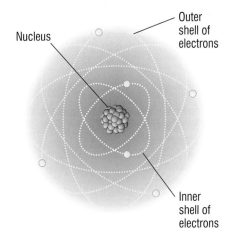

Figure 4 The structure of the atom

⚙ How Science Works

Comparing nuclear power and fossil fuel power

	Nuclear power station	Fossil fuel power station
Fuel	Uranium or plutonium	Coal, oil or gas
Energy released per kg of fuel	1 000 000 kWh (= about 10 000 × energy released per kg of fossil fuel)	100 kWh
Waste	Radioactive waste that needs to be stored for many years	Non-radioactive waste
Greenhouse gases	No – because uranium releases energy without burning	Yes – because fossil fuels produce gases such as carbon dioxide when they burn

Summary questions

1 Copy and complete **a** to **c** using the words below:

coal gas oil uranium

   a The fuel that is not a fossil fuel is ....
   b Power stations that use .... as the fuel can be switched on very quickly.
   c Greenhouse gases are produced in a power station that uses coal, gas or .... as fuel.

2 **a** State one advantage and one disadvantage of:
   i an oil-fired power station compared with a nuclear power station
   ii a gas-fired power station compared with a coal-fired power station.
   b Look at the table above.
      How many kilograms of fossil fuel would give the same amount of energy as 1 kilogram of uranium fuel?

3 **a** Explain why ethanol is described as a biofuel.
   b Ethanol is also described as carbon-neutral. What is a carbon-neutral fuel?

Key points

● Electricity generators in power stations are driven by turbines.

● Coal, oil and natural gas are burned in fossil fuel power stations.

● Uranium or plutonium are used as the fuel in a nuclear power station. Much more energy is released per kg from uranium or plutonium than from fossil fuel.

● Biofuels are renewable sources of energy. Biofuels such as methane and ethanol can be used to generate electricity.

P1 4.2 | Energy from wind and water

Learning objectives

● What does a wind turbine consist of?

● How do we use waves to generate electricity?

● What type of power station uses water running downhill to generate electricity?

● How can we use the tides to generate electricity?

Strong winds can cause lots of damage on a very stormy day. Even when the wind is much weaker, it can still turn a wind turbine. Energy from the wind and other natural sources such as **waves** and **tides** is called **renewable energy**. That's because such natural sources of energy can never be used up.

In addition, no fuel is needed to produce electricity from these natural sources so they are carbon-free to run.

Wind power

A wind turbine is an electricity generator at the top of a narrow tower. The force of the wind drives the turbine's blades around. This turns a generator. The power generated increases as the wind speed increases.

a   What happens if the wind stops blowing?

Wave power

A wave generator uses the waves to make a floating generator move up and down. This motion turns the generator so it generates electricity. A cable between the generator and the shore delivers electricity to the grid system.

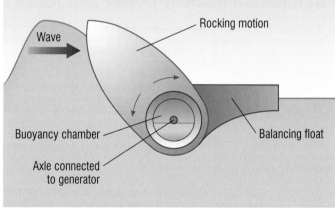

Figure 2 Energy from waves

Wave generators need to withstand storms and they don't produce a constant supply of electricity. Also, lots of cables (and buildings) are needed along the coast to connect the wave generators to the electricity grid. This can spoil areas of coastline. Tidal flow patterns might also change, affecting the habitats of marine life and birds.

b   What could happen if the waves get too high?

Hydroelectric power

We can generate hydroelectricity when rainwater collected in a reservoir (or water in a pumped storage scheme) flows downhill. The flowing water drives turbines that turn electricity generators at the foot of the hill.

c   Where does the energy for hydroelectricity come from?

Figure 1 A wind farm – why do some people oppose these developments?

How Science Works

When electricity demand is low, we can use electricity from wind turbines, wave generators and other electricity generators to pump water uphill into a reservoir. When demand is high, we can let the water run downhill through a hydroelectric generator.

Tidal power

A tidal power station traps water from each high tide behind a barrage. We can then release the high tide into the sea through turbines. The turbines drive generators in the barrage.

One of the most promising sites in Britain is the Severn estuary. This is because the estuary rapidly becomes narrower as you move up-river away from the open sea. So it funnels the incoming tide and makes it higher.

d Why is tidal power more reliable than wind power?

Figure 3 A hydroelectric scheme

Figure 4 A tidal power station

Summary questions

1 Copy and complete **a** to **d** using the words below:

hydroelectric   tidal   wave   wind

a .... power does not need water.

b .... power does not need energy from the Sun.

c .... power is obtained from water running downhill.

d .... power is obtained from water moving up and down.

2 **a** Use the table below for this question. The output of each source is given in millions of watts (MW).
   i How many wind turbines would give the same total power output as a tidal power station?
   ii How many kilometres of wave generators would give the same total output as a hydroelectric power station?

b Use the words below to fill in the location column in the table.

coastline   estuaries   hilly or coastal areas   mountain areas

	Output	Location	Total cost in £ per MW
Hydroelectric power station	500 MW per station		50
Tidal power station	2000 MW per station		300
Wave power generators	20 MW per kilometre of coastline		100
Wind turbines	2 MW per wind turbine		90

3 The last column of the table above shows an estimate of the total cost per MW of generating electricity using different renewable energy sources. The total cost for each includes its running costs and the capital costs to set it up.
  a The capital cost per MW of a tidal power station is much higher than that of a hydroelectric power station. Give one reason for this difference.
  b **i** Which energy resource has the lowest total cost per MW?
    ii Give two reasons why this resource might be unsuitable in many areas.

Key points

- A wind turbine is an electricity generator on top of a tall tower.
- Waves generate electricity by turning a floating generator.
- Hydroelectricity generators are turned by water running downhill.
- A tidal power station traps each high tide and uses it to turn generators.

P1 4.3 — Power from the Sun and the Earth

Learning objectives

- What are solar cells and how do we use them?
- What is the difference between a panel of solar cells and a solar heating panel?
- What is geothermal energy?
- How can we use geothermal energy to generate electricity?

Figure 2 A solar-powered vehicle. Think of some advantages and disadvantages of this car.

∞ links

For more information on solar heating panels, look back at P1 1.9 Heating and insulating buildings.

Solar radiation transfers energy to you from the Sun. That can sometimes be more energy than you want if you get sunburnt. But we can use the Sun's energy to generate electricity using **solar cells**. We can also use the Sun's energy to heat water directly in solar heating panels.

a Which generates electricity – a solar cell or a solar heating panel?

Practical

Solar cells

Use a solar cell panel to drive a small electric motor.

- See what happens if you gradually cover the solar cells with a card.

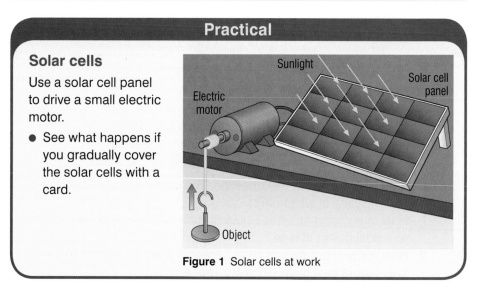

Figure 1 Solar cells at work

Solar cells at present convert less than 10% of the solar energy they absorb into electrical energy. We can connect them together to make solar cell panels.

- They are useful where we only need small amounts of electricity (e.g. in watches and calculators) or in remote places (e.g. on small islands in the middle of an ocean).

- They are very expensive to buy even though they cost nothing to run.

- We need lots of them – and plenty of sunshine – to generate enough power to be useful.

A **solar heating panel** heats water that flows through it. Even on a cloudy day in Britain, a solar heating panel on a house roof can supply plenty of hot water.

b If the water stopped flowing through a solar heating panel, what would happen?

A **solar power tower** uses thousands of flat mirrors to reflect sunlight on to a large water tank at the top of a tower. The mirrors on the ground surround the base of the tower.

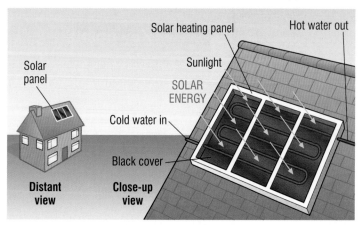

Figure 3 Solar water heating

- The water in the tank is turned to steam by the heating effect of the solar radiation directed at the water tank.
- The steam is piped down to ground level where it turns electricity generators.
- The mirrors are controlled by a computer so they track the Sun.

A solar power tower in a hot dry climate can generate more than 20 MW of electrical power.

Figure 4 A solar power tower

> **c** The solar furnace shown in Figure 3 in P1 1.1 uses 63 flat tracking mirrors to reflect solar radiation on to the giant reflector. Why does the solar power tower in Figure 4 opposite collect much more solar radiation than this solar furnace?

Study tip

Make sure you know the difference between a solar cell panel (in which sunlight is used to make electricity) and a solar heating panel (in which sunlight is used to heat water).

Geothermal energy

Geothermal energy comes from energy released by radioactive substances, deep within the Earth.

- The energy released by these radioactive substances heats the surrounding rock.
- As a result, energy is transferred by heating towards the Earth's surface.

We can build **geothermal power stations** in volcanic areas or where there are hot rocks deep below the surface. Water gets pumped down to these rocks to produce steam. Then the steam produced drives electricity turbines at ground level.

In some areas, we can heat buildings using geothermal energy directly. Heat flow from underground is called **ground heat.** It can be used to heat water in long lengths of underground pipes. The hot water is then pumped round the building. Ground heat is used as under-floor heating in some large 'eco-buildings'.

> **d** Why do geothermal power stations not need energy from the Sun?

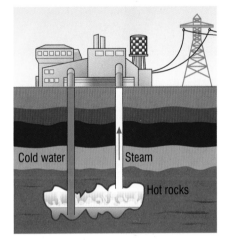

Figure 5 A geothermal power station

Summary questions

1 Copy and complete **a** to **c** using the words below:

   geothermal   solar   radiation   radioactivity

   a A suitable energy resource for a calculator is .... energy.
   b .... inside the Earth releases .... energy.
   c .... from the Sun generates electricity in a solar cell.

2 A satellite in space uses a solar cell panel for electricity. The panel generates 300 W of electrical power and has an area of 10m^2.
   a Each cell generates 0.2 W. How many cells are in the panel?
   b The satellite carries batteries that are charged by electricity from the solar cell panels. Why are batteries carried as well as solar cell panels?

3 A certain geothermal power station has a power output of 200 000 W.
   a How many kilowatt-hours of electrical energy does the power station generate in 24 hours?
   b State one advantage and one disadvantage of a geothermal power station compared with a wind turbine.

Key points

- Solar cells are flat solid cells that convert solar energy directly into electricity.
- Solar heating panels use the Sun's energy to heat water directly.
- Geothermal energy comes from the energy released by radioactive substances deep inside the Earth.
- Water pumped into hot rocks underground produces steam to drive turbines that generate electricity.

P1 4.4 Energy and the environment

Learning objectives

- What do fossil fuels do to our environment?

- Why are people concerned about nuclear power?

- How do renewable energy resources affect our environment?

Can we get energy without creating any problems? Look at the pie chart in Figure 1.

It shows the energy sources we use at present to generate electricity. What effect does each one have on our environment?

How Science Works

When a popular TV programme ends, lots of people decide to put the kettle on. The national demand for electricity leaps as a result. Engineers meet these surges in demand by switching gas turbine engines on in gas-fired power stations.

Fossil fuel problems

- When we burn coal, oil or gas, greenhouse gases such as carbon dioxide are released. We think that these gases cause global warming. We get some of our electricity from oil-fired power stations. We use much more oil to produce fuels for transport.

- Burning fossil fuels can also produce sulfur dioxide. This gas causes **acid rain**. We can remove the sulfur from a fuel before burning it to stop acid rain. For example, natural gas has its sulfur impurities removed before we use it.

- Fossil fuels are non-renewable. Sooner or later, we will have used up the Earth's reserves of fossil fuels. We will then have to find alternative sources of energy. But how soon? Oil and gas reserves could be used up within the next 50 years. Coal reserves will last much longer.

- **Carbon capture and storage** (CCS) could be used to stop carbon dioxide emissions into the atmosphere from fossil fuel power stations. Old oil and gas fields could be used for storage.

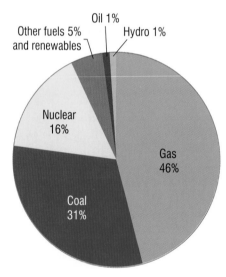

Figure 1 Energy sources for electricity

a Burning fossil fuels in power stations pollutes our atmosphere. Which gas contributes towards:
  i global warming?
  ii acid rain?

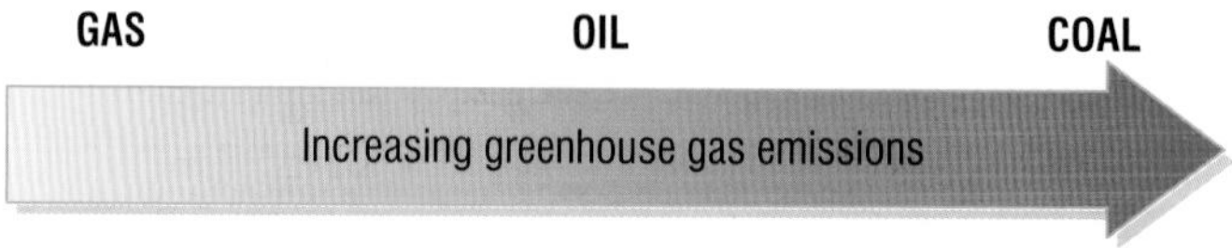

GAS   OIL   COAL

Increasing greenhouse gas emissions

Figure 2 Greenhouse gases from fossil fuels

links

For more information on pollution from fuels, see C1 4.3 Burning fuels, and C1 4.4 Cleaner fuels.

Nuclear v. renewable

We need to cut back on our use of fossil fuels to stop global warming. Should we rely on nuclear power or on renewable energy in the future?

Nuclear power

Advantages

- No greenhouse gases (unlike fossil fuel).

- Much more energy from each kilogram of uranium (or plutonium) fuel than from fossil fuel.

Did you know ...?

The Gobi Desert is one of the most remote regions on Earth. Many areas do not have mains electricity. Yet people who live there can watch TV programmes – just as you can. All they need is a solar panel and satellite TV.

Disadvantages

● Used fuel rods contain radioactive waste, which has to be stored safely for centuries.

● Nuclear reactors are safe in normal operation. However, an explosion at one could release radioactive material over a wide area. This would affect these areas for many years.

b Why is nuclear fuel non-renewable?

Renewable energy sources and the environment

Advantages

● They will never run out.

● They do not produce greenhouse gases or acid rain.

● They do not create radioactive waste products.

● They can be used where connection to the National Grid is uneconomic. For example, solar cells can be used for road signs and hydroelectricity can be used in remote areas.

Disadvantages

● Wind turbines create a whining noise that can upset people nearby and some people consider them unsightly.

● Tidal barrages affect river estuaries and the habitats of creatures and plants there.

● Hydroelectric schemes need large reservoirs of water, which can affect nearby plant and animal life. Habitats are often flooded to create dams.

● Solar cells would need to cover large areas to generate large amounts of power.

c Do wind turbines affect plant and animal life?

Summary questions

1 Copy and complete **a** to **c** using the words below:

acid rain   fossil fuels   greenhouse gas   plant and animal life
radioactive waste

a Most of Britain's electricity is produced by power stations that burn

....

b A gas-fired power station does not produce .... or much ....

c A tidal power station does not produce .... as a nuclear power station does but it does affect .... locally.

2 Match each energy source with a problem it causes.

Energy source	Problem
i Coal	A Noise
ii Hydroelectricity	B Acid rain
iii Uranium	C Radioactive waste
iv Wind power	D Takes up land

3 **a** List three possible renewable energy resources that could be used to generate electricity for people on a remote flat island in a hot climate.

b List three types of power stations that do not release greenhouse gases into the atmosphere.

Did you know …?

In 1986, some nuclear reactors at Chernobyl in Ukraine overheated and exploded. Radioactive substances were thrown high into the atmosphere. Chernobyl and the surrounding towns were evacuated. Radioactive material from Chernobyl was also deposited on parts of Britain.

Figure 3 Chernobyl, the site of the world's most serious accident at a nuclear power station

Figure 4 The effects of acid rain

Key points

● Fossil fuels produce increased levels of greenhouse gases which could cause global warming.

● Nuclear fuels produce radioactive waste.

● Renewable energy resources can affect plant and animal life.

P1 4.5 — The National Grid ⓚ

Learning objectives

- What is the National Grid?
- What do the transformers do in the National Grid?
- Why do we use high voltages in the National Grid?

Your electricity supply at home reaches you through the **National Grid**. This is a network of cables that distributes electricity from power stations to homes and other buildings. The network also contains **transformers**. Step-up transformers are used at power stations. Step-down transformers are used at substations near homes.

The National Grid's voltage is 132 000 V or more. This is because transmitting electricity at a high voltage reduces power loss, making the system more efficient.

Power stations produce electricity at a voltage of 25 000 V.

- We use **step-up transformers** to step this voltage up to the grid voltage.
- We use **step-down transformers** at local substations to step the grid voltage down to 230 V for use in homes and offices.

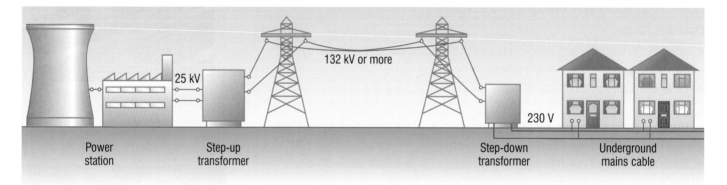

Figure 1 The National Grid

Figure 2 Electricity pylons carry the high voltage cables of the National Grid

Demonstration

Modelling the National Grid

Watch a demonstration of the effect of a transformer using this apparatus.

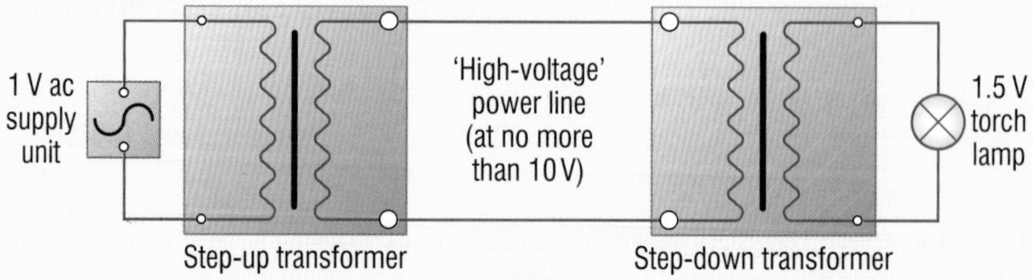

Figure 3 A model power line

Study tip

Remember that step-up transformers are used at power stations and step-down transformers are used at sub-stations near homes.

??? Did you know … ?

The National Grid was set up in 1926. The UK government decided electricity would be supplied to homes at 240 V. This was lowered to 230 V in 1994.

Power and the grid voltage

The electrical power supplied to any appliance depends on the appliance's current and its voltage. To supply a certain amount of power, we can lower the current if we raise the voltage. This is what a step-up transformer does in the grid system.

A step-up transformer raises the voltage, so less current is needed to transfer the same amount of power. A lower current passes through the grid cables. So energy losses due to the heating effect of the current are reduced to almost zero. But we need to lower the voltage at the end of the grid cables before we can use mains electricity at home.

a What difference would it make if we didn't step-up the grid voltage?

How Science Works

Underground or overhead?

Lots of people object to electricity pylons. They say they spoil the landscape or they affect their health. Electric currents produce electric and magnetic fields that might affect people.

Why don't we bury all cables underground?

Underground cables would be much more expensive, much more difficult to repair, and difficult to bury where they cross canals, rivers and roads.

What's more, overhead cables are high above the ground. Underground cables could affect people more because the cables wouldn't be very deep.

b Suggest two reasons why underground cables are more difficult to repair than overhead ones.

Summary questions

1 Copy and complete **a** and **b** using the words below:

higher   down   lower   up

a Power stations are connected to the National Grid using step-............ transformers. This type of transformer makes the voltage ....

b Homes are connected to the National Grid using step-............ transformers. This type of transformer makes the voltage ....

2 a Why is electrical energy transferred through the National Grid at a much higher voltage than the voltage generated in a power station?

b Why are transformers needed to connect local substations to the National Grid?

3 A step-up transformer connects a power station to the cables of the National Grid.

a What does the transformer do to
   i the voltage
   ii the current?

b Why are step-down transformers used between the end of the grid cables and the mains cables that supply mains electricity to our homes?

Study tip

You need to remember that:

- step-up transformers raise the voltage and lower the current
- step-down transformers lower the voltage and raise the current.

??? Did you know ... ?

The insulators used on electricity pylons need to be very effective or else the electricity would short-circuit to the ground. In winter, ice on the cables can cause them to snap. Teams of electrical engineers are always on standby to deal with sudden emergencies.

Figure 4 Engineers at work on the Grid. They certainly need a head for heights!

Key points

- The National Grid is a network of cables and transformers that distributes electricity to our homes from distant power stations and renewable energy generators.

- Step-up transformers are used to step up power station voltages to the grid voltage. Step-down transformers are used to step the grid voltage down for use in our homes.

- A high grid voltage reduces energy loss and makes the system more efficient.

P1 4.6

Big energy issues

Learning objectives

- How do we best use our electricity supplies to meet variations in demand?

- How do we best use our electricity supplies to meet base-load demand?

- Which energy resources need to be developed to meet our energy needs in future?

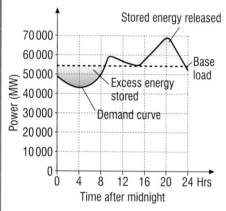

Figure 1 Example of electricity demand

Supply and demand

The demand for electricity varies during each day. It is also higher in winter than in summer. Our electricity generators need to match these changes in demand.

Power stations can't just 'start up' instantly. The **start-up time** depends on the type of power station.

NATURAL GAS   OIL   COAL   NUCLEAR

Shortest start-up time   Longest start-up time

a Which type of power station takes longest to start up?

Renewable energy resources are unreliable. The amount of electricity they generate depends on the conditions.

Table 1

Hydroelectric	Upland reservoir could run dry
Wind, waves	Wind and waves too weak on very calm days
Tidal	Height of tide varies both on a monthly and yearly cycle
Solar	No solar energy at night and variable during the day.

The variable demand for electricity is met by:

- using nuclear, coal- and oil-fired power stations to provide a constant amount of electricity (the **base load** demand)

- using gas-fired power stations and pumped-storage schemes to meet daily variations in demand and extra demand in winter

- using renewable energy sources when demand is high and renewables are in operation (e.g. use of wind turbines in winter when wind speeds are suitable)

- using renewable energy sources when demand is low to store energy in pumped storage schemes.

b Which type of power station can be used to satisfy sudden high demands for electricity which occur every day?

Figure 2 A nuclear power station

Activity

The big energy debate

A big energy debate is taking place at your school. Is it possible to generate enough electricity and to reduce the release of greenhouse gases? Your teacher will chair the debate.

Professor Jenny Jones has already spoken in favour of nuclear power and carbon capture. Here is a summary of what she said: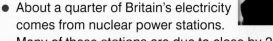

- About a quarter of Britain's electricity comes from nuclear power stations. Many of these stations are due to close by 2020. A new nuclear power station takes several years to build. We need to build more new nuclear power stations – or the lights will go out!

- We can't rely on wind power because when there is no wind, the wind turbines would not generate electricity. We can't rely on solar power at night or in winter. Nuclear power on its own won't give us enough electricity. We have to continue to burn fossil fuels but we can capture and store the greenhouse gases they produce in old oil or gas fields.

The leader of GoGreenUK, Peter Potts, has just finished speaking in favour of renewable energy and energy saving. Here is his summary: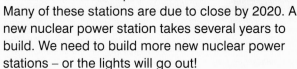

- We need to reduce our greenhouse gas emissions so we have to stop burning fossil fuels. We need to develop renewable energy resources on a much larger scale. We think that we can get most of our electricity from renewable energy devices like wind turbines and solar panels fitted to buildings. We should use public transport more to cut down on how much oil we need.

- If we insulate our homes better and make domestic appliances like fridges more efficient, we wouldn't need as much electricity. We need to use energy more efficiently. Then we wouldn't need new nuclear power stations.

Debate

Now it's your turn to raise points and ask questions. Choose which side of the debate you are on – for, against or undecided!

Some possible points that could be raised are listed below. Add some more points if you think they are reasonable. Your teacher will invite people to ask questions.

- The cost of building and running a nuclear power station is very high. So is the cost of decommissioning it (i.e. taking it out of use).

- Radioactive waste products are dangerous. No one wants a nuclear reactor to be built where they live.

- The capital costs of setting up renewable energy resources are high because lots of expensive equipment is needed to 'collect' large quantities of renewable energy.

- Carbon capture and storage is a new technology and likely to be expensive.

- Most home owners are unlikely to buy energy-saving improvements until energy bills go up even more.

Summary questions

1 Copy and complete using the words below:

coal gas nuclear oil

A .... power station can be started faster than any other type of power station. A .... power station does not produce greenhouse gases. The reserves of .... are likely to last longer than any other fossil fuel reserves. More public transport would reduce our use of ....

2 We need to cut back on fossil fuels to reduce the production of greenhouse gases. What could happen if the only energy we used was:
   a renewable energy
   b nuclear power?

3 a Why are nuclear power stations unsuitable for meeting daily variations in the demand for electricity?
   b What are pumped storage schemes and why are they useful?

Key points

- Gas-fired power stations and pumped-storage stations can meet variations in demand.

- Nuclear, coal and oil power stations can meet base-load demand.

- Nuclear power stations, fossil-fuel power stations using carbon capture and renewable energy are all likely to contribute to future energy supplies.

Summary questions (k)

1 Answer **a** to **e** using the list of fuels below:

coal   natural gas   oil   uranium   wood

  a Which fuels from the list are fossil fuels?

  b Which fuels from the list cause acid rain?

  c Which fuels release chemical energy when they are used?

  d Which fuel releases the most energy per kilogram?

  e Which fuel produces radioactive waste?

2 a Copy and complete **i** to **iv** using the words below:

hydroelectric   tidal   wave   wind

    i .... power stations trap sea water.

    ii .... power stations trap rain water.

    iii .... generators must be located along the coastline.

    iv .... turbines can be located on hills or offshore.

  b Which renewable energy resource transfers:

    i the kinetic energy of moving air to electrical energy

    ii the gravitational potential energy of water running downhill into electrical energy

    iii the kinetic energy of water moving up and down to electrical energy?

3 a Copy and complete **i** to **iv** using the words below:

coal-fired   geothermal   hydroelectric   nuclear

    i A .... power station does not produce greenhouse gases and uses energy which is from inside the Earth.

    ii A .... power station uses running water and does not produce greenhouse gases.

    iii A .... power station releases greenhouse gases.

    iv A .... power station does not release greenhouse gases but does produce waste products that need to be stored for many years.

  b Wood can be used as a fuel. State whether it is

    i renewable or non-renewable

    ii a fossil fuel or a non-fossil fuel.

4 a Figure 1 shows a landscape showing three different renewable energy resources, numbered 1 to 3. Match each type of energy resource with one of the labels below.

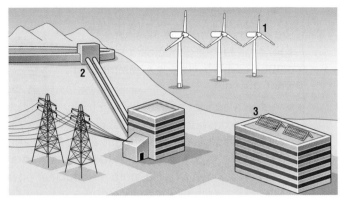

Figure 1 Renewable energy

Hydroelectricity   Solar energy   Wind energy

  b Which of the three resources shown is not likely to produce as much energy as the others if the area is

    i hot, dry and windy

    ii wet and windy?

5 Copy and complete **a** to **d** using the words below. Each word or phrase can be used more than once.

cheaper   more expensive   longer   shorter

  a Wind turbines are .... to build than nuclear power stations and .... to run.

  b Nuclear power stations take .... to decommission than fossil fuel power stations.

  c Solar cells are .... to install than solar heating panels.

  d A gas-fired power station has a .... start-up time compared to a nuclear power station.

6 a i What are transformers used for in the National Grid?

    ii What type of transformer is connected between the generators in the power station and the cables of the grid system?

  b i What can you say about the voltage of the cables in the grid system compared with the voltages at the power station generator and at the mains cables into the home?

    ii What can you say about the current through the grid cables compared with the current from the power station generator?

    iii What is the reason for making the grid voltage different from the generator voltage?

Practice questions ⓚ

1 Electricity may be generated in a coal-fired power station.

Copy and complete the following sentences using words from the list below. Each word can be used once, more than once or not at all.

electricity fuel generator steam turbine water wood

In a coal-fired power station, ... is burned to heat .... This produces .... at high pressure which makes a .... spin round. This then drives a .... that produces .... (6)

2 Various power sources can be used to generate electricity.

Match the power sources in the list with the statements 1 to 4 in the table.

A falling water
B tides
C waves
D wind

	Statement
1	the source of hydroelectric power
2	used with a floating generator
3	very unpredictable and at times may stop altogether
4	will produce a predictable cycle of power generation during the day

(4)

3 A solar cell panel and a solar heating panel work in different ways.

Which statement below is correct?

A A solar cell produces light when it is supplied with electricity.
B A solar cell generates electricity when it is supplied with light.
C A solar heating panel produces heat when it is supplied with electricity.
D A solar heating panel produces electricity when it is supplied with heat. (1)

4 Gas-fired power stations have a shorter start-up time than other power stations. Give **one** reason why is it important to have power stations with a short start-up time. (2)

5 During the night, when demand for electricity is low, a wind farm may be generating a large amount of power. Explain how, by using another type of power station, this power could be stored and used when it is needed. (3)

6 Explain why step-up transformers are used in the National Grid. (2)

7 Palm oil can be used to make a biofuel called biodiesel. Biodiesel can be used instead of the normal type of diesel obtained by refining crude oil.

a Suggest **two** advantages of using biodiesel rather than normal diesel. (2)

b Suggest **two** disadvantages of using biodiesel rather than normal diesel. (2)

8 The pie chart shows the main sources of energy used in power stations in a country last year.

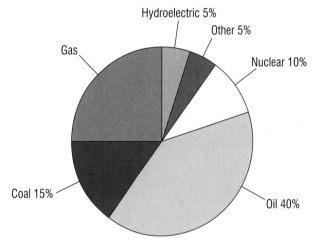

a What fraction of the energy used in power stations was obtained from gas? (2)

b Name **one** source of energy shown that is a fossil fuel. (1)

c Name **one** source of energy shown that is renewable. (1)

d Name **one** source of energy that could be included in the label 'other'. (1)

e Name **one** source of energy that does not cause carbon dioxide to be released when it is used. (1)

9 *In this question you will be assessed on using good English, organising information clearly and using specialist terms where appropriate.*

Power stations that burn fossil fuels produce waste gases that can cause pollution.

Describe the effect that these gases could have on the environment and what could be done to reduce the amount of these gases emitted by power stations. (6)

P1 5.1

The nature of waves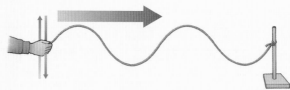

Learning objectives

- What can we use waves for?
- What are transverse waves?
- What are longitudinal waves?
- Which types of waves are transverse and which are longitudinal?

Figure 1 Big waves

We use waves to transfer information and we can use them to transfer energy. We can use information transferred by waves in communications, for example when you use a mobile phone or listen to the radio.

There are different types of waves. These include:

- sound waves, water waves, waves on springs and ropes and seismic waves produced by earthquakes. These are examples of **mechanical waves**, which are vibrations that travel through a medium (substance).
- light waves, radio waves and microwaves. These are examples of **electromagnetic waves** which can all travel through a vacuum at the same speed of 300000 kilometres per second. No medium is needed.

Practical

Observing mechanical waves

Figure 2 shows how we can make waves on a rope by moving one end up and down.

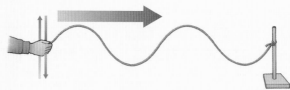

Figure 2 Transverse waves

Tie a ribbon to the middle of the rope. Move one end of the rope up and down. You will see that the waves move along the rope but the ribbon doesn't move along the rope – it just moves up and down. This type of wave is known as a **transverse wave.** We say the ribbon **vibrates** or **oscillates.** This means it moves repeatedly between two positions. When the ribbon is at the top of a wave, we say it is at the **peak** (or crest) of the wave.

Repeat the test with the slinky. You should observe the same effects if you move one end of the slinky up and down.

However, if you push and pull the end of the slinky as shown in Figure 3, you will see a different type of wave, known as a **longitudinal wave.** Notice that there are areas of **compression** (coils squashed together) and areas of **rarefaction** (coils spread further apart) moving along the slinky.

Hand moved backwards and forwards along the line of the slinky

Figure 3 Making longitudinal waves on a slinky

- How does the ribbon move when you send **longitudinal** waves along the slinky?

⊂⊃ links

For more information on electromagnetic waves, see P1 6.1 The electromagnetic spectrum.

Study tip

You are **not** required to recall the value of the speed of electromagnetic waves through a vacuum. If you need it to answer a question, it will be provided for you.

Transverse waves

Imagine we send waves along a rope which has a white spot painted on it. The spot would be seen to move up and down without moving along the rope. In other words, the spot would vibrate **perpendicular** (at right angles) to the direction which the waves are moving. The waves on a rope are called **transverse waves** because the vibrations are up and down or from side to side. All electromagnetic waves are transverse waves.

The vibrations of a transverse wave are perpendicular to the direction in which the waves transfer energy.

a State one type of wave that is mechanical and transverse.

Longitudinal waves

The slinky spring in Figure 3 is useful to demonstrate how sound waves travel. When one end of the slinky is pushed in and out repeatedly, vibrations travel along the spring. The vibrations are parallel to the direction in which the waves transfer energy along the spring. Waves that travel in this way are called **longitudinal waves**.

Sound waves are longitudinal waves. When an object vibrates in air, it makes the air around it vibrate as it pushes and pulls on the air. The vibrations (**compressions** and **rarefactions**) which travel through the air are sound waves. The vibrations are along the direction in which the wave travels.

The vibrations of a longitudinal wave are parallel to the direction in which the waves are travelling.

Therefore mechanical waves can be transverse or longitudinal.

b When a sound wave passes through air, what happens to the air particles at a compression?

Summary questions

1 Copy and complete **a** to **d** using the words below:

longitudinal parallel perpendicular transverse

a Sound waves are .... waves.
b Light waves are .... waves.
c Transverse waves vibrate .... to the direction of energy transfer of the waves.
d Longitudinal waves vibrate .... to the direction of energy transfer of the waves.

2 A long rope with a knot tied in the middle lies straight along a smooth floor. A student picks up one end of the rope. This sends waves along the rope.
   a Are the waves on the rope transverse or longitudinal waves?
   b What can you say about:
      i the direction of energy transfer along the rope?
      ii the movement of the knot?

3 Describe how to use a slinky spring to demonstrate to a friend the difference between longitudinal waves and transverse waves.

⚭ links

For more information on sound, see P1 5.5 Wave properties: diffraction, and P1 5.6 Sound.

Study tip

Make sure that you understand the difference between transverse waves and longitudinal waves.

Did you know …?

When we pluck a guitar string, it vibrates because we send transverse waves along the string. The vibrating string sends sound waves into the surrounding air. The sound waves are longitudinal.

Key points

- We use waves to transfer energy and transfer information.

- Transverse waves vibrate at right angles to the direction of energy transfer of the waves. All electromagnetic waves are transverse waves.

- Longitudinal waves vibrate parallel to the direction of energy transfer of the waves. A sound wave is an example of a longitudinal wave.

- Mechanical waves, which need a medium (substance) to travel through, may be transverse or longitudinal waves.

277

P1 5.2 — Measuring waves

Learning objectives

- What do we mean by the amplitude of a wave?
- What do we mean by the frequency of a wave?
- What do we mean by the wavelength of a wave?
- What is the relationship between the speed, wavelength and frequency of a wave?

We need to measure waves if we want to find out how much energy or information they carry. Figure 1 shows a snapshot of waves on a rope. The **crests** or peaks are at the top of the wave. The **troughs** are at the bottom. They are equally spaced.

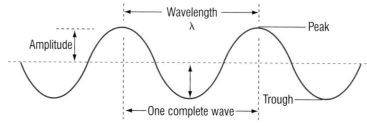

Figure 1 Waves on a rope

- The **amplitude** of the waves is the height of the wave crest or the depth of the wave trough from the middle, which is the position of the rope at rest.
  The bigger the amplitude of the waves, the more energy the waves carry.
- The **wavelength** of the waves is the distance from one wave crest to the next crest.

> **a** Use a millimetre rule to measure the amplitude and the wavelength of the waves in Figure 1.

Frequency

If we made a video of the waves on the rope, we would see the waves moving steadily across the screen. The number of wavecrests passing a fixed point every second is the **frequency** of the waves.

The unit of frequency is the **hertz** (Hz). One wave crest passing each second is a frequency of 1 Hz.

Wave speed

Figure 2 shows a ripple tank, which is used to study water waves in controlled conditions. We can make straight waves by moving a ruler up and down on the water surface in a ripple tank. Straight waves are called **plane** waves. The waves all move at the same speed and keep the same distance apart.

The **speed** of the waves is the distance travelled by a wave crest or a wave trough every second.

For example, sound waves in air travel at a speed of 340 m/s. In 5 seconds, sound waves travel a distance of 1700 m (= 340 m/s × 5 s).

For waves of constant frequency, the speed of the waves depends on the frequency and the wavelength as follows:

$$\text{wave speed} = \text{frequency} \times \text{wavelength}$$
$$\text{(metre/second, m/s)} \quad \text{(hertz, Hz)} \quad \text{(metre, m)}$$

Study tip

A common error is to think that the amplitude is the distance from the top of the crest to the bottom of the trough.

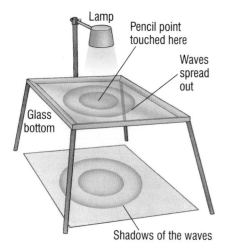

Figure 2 The ripple tank

Maths skills

We can write the wave speed equation as $v = f \times \lambda$

where v = speed, f = frequency, λ = wavelength.

Note: λ is pronounced 'lambda'.

Practical

Making straight (plane) waves

To measure the speed of the waves:

Use a stopwatch to measure the time it takes for a wave to travel from the ruler to the side of the ripple tank.

Measure the distance the waves travel in this time.

Use the equation speed $= \dfrac{\text{distance}}{\text{time}}$ to calculate the speed of the waves.

Observe the effect on the waves of moving the ruler up and down faster. More waves are produced every second and they are closer together.

- Find out if the speed of the waves has changed.

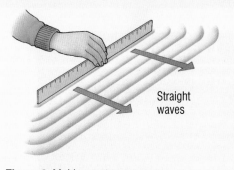

Straight waves

Figure 3 Making water waves

To understand what the wave speed equation means, look at Figure 4. The surfer is riding on the crest of some unusually fast waves.

Suppose the frequency of the waves is 3 Hz and the wavelength of the waves is 4.0 m.

- At this frequency, 3 wave crests pass a fixed point once every second (because the frequency is 3 Hz).
- The surfer therefore moves forward a distance of 3 wavelengths every second or 12 m (= 3 × 4.0 m).

The speed of the surfer is therefore 12 m/s.

This speed is equal to the frequency × the wavelength of the waves: $v = f \times \lambda$.

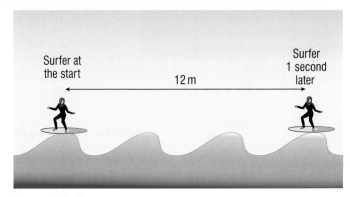

Surfer at the start

12 m

Surfer 1 second later

Figure 4 Surfing

Summary questions

1 Copy and complete **a** to **d** using the words below. Each word can be used more than once.

amplitude frequency speed wavelength

 a The hertz is the unit of ....
 b The distance from one wave crest to the next is the .... of a wave.
 c For water waves, the height of a wave crest above the undisturbed water surface is the .... of the wave.
 d .... × frequency = ...

2 Figure 5 shows a snapshot of a wave travelling from left to right along a rope.
 a Copy Figure 5 and mark on your diagram
  i one wavelength
  ii the amplitude of the waves.

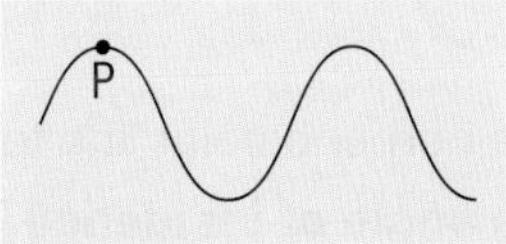

Figure 5 A wave on a rope

 b Describe the motion of point P on the rope when the wave crest at P moves along by a distance of one wavelength.

3 **a** A speedboat on a lake sends waves travelling across a lake at a frequency of 2.0 Hz and a wavelength of 3.0 m. Calculate the speed of the waves.
 b If the waves had been produced at a frequency of 1.0 Hz and travelled at the speed calculated in **a**, what would be their wavelength? **[H]**

Key points

- For any wave, its amplitude is the height of the wave crest or the depth of the wave trough from the position at rest.

- For any wave, its frequency is the number of wave crests passing a point in one second.

- For any wave, its wavelength is the distance from one wave crest to the next wave crest. This is the same as the distance from one wave trough to the next wave trough.

- $v = f \times \lambda$

P1 5.3

Wave properties: reflection

Learning objectives

- What is the normal in a diagram showing light rays?

- What is an angle of incidence?

- What can we say about the reflection of a light ray at a plane mirror?

- How is an image formed by a plane mirror?

If you visit a Hall of Mirrors at a funfair, you will see some strange images of yourself. A tall, thin image or a short, broad image of yourself means you are looking into a mirror that is curved. If you want to see a normal image of yourself, look in a **plane mirror**. Such a mirror is perfectly flat. You see an exact mirror **image** of yourself.

Figure 1 A good image

Investigating the reflection of waves using a ripple tank

Light consists of waves. Figure 2 shows how we can investigate the reflection of waves using a ripple tank. The investigations show that when plane (straight) waves reflect from a flat reflector, the reflected waves are at the same angle to the reflector as the incident waves.

The law of reflection

We use light rays to show us the direction light waves are moving in. Figure 3 shows how we can investigate the reflection of a light ray from a ray box using a plane mirror.

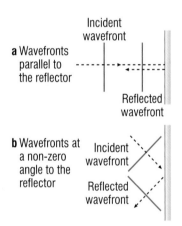

a Wavefronts parallel to the reflector

Incident wavefront

Reflected wavefront

b Wavefronts at a non-zero angle to the reflector

Incident wavefront

Reflected wavefront

Figure 2 Reflection of plane waves

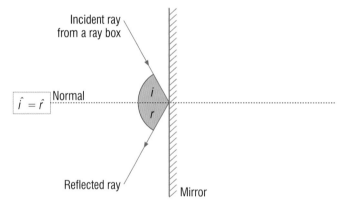

Incident ray from a ray box

$\hat{i} = \hat{r}$ Normal

i

r

Reflected ray

Mirror

Figure 3 The law of reflection

- The perpendicular line to the mirror is called the **normal**.
- The **angle of incidence** is the angle between the incident ray and the normal.
- The **angle of reflection** is the angle between the reflected ray and the normal.

Measurements show that for any light ray reflected by a plane mirror:

the angle of incidence = the angle of reflection

Practical

A reflection test

Use a ray box and a plane mirror as shown in Figure 3 to test the law of reflection for different angles of incidence.

a If the angle of reflection of a light ray from a plane mirror is 20° what is:
  i the angle of incidence?
  ii the angle between the incident ray and the reflected ray?

Image formation by a plane mirror

Figure 4 shows how an image is formed by a plane mirror. This ray diagram shows the path of two light rays from a point object that reflect off the mirror. The image and the object in Figure 4 are at equal distances from the mirror.

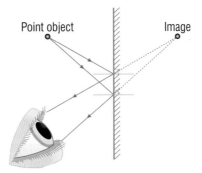

Figure 4 Image formation by a plane mirror

Real and virtual images

The image formed by a plane mirror is virtual, upright (the same way up as the object) and laterally inverted (back to front but not upside down). A **virtual image** can't be projected on to a screen like the movie images that you see at a cinema. An image on a screen is described as a **real image** because it is formed by focusing light rays on to the screen.

b When you use a mirror, is the image real or virtual?

?? ? Did you know ...?

Ambulances and police cars often carry a 'mirror image' sign at the front. This is so a driver in a vehicle in front looking at their rear-view mirror can read the sign as it gets 'laterally inverted' (back to front but not upside down).

Figure 5 A mirror sign on an ambulance

Summary questions

1 Copy and complete **a** to **c** using the words below. Each word can be used once, more than once, or not at all.

equal to   greater than   less than

a The angle of incidence of a light ray at a plane mirror is always .... 90 degrees.

b The angle between the normal and the mirror is always .... 90 degrees.

c The angle of incidence of a light ray at a plane mirror is always .... the angle of reflection of the light ray.

2 A point object O is placed in front of a plane mirror, as shown.

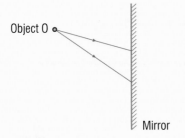

Figure 6

a Complete the path of the two rays shown from O after they have reflected off the mirror.

b i Use the reflected rays to locate the image of O.

ii Show that the image and the object are the same distance from the mirror.

3 Two plane mirrors are placed perpendicular to each other. Draw a ray diagram to show the path of a light ray at an angle of incidence of 60° that reflects off both mirrors.

Key points

- The normal at a point on a mirror is a line drawn perpendicular to the mirror.

- For a light ray reflected by a plane mirror:

  1 The angle of incidence is the angle between the incident ray and the normal.

  2 The angle of reflection is the angle between the reflected ray and the normal.

- The law of reflection states that:
  the angle of incidence = the angle of reflection.

P1 5.4

Wave properties: refraction

Learning objectives

- What is refraction?

- What can we say about the refraction of a light ray when it goes from air into glass?

- What can we say about the refraction of a light ray when it goes from glass into air?

When you have your eyes tested, the optician might test different lenses in front of each of your eyes. Each lens changes the direction of light passing through it. This change of direction is known as **refraction**.

Refraction is a property of all forms of waves including light and sound. Figure 1 shows how we can see refraction of waves in a ripple tank.

A glass plate is submerged in a ripple tank. The water above the glass plate is shallower than the water in the rest of the tank. The waves are slower in shallow water than in deep water. If the waves are not parallel to the **boundary**, they change direction when they cross the boundary:

- towards the normal when they cross from deep to shallow water

- away from the normal when they cross from shallow to deep water.

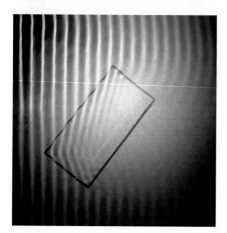

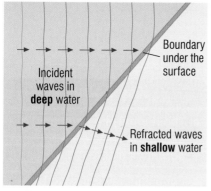

Figure 1 Refraction of water waves

Practical

Investigating refraction of light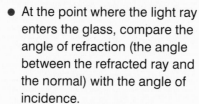

Figure 2 shows how you can use a ray box and a rectangular glass block to investigate the refraction of a light ray when it enters glass. The ray changes direction at the boundary between air and glass (unless it is along the normal).

- At the point where the light ray enters the glass, compare the angle of refraction (the angle between the refracted ray and the normal) with the angle of incidence.

You should find that the angle of refraction at the point of entry is always less than the angle of incidence.

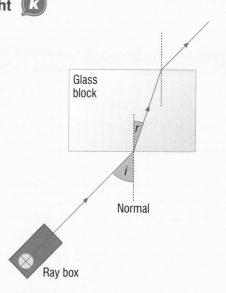

Figure 2 Refraction of light

Refraction rules

Your investigation should show that a light ray:

- changes direction towards the normal when it travels from air into glass. The angle of refraction (r) is smaller than the angle of incidence (i).

- changes direction away from the normal when it travels from glass into air. The angle of refraction (r) is greater than the angle of incidence (i).

> **a** If a light ray enters a rectangular glass block along the normal, does it leave the block along the normal?

Study tip

Remember that angles of incidence, reflection and refraction are always measured between the ray and the normal.

Refraction by a prism

Figure 3 shows what happens when a narrow beam of white light passes through a triangular glass prism. The beam comes out of the prism in a different direction to the incident ray and is split into the colours of the spectrum.

White light contains all the colours of the spectrum. Each colour of light is refracted slightly differently. So the prism splits the light into colours.

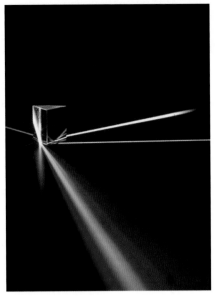

Figure 3 Refraction by a prism

?? Did you know ... ?

A swimming pool always appears shallower than it really is. The next time you jump into water, make sure you know how deep it is. Light from the bottom refracts at the surface. This makes the water appear shallower than it is.

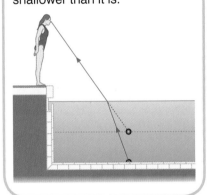

Summary questions

1. Copy and complete **a** to **d** using the words below:

   away from   greater than   less than   towards

   a When a light ray travels from air into glass, it refracts .... the normal.

   b When a light ray travels from glass into air, it refracts .... the normal.

   c When a light ray travels from air into glass, the angle of refraction is .... the angle of incidence.

   d When a light ray travels from glass into air, the angle of refraction is .... the angle of incidence.

2. Copy and complete the path of the light ray through each glass object below.

   a    **b**

   Figure 4

3. A light ray from the bottom of a swimming pool refracts at the water surface. Its angle of incidence is 40 degrees and its angle of refraction is 75 degrees.

   a Draw a diagram to show the path of this light ray from the bottom of the swimming pool into the air above the pool.

   b Use your diagram to explain why the swimming pool appears shallower than it really is when viewed from above.

?? Did you know ... ?

A **rainbow** is caused by refraction of light when sunlight shines on rain droplets. The droplets refract sunlight and split it into the colours of the spectrum.

Key points

- Refraction is the change of direction of waves when they travel across a boundary.

- When a light ray refracts as it travels from air into glass, the angle of refraction is less than the angle of incidence.

- When a light ray refracts as it travels from glass into air, the angle of refraction is more than the angle of incidence.

P1 5.5 Wave properties: diffraction ⓚ

Learning objectives

- What do we mean by diffraction?
- What is the effect of gap width on the diffraction of waves?
- Why is radio and TV reception often poor in hilly areas?

Diffraction is the spreading of waves when they pass through a gap or move past an obstacle. The waves that pass through the gap or past the edges of the obstacle can spread out. Figure 1 shows waves in a ripple tank spreading out after they pass through two gaps. The effect is most noticeable if the wavelength of the waves is similar to the width of the gap. You can see from Figure 1 that

- the narrower the gap, the more the waves spread out
- the wider the gap, the less the waves spread out.

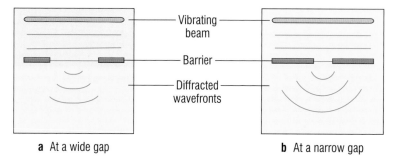

a At a wide gap   b At a narrow gap

Figure 1 Diffraction of waves by a gap: **a** A wide gap **b** A narrow gap

??? Did you know ... ?

Sea waves entering a harbour through a narrow entrance spread out after passing through the entrance. Look out for this diffraction effect the next time you visit a harbour.

Figure 2 Image of two colliding galaxies taken by the Hubble Space Telescope

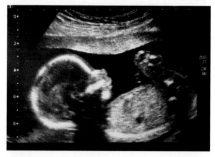

Figure 3 An ultrasonic scan of a baby in the womb

Practical

Investigating diffraction

Use a ripple tank as in Figure 1 to direct plane waves continuously at a gap between two metal barriers. Notice that the waves spread out after they pass through the gap. In other words, they are diffracted by the gap.

Change the gap spacing and observe the effect on the diffraction of the waves that pass through the gap. You should find that the diffraction of the waves increases as the gap is made narrower, as shown in Figure 1.

Diffraction details

Diffraction of light is important in any optical instrument. The Hubble Space Telescope in its orbit above the Earth has provided amazing images of objects far away in space. Its focusing mirror is 2.4 m in diameter. When it is used, astronomers can see separate images of objects which are far too close to be seen separately using a narrower telescope. Little diffraction occurs when light passes through the Hubble Space Telescope because it is so wide. So its images are very clear and very detailed.

Diffraction of ultrasonic waves is an important factor in the design of an ultrasonic scanner. Ultrasonic waves are sound waves at frequencies above the range of the human ear. An ultrasonic scan can be made of a baby in the womb. The ultrasonic waves spread out from a hand-held transmitter and then reflect from the tissue boundaries inside the womb. If the transmitter is too narrow, the waves spread out too much and the image is not very clear.

a The two examples of diffraction above show that both transverse and longitudinal waves can be diffracted. Which is which?

Demonstration

Tests using microwaves

A microwave transmitter and a detector can be used to demonstrate diffraction of microwaves. The transmitter produces microwaves of wavelength 3.0 cm.

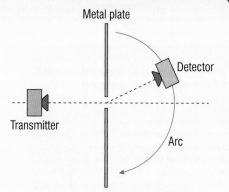

Figure 4 Using microwaves (top view)

1 Place a metal plate between the transmitter and the detector across the path of the microwaves. Microwaves can still be detected behind the metal plate. This is because some microwaves diffract round the edge of the plate.

● Why do the microwaves not go through the metal plates?

2 Place two metal plates separated by a gap across the path of the microwaves, as shown in Figure 4. The microwaves pass through the gap but not through the plates. When the detector is moved along an arc centred on the gap, it detects microwaves that have spread out from the gap.

When the gap is made wider, the microwaves passing through the gap spread out less. The detector needs to be nearer the centre of the arc to detect the microwaves.

Signal problems

People in hilly areas often have poor TV reception. The signal from a TV transmitter mast is carried by radio waves. If there are hills between a TV receiver and the transmitter mast, the signal may not reach the receiver. The radio waves passing the top of a hill are diffracted by the hill but they do not spread enough behind the hill.

b What type of waves carry TV signals?

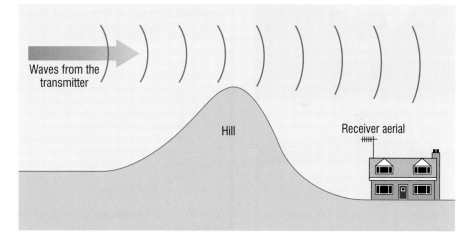

Figure 5 Poor reception

Summary questions

1 Copy and complete **a** and **b** using the words below. Each term can be used once, twice, or not at all.

more than   less than   the same as

a Diffracted waves spread out from a narrow gap .... they would from a wider gap.

b When waves pass through a gap, their wavelength is .... it was before it passed through the gap.

2 a State what is meant by diffraction.

b Explain why the TV reception from a transmitter mast can be poor in hilly areas.

3 A small portable radio inside a room can be heard all along a corridor that runs past the room when its door is open. Explain why it can be heard by someone in the corridor who is not near the door.

Key points

● Diffraction is the spreading out of waves when they pass through a gap or round the edge of an obstacle.

● The narrower a gap is, the greater the diffraction is.

● If radio waves do not diffract enough when they go over hills, radio and TV reception will be poor.

P1 5.6   Sound

- What range of frequencies can be detected by the human ear?
- What are sound waves?
- What are echoes?

Figure 1 Making sound waves

??? Did you know ...?

When you blow a round whistle, you force a small ball inside the whistle to go round and round inside. Each time it goes round, its movement draws air in then pushes it out. Sound waves are produced as a result.

Investigating sound waves

Sound waves are easy to produce. Your vocal cords vibrate and produce sound waves every time you speak. Any object vibrating in air makes the layers of air near the object vibrate. These layers make the layers of air further away vibrate. The vibrating object pushes and pulls repeatedly on the air. This sends out the vibrations of the air in waves of compressions and rarefactions. When the waves reach your ears, they make your eardrums vibrate in and out so you hear sound as a result.

The vibrations travelling through the air are sound waves. The waves are longitudinal because the air particles vibrate along the direction in which the waves transfer energy.

Practical

Investigating sound waves

You can use a loudspeaker to produce sound waves by passing alternating current through it. Figure 2 shows how to do this using a signal generator. This is an alternating current supply unit with a variable frequency dial.

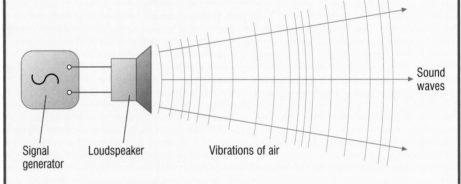

Figure 2 Using a loudspeaker

- If you observe the loudspeaker closely, you can see it vibrating. It produces sound waves as it pushes the surrounding air backwards and forwards.
- If you alter the frequency dial of the signal generator, you can change the frequency of the sound waves.

Find out the lowest and the highest frequency you can hear. Young people can usually hear sound frequencies from about 20 Hz to about 20 000 Hz. Older people in general can't hear frequencies at the higher end of this range.

a   Which animal produces sound waves at a higher frequency, an elephant or a mouse?

Sound waves cannot travel through a vacuum. You can test this by listening to an electric bell in a bell jar. As the air is pumped out of the bell jar, the ringing sound fades away.

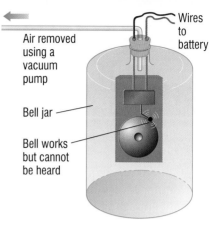

Figure 3 A sound test

b What would you notice if the air is let back into the bell jar?

Reflection of sound

Have you ever created an echo? An **echo** is an example of reflection of sound. Echoes can be heard in a large hall or gallery which has bare, smooth walls.

● If the walls are covered in soft fabric, the fabric will absorb sound instead of reflecting it. No echoes will be heard.

● If the wall surface is uneven (not smooth), echoes will not be heard because the reflected sound is 'broken up' and scattered.

c What happens to the energy of the sound waves when they are absorbed by a fabric?

Refraction of sound

Sound travels through air at a speed of about 340 m/s. The warmer the air is, the greater the speed of sound. At night you can hear sound a long way from its source. This is because sound waves refract back to the ground instead of travelling away from the ground. Refraction takes place at the boundaries between layers of air at different temperatures. In the daytime, sound refracts upwards, not downwards, because the air near the ground is warmer than air higher up.

Figure 4 Refraction of sound

Summary questions

1 Copy and complete **a** and **b** using the words below:

absorbed   reflected   scattered

a An echo is heard when sound is .... from a bare, smooth wall.
b Sound waves are .... by a rough wall and .... by soft fabric.

2 **a** What is the highest frequency of sound the human ear can hear?
b Why does a round whistle produce sound at a constant frequency when you blow steadily into it?

3 **a** A boat is at sea in a mist. The captain wants to know if the boat is near any cliffs so he sounds the horn and listens for an echo. Why would hearing an echo tell him he is near the cliffs?
b Explain why someone in a large cavern can sometimes hear more than one echo of a sound.

Key points

● The frequency range of the normal human ear is from about 20 Hz to about 20 000 Hz.

● Sound waves are vibrations that travel through a medium (substance). They cannot travel through a vacuum (as in space).

● Echoes are due to sound waves reflected from a smooth, hard surface.

P1 5.7

Musical sounds ⓚ

Learning objectives

- What determines the pitch of a note?

- What happens to the loudness of a note as the amplitude increases?

- How are sound waves created by musical instruments?

Figure 1 Making music

What type of music do you like? Whatever your taste in music is, when you listen to it you usually hear sounds produced by specially-designed instruments. Even your voice is produced by a biological organ that has the job of producing sound.

- Musical notes are easy to listen to because they are rhythmic. The sound waves change smoothly and the wave pattern repeats itself regularly.

- Noise consists of sound waves that vary in frequency without any pattern.

a Name four different vehicles that produce sound through a loudspeaker or a siren.

Practical

Investigating different sounds

Use a microphone connected to an oscilloscope to display the waveforms of different sounds.

Figure 2 Investigating different sound waves

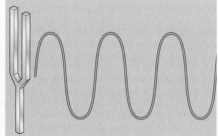

Figure 3 Tuning fork waves

1 Test a tuning fork to see the waveform of a sound of constant frequency.

2 Compare the pure waveform of a tuning fork with the sound you produce when you talk or sing or whistle. You may be able to produce a pure waveform when you whistle or sing but not when you talk.

3 Use a signal generator connected to a loudspeaker to produce sound waves. The waveform on the oscilloscope screen should be a pure waveform.

b What can you say about the waveform of a sound when you make the sound quieter?

Your investigations should show you that:

- **increasing the loudness** of a sound increases the **amplitude** of the waves. So the waves on the screen become taller.

- **increasing the frequency of a sound** (the number of waves per second) increases its **pitch**. This makes more waves appear on the screen.

Figure 4 shows the waveforms for different sounds from the loudspeaker.

c How would the waveform in Figure 4a change if the loudness and the pitch are both reduced?

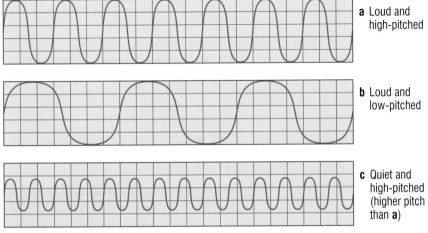

a Loud and high-pitched

b Loud and low-pitched

c Quiet and high-pitched (higher pitch than **a**)

Figure 4 Investigating sounds

Musical instruments

When you play a musical instrument, you create sound waves by making the instrument and the air inside it vibrate. Each new cycle of vibrations makes the vibrations stronger at certain frequencies. We say the instrument **resonates** at these frequencies. Because the instrument and the air inside it vibrate strongly at these frequencies when it is played, we hear recognisable notes of sound from the instrument.

- A wind instrument such as a flute is designed so that the air inside resonates when it is played. You can make the air in an empty bottle resonate by blowing across the top gently.
- A string instrument such as a guitar produces sound when the strings vibrate. The vibrating strings make the surfaces of the instrument vibrate and produce sound waves in the air. In an acoustic guitar, the air inside the hollow body of the guitar (the sound box) vibrates too.
- A percussion instrument such as a drum vibrates and produces sound when it is struck.

Study tip

Be sure you know the meaning of the words **frequency** and **amplitude**.

Practical

Musical instruments

Investigate the waveform produced by a musical instrument, such as a flute.

You should find its waveform changes smoothly, like the one in Figure 5 – but only if you can play it correctly. The waveform is a mixture of frequencies rather than a single frequency waveform like Figure 3.

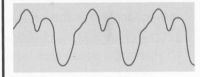

Figure 5 Flute wave pattern

Summary questions

1 Copy and complete **a** to **c** using the words below:

   amplitude   frequency   vibrations

   a When a drum is struck, sound waves are created by the .... of the drumskin.
   b The loudness of a sound is increased by increasing the .... of the sound waves.
   c The pitch of a sound is increased by increasing the .... of the sound waves.

2 A microphone and an oscilloscope are used to investigate sound from a loudspeaker connected to a signal generator. What change would you expect to see on the oscilloscope screen if the sound is:
   a made louder at the same frequency
   b made lower in frequency at the same loudness?

3 **a** How does the note produced by a guitar string change if the string is
      i shortened   **ii** tightened?
   b Compare the sound produced by a violin with the sound produced by a drum.

Key points

- The pitch of a note increases if the frequency of the sound waves increases.

- The loudness of a note increases if the amplitude of the sound waves increases.

- Vibrations created in an instrument when it is played produce sound waves.

Summary questions 🅚

1 Figure 1 shows an incomplete ray diagram of image formation by a plane mirror.

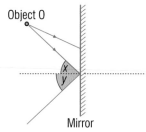

Figure 1

a What can you say about the angles *x* and *y* in the diagram?

b Complete the ray diagram to locate the image.

c What can you say about the distance from the image to the mirror compared with the distance from the object to the mirror?

2 a Figure 2 shows a light ray directed into a glass block.

Figure 2

i Sketch the path of the light ray through the block.

ii Describe how the direction of the light ray changes as it passes into and out of the block.

b Copy and complete **i** to **iii** using the words below:

diffraction   reflection   refraction

i The change of the direction of a light ray when it enters a glass block from air is an example of ....

ii The spreading of waves when they pass through a gap is an example of ....

iii The image of an object seen in a mirror is formed because the mirror causes light from the object to undergo ....

3 Copy and complete **a** to **c** using the words below. Each word can be used more than once.

light   radio   sound

a .... waves and .... waves travel at the same speed through air.

b .... waves are longitudinal waves.

c .... waves cannot travel through a vacuum.

4 Waves travel a distance of 30 m across a pond in 10 seconds. The waves have a wavelength of 1.5 m.

a Calculate the speed of the waves.

b Show that the frequency of the waves is 2.0 Hz.

5 a A loudspeaker is used to produce sound waves. In terms of the amplitude of the sound waves, explain why the sound is fainter further away from the loudspeaker.

b A microphone is connected to an oscilloscope. Figure 3 shows the display on the screen of the oscilloscope when the microphone detects sound waves from a loudspeaker.

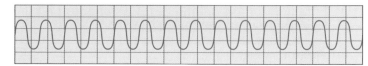

Figure 3

Describe how the waveform displayed on the oscilloscope screen changes if the sound from the loudspeaker is

i made louder

ii reduced in pitch.

6 Copy and complete **a** to **c** using the words below.

absorbed   reflected   scattered   smooth   soft   rough

a An echo is due to sound waves that are .... from a .... wall.

b When sound waves are directed at a .... surface, they are broken up and ....

c When sound waves are directed at a wall covered with a .... material, they are .... and not reflected.

7 a What is the highest frequency the human ear can hear?

b A sound meter is used to measure the loudness of the sound reflected from an object. Describe how you would use the meter and the arrangement shown in Figure 4 to test if more sound is reflected from a board than from a cushion in place of the board. The control knob and a frequency dial can be used to change the loudness and the frequency of the sound from the loudspeaker. List the variables that you would need to keep constant in your test.

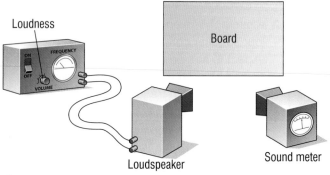

Figure 4

Practice questions

1 Draw labelled diagrams to explain what is meant by

  a a transverse wave (2)

  b a longitudinal wave. (2)

2 Match the words in the list with the descriptions **1** to **4** in the table.

  A amplitude
  B frequency
  C wave speed
  D wavelength

	Description
1	The distance travelled by a wave crest every second.
2	The distance from one crest to the next.
3	The height of the wave crest from the rest position.
4	The number of crests passing a fixed point every second.

  (4)

3 Which of the following is a correct description of the image in a plane mirror?

  A It is a virtual image
  B It can be focused on to a screen
  C It is on the surface of the mirror
  D It is upside down (1)

4 When a ray of light passes from air into glass it usually changes direction.

  a What is the name given to this effect? (1)

  b Which diagram correctly shows what happens to a ray of light as it passes through a glass block?

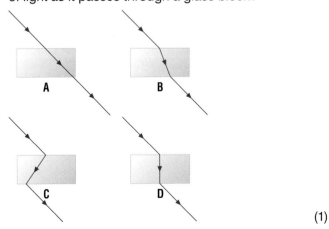

  (1)

5 The diagram represents some water waves passing through a narrow gap.

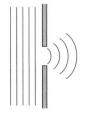

  Give the name of the effect being shown by the waves. When is it most significant? (2)

6 Give one similarity and one difference between a sound wave and a light wave. (2)

7 A sound wave in air has a frequency of 256 Hz. The wavelength of the wave is 1.3 m.

  Calculate the speed of sound in air. Write down the equation you use. Show clearly how you work out your answer and give the unit. (2)

8 a Give **one** example of each of the following from everyday life.

    i reflection of light (1)
    ii reflection of sound (1)
    iii refraction of light (1)
    iv diffraction of sound (1)

  b We do not normally see diffraction of light in everyday life.

    Suggest a reason for this. (2)

9 Electromagnetic waves travel at a speed of 300 000 000 m/s.

  BBC Radio 4 is transmitted using a wavelength of 1500 metres.

  Calculate the frequency of these waves.

  Write down the equation you use. Show clearly how you work out your answer and give the unit. [H] (3)

10 *In this question you will be assessed on using good English, organising information clearly and using specialist terms where appropriate.*

  The diagram shows an oscilloscope trace of the sound wave produced by a musical instrument.

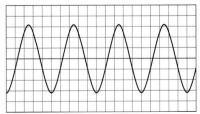

  Explain, in detail, how the waveform would change if the instrument produced a sound which was louder and at a higher pitch. (6)

P1 6.1

The electromagnetic spectrum

Learning objectives

- What are the parts of the electromagnetic spectrum?

- How can we calculate the frequency or wavelength of electromagnetic waves?

We all use waves from different parts of the **electromagnetic spectrum**. Figure 1 shows the spectrum and some of its uses.

Electromagnetic waves are electric and magnetic disturbances that transfer energy from one place to another.

Electromagnetic waves do not transfer matter. The energy they transfer depends on the **wavelength** of the waves. This is why waves of different wavelengths have different effects. Figure 1 shows some of the uses of each part of the electromagnetic spectrum.

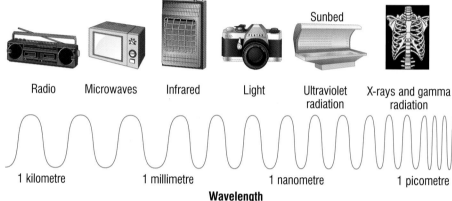

| Radio | Microwaves | Infrared | Light | Ultraviolet radiation | X-rays and gamma radiation |

1 kilometre   1 millimetre   1 nanometre   1 picometre

Wavelength

(1 nanometre = 0.000 001 millimetres, 1 picometre = 0.001 nanometres)

Figure 1 The spectrum is continuous. The frequencies and wavelengths at the boundaries are approximate as the different parts of the spectrum are not precisely defined.

Waves from different parts of the electromagnetic spectrum have different wavelengths.

- Long-wave radio waves have wavelengths as long as 10 km.
- X-rays and gamma rays have wavelengths as short as a millionth of a millionth of a millimetre (= 0.000 000 000 001 mm).

a Where in the electromagnetic spectrum would you find waves of wavelength 10 millimetres?

Study tip

The spectrum of visible light covers just a very tiny part of the electromagnetic spectrum. The wavelength decreases from radio waves to gamma rays.

The speed of electromagnetic waves

All electromagnetic waves travel at a speed of 300 million m/s through space or in a vacuum. This is the distance the waves travel each second.

We can link the speed of the waves to their frequency and wavelength using the **wave speed** equation:

$$v = f \times \lambda$$

Where:
v = wave speed in metres per second, m/s
f = frequency in hertz, Hz
λ = wavelength in metres, m

⬭ links

For more information on the wave speed equation, look back at P1 5.2 Measuring waves.

b Work out the wavelength of electromagnetic waves of frequency 200 million Hz.

c Work out the frequency of electromagnetic waves of wavelength 1500 m.

Higher

 Maths skills

We can work out the wavelength if we know the frequency and the wave speed. To do this, we rearrange the equation into:

$$\lambda = \frac{v}{f}$$

We can work out the frequency if we know the wavelength and the wave speed. To do this, we rearrange the equation into:

$$f = \frac{v}{\lambda}$$

Where:
v = speed in metres per second, m/s
f = frequency in hertz, Hz
λ = wavelength in metres, m.

Worked example

A mobile phone gives out electromagnetic waves of frequency 900 million Hz. Calculate the wavelength of these waves.
The speed of electromagnetic waves in air = 300 million m/s.

Solution

$$\text{wavelength } \lambda \text{ (in metres)} = \frac{\text{wave speed } v \text{ (in m/s)}}{\text{frequency } f \text{ (in Hz)}} =$$

$$\frac{300\,000\,000 \text{ m/s}}{900\,000\,000 \text{ Hz}} = 0.33 \text{ m}$$

Energy and frequency

The wave speed equation shows us that the shorter the wavelength of the waves, the higher their frequency is. The energy of the waves increases as the frequency increases. The energy and frequency of the waves therefore increases from radio waves to gamma rays as the wavelength decreases.

Summary questions

1 Copy and complete **a** to **c** using the words below:

greater than smaller than the same as

   a The wavelength of light waves is .... the wavelength of radio waves.

   b The speed of radio waves in a vacuum is .... the speed of gamma rays.

   c The frequency of X-rays is .... the frequency of infrared radiation.

2 Fill in the missing parts of the electromagnetic spectrum in the list below.

   radio ...a... infrared visible ...b... X-rays ...c...

3 Electromagnetic waves travel through space at a speed of 300 million metres per second. Calculate:
   a the wavelength of radio waves of frequency 600 million Hz
   b the frequency of microwaves of wavelength 0.30 m.

4 A distant star explodes and emits light and gamma rays simultaneously. Explain why the gamma rays and the light waves reach the Earth at the same time.

Key points

- The electromagnetic spectrum (in order of decreasing wavelength, increasing frequency and energy) is:
  – radio waves
  – microwaves
  – infrared radiation
  – light
  – ultraviolet radiation
  – gamma radiation and X-rays.

- The wave speed equation is used to calculate the frequency or wavelength of electromagnetic waves.

P1 6.2

Light, infrared, microwaves and radio waves

Learning objectives

- What is white light?
- What do we use infrared radiation, microwaves and radio waves for?
- What are the hazards of these types of electromagnetic radiation?

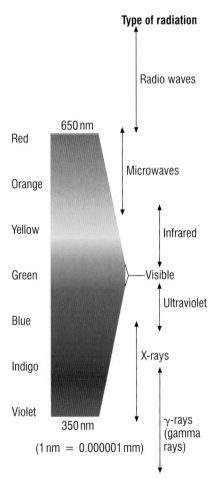

Figure 1 The electromagnetic spectrum with an expanded view of the visible range

Type of radiation

Radio waves

650 nm

Red

Orange

Microwaves

Yellow

Infrared

Green
Visible

Ultraviolet

Blue

X-rays

Indigo

Violet

350 nm
γ-rays (gamma rays)

(1 nm = 0.000001 mm)

Light and colour

Light from ordinary lamps and from the Sun is called **white light**. This is because it has all the colours of the visible spectrum in it. The wavelength increases across the spectrum as you go from violet to red.

You see the colours of the spectrum when you look at a rainbow. You can also see them if you use a glass prism to split a beam of white light.

Photographers need to know how shades and colours of light affect the photographs they take.

1 **In a film camera**, the light is focused by the camera lens on to a light-sensitive film. The film then needs to be developed to see the image of the objects that were photographed.

2 **In a digital camera**, the light is focused by the lens on to a sensor. This consists of thousands of tiny light-sensitive cells called **pixels**. Each pixel gives a dot of the image. The image can be seen on a small screen at the back of the camera. When a photograph is taken, the image is stored electronically on a memory card.

   a Why is a 10 million pixel camera better than a 2 million pixel camera?

Infrared radiation

All objects emit infrared radiation.

- The hotter an object is, the more infrared radiation it emits.
- Infrared radiation is absorbed by the skin. It damages or kills skin cells because it heats up the cells.

   b Where does infrared radiation lie in the electromagnetic spectrum?

Infrared devices

- **Optical fibres** in communications systems use infrared radiation instead of light. This is because infrared radiation is absorbed less than light in the glass fibres.
- **Remote control handsets** for TV and video equipment transmit signals carried by infrared radiation. When you press a button on the handset, it sends out a sequence of infrared pulses.
- **Infrared scanners** are used in medicine to detect 'hot spots' on the body surface. These hot areas can mean the underlying tissue is unhealthy.
- You can use **infrared cameras** to see people and animals in darkness.

   c Does infrared radiation pass through a thin sheet of paper?

⚭ links

For more information on infrared radiation, look back at P1 1.1 Infrared radiation.

Microwaves

Microwaves lie between radio waves and infrared radiation in the electromagnetic spectrum. They are called '**micro**waves' because they are shorter in wavelength than radio waves.

We use microwaves for communications, e.g. **satellite TV**, because they can pass through the atmosphere and reach satellites above the Earth. We also use them to beam signals from one place to another. That's because microwaves don't spread out as much as radio waves. Microwaves (as well as radio waves) are used to carry **mobile phone** signals.

Radio waves

Radio wave frequencies range from about 300 000 Hz to 3000 million Hz (where microwave frequencies start). Radio waves are longer in wavelength and lower in frequency than microwaves.

As explained in P1 6.3, we use radio waves to carry **radio, TV and mobile phone** signals.

We can also use radio waves instead of cables to connect a computer to other devices such as a printer or a 'mouse'. For example, Bluetooth-enabled devices can communicate with each other over a range of about 10 metres. No cables are needed – just a Bluetooth radio in each device and the necessary software. Such wireless connections work at frequencies of about 2400 million hertz, and they operate at low power.

Bluetooth was set up by the electronics manufacturers. They realised the need to agree on the radio frequencies to be used for common software.

> **d** If wireless-enabled devices operated at higher power, how would their range be affected?

Summary questions

1 Copy and complete **a** and **b** using the words below:

infrared radiation   visible light   microwaves   radio waves

  a In a TV set, the aerial detects .... and the screen emits ....
  b A satellite TV receiver detects ...., which pass through the atmosphere, unlike ...., which have a shorter wavelength.

2 Mobile phones use electromagnetic waves in a wavelength range that includes short-wave radio waves and microwaves.
  a What would be the effect on mobile phone users if remote control handsets operated in this range as well?
  b Why do our emergency services use radio waves in a wavelength range that no else is allowed to use?

3 The four devices listed below each emit a different type of electromagnetic radiation. State the type of radiation each one emits.
  a A TV transmitter mast.
  b A TV satellite.
  c A TV remote handset.
  d A TV receiver.

Practical

Testing infrared radiation

Can infrared radiation pass through paper? Use a remote handset to find out.

Demonstration

Demonstrating microwaves

Look at the demonstration.
● What does this show?

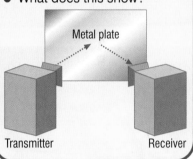

Metal plate

Transmitter   Receiver

Key points

● White light contains all the colours of the visible spectrum.

● Infrared radiation is used for carrying signals from remote handsets and inside optical fibres.
We use microwaves to carry satellite TV programmes and mobile phone calls.
Radio waves are used for radio and TV broadcasting, radio communications and mobile phone calls.

● Different types of electromagnetic radiation are hazardous in different ways. Microwaves and radio waves can cause internal heating. Infrared radiation can cause skin burns.

P1 6.3

Communications

Learning objectives

- Why do we use radio waves of different frequencies for different purposes?

- Which waves do we use for satellite TV?

- How can we evaluate whether or not mobile phones are safe to use?

- What are optical fibres?

Figure 1 Sending microwave signals to a satellite

??? Did you know ... ?

Satellite TV signals are carried by microwaves. We can detect the signals on the ground because they pass straight through a layer of ionised gas in the upper atmosphere. This layer reflects lower-frequency radio waves.

Figure 2 A mobile phone mast

Radio communications

Radio waves are emitted from an aerial when we apply an alternating voltage to the aerial. The frequency of the radio waves produced is the same as the frequency of the alternating voltage.

When the radio waves pass across a receiver aerial, they cause a tiny alternating voltage in the aerial. The frequency of the alternating voltage is the same as the frequency of the radio waves received. The aerial is connected to a loudspeaker. The alternating voltage from the aerial is used to make the loudspeaker send out sound waves.

The radio and microwave spectrum is divided into **bands** of different wavelength ranges. This is because the shorter the wavelength of the waves:

- the more information they can carry
- the shorter their range (due to increasing absorption by the atmosphere)
- the less they spread out (because they diffract less).

Radio wavelengths

Microwaves and radio waves of different wavelengths are used for different communications purposes. Examples are given below.

- **Microwaves** are used for satellite phone and TV links and satellite TV broadcasting. This is because microwaves can travel between satellites in space and the ground. Also, they spread out less than radio waves do so the signal doesn't weaken as much.

- **Radio waves of wavelengths less than about 1 metre** are used for TV broadcasting from TV masts because they can carry more information than longer radio waves.

- **Radio waves of wavelengths from about 1 metre up to about 100 m** are used by local radio stations (and for the emergency services) because their range is limited to the area round the transmitter.

- **Radio waves of wavelengths greater than 100 m** are used by national and international radio stations because they have a much longer range than shorter wavelength radio waves.

a Why do microwaves spread out less than radio waves do?

Mobile phone radiation

A mobile phone sends a radio signal from your phone. The signal is picked up by a local mobile phone mast and is sent through the phone network to the other phone. The 'return' signal goes through the phone network back to the mobile phone mast near you and then on to you. The signals to and from your local mast are carried by radio waves of different frequencies.

The radio waves to and from a mobile phone have a wavelength of about 30 cm. Radio waves at this wavelength are not quite in the microwave range but they do have a similar heating effect to microwaves. So they are usually referred to as microwaves.

b Why should signals to and from a mobile phone be at different frequencies?

How Science Works

Is mobile phone radiation dangerous?

The radiation is much weaker than the microwave radiation in an oven. But when you use a mobile phone, it is very close to your brain. Some scientists think the radiation might affect the brain. As children have thinner skulls than adults, their brains might be more affected by mobile phone radiation. A UK government report published in May 2000 recommended that the use of mobile phones by children should be limited.

Mobile phone hazards

Here are some findings by different groups of scientists:

The short-term memory of volunteers using a mobile phone was found to be unaffected by whether the phone was on or off.

The brains of rats exposed to microwaves were found to respond less to electrical impulses than the brains of unexposed rats.

Mice exposed to microwaves by some scientists developed more cancers than unexposed mice. Other scientists were unable to confirm this effect.

A survey of mobile phone users in Norway and Sweden found they experienced headaches and fatigue. No control group of people who did not use a mobile phone was surveyed.

- What conclusions do you draw from the evidence above?
- Suggest how researchers could improve the validity of any conclusions we can draw.

Optical fibre communications

Optical fibres are very thin glass fibres. We use them to transmit signals carried by light or infrared radiation. The light rays can't escape from the fibre. When they reach the surface of the fibre, they are reflected back into the fibre.

In comparison with radio waves and microwaves:

- optical fibres can carry much more information – this is because light has a much smaller wavelength than radio waves so can carry more pulses of waves
- optical fibres are more secure because the signals stay in the fibre.

c   Why are signals in an optical fibre more secure than radio signals?

Summary questions

1   Copy and complete **a** to **c** using the words below. Each term can be used more than once.

infrared radiation   microwaves   radio waves

   a Mobile phone signals are carried by ....
   b Optical fibre signals are carried by ....
   c A beam of .... can travel from the ground to a satellite but a beam of .... cannot if its frequency is below 30 MHz.

2   **a** Why could children be more affected by mobile phone radiation than adults?
   b Why can light waves carry more information than radio waves?

3   Explain why microwaves are used for satellite TV and radio waves for terrestrial TV.

Demonstration

Demonstrating an optical fibre

Observe light shone into an optical fibre. You should see the reflection of light inside an optical fibre. This is known as total internal reflection.

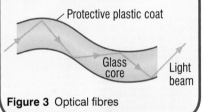

Figure 3 Optical fibres

Key points

- Radio waves of different frequencies are used for different purposes because the wavelength (and therefore frequency) of waves affects:
  - how far they can go
  - how much they spread
  - how much information they can carry.

- Microwaves are used for satellite TV signals.

- Further research is needed to evaluate whether or not mobile phones are safe to use.

- Optical fibres are very thin transparent fibres that are used to transmit signals by light and infrared radiation.

P1 6.4

The expanding universe

Learning objectives

- What do we mean by red-shift of a light source?

- How does red-shift depend on speed?

- How do we know the distant galaxies are moving away from us?

- Why do we think the universe is expanding?

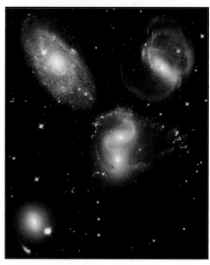

Figure 1 Galaxies

The Doppler effect ⓒ

The **Doppler effect** is the change in the observed wavelength (and frequency) of waves due to the motion of the source of the waves. Christian Doppler discovered the effect in 1842 using sound waves. He demonstrated it by using an open railway carriage filled with trumpeters. The spectators had to listen to the pitch of the trumpets as they sped past. Another example, explained below, is the red-shift of the light from a distant galaxy moving away from us.

Red-shift

We live on the third rock out from a middle-aged star on the outskirts of a big galaxy we call the Milky Way. The galaxy contains about 100 000 million stars. Its size is about 100 000 light years across. This means that light takes 100 000 years to travel across it. But it's just one of billions of galaxies in the universe. The furthest galaxies are about 13 000 million light years away!

a Why do stars appear as points of light?

We can find out lots of things about stars and galaxies by studying the light from them. We can use a prism to split the light into a spectrum. The wavelength of light increases across the spectrum from blue to red. We can tell from its spectrum if a star or galaxy is moving towards us or away from us. This is because:

- the light waves are stretched out if the star or galaxy is moving away from us. The wavelength of the waves is increased. We call this a **red-shift** because the spectrum of light is shifted towards the red part of the spectrum.
- the light waves are squashed together if the star or galaxy is moving towards us. The wavelength of the waves is reduced. We call this a **blue-shift** because the spectrum of light is shifted towards the blue part of the spectrum.

The dark spectral lines shown in Figure 2 are caused by absorption of light by certain atoms such as hydrogen that make up a star or galaxy. The position of these lines tells us if there is a shift and if so, whether it is a red-shift or a blue-shift.

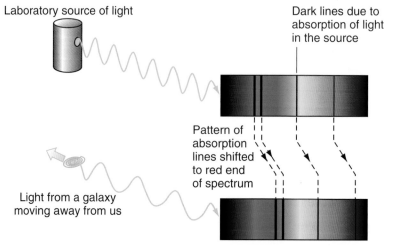

Figure 2 Red-shift

The bigger the shift, the more the waves are squashed together or stretched out. So the faster the star or galaxy must be moving towards or away from us. In other words:

the faster a star or galaxy is moving (relative to us), the bigger the shift is.

> **b** What do you think happens to the wavelength of the light from a star that is moving towards us?

Expanding universe

In 1929, Edwin Hubble discovered that:

1 the light from distant galaxies was red-shifted

2 the further a galaxy is from us, the bigger its red-shift is.

He concluded that:

- the distant galaxies are moving away from us (i.e. receding)
- the greater the distance a galaxy is from us, the greater the speed is at which it is moving away from us (its speed of recession).

Why should the distant galaxies be moving away from us? We have no special place in the universe, so all the distant galaxies must be moving away from each other. In other words, **the whole universe is expanding**.

> **c** Galaxy X is 2000 million light years away. Galaxy Y is 4000 million light years away. Which galaxy, X or Y, has the bigger red-shift?

??? Did you know …?

You can hear the Doppler effect when an ambulance with its siren on goes speeding past.

- As it approaches, the sound waves it sends out are squashed up so their frequency is higher (and the wavelength shorter) than if the siren was stationary. So you hear a higher pitch.

- As it travels away from you, the sound waves it sends out are stretched out so their frequency is lower (and the wavelength longer) than if the siren was stationary. So you hear a lower pitch.

Summary questions

1 Copy and complete **a** to **d** using the words below:

approaching expanding orbiting receding

  a The Earth is .... the Sun.

  b The universe is ....

  c The distant galaxies are ....

  d A blue-shift in the light from a star would tell us it is ....

2 a Put these objects in order of increasing size:

    Andromeda galaxy Earth Sun universe

  b Copy and complete **i** and **ii** using the words below:

    galaxy star red-shift planet

    i The Earth is a .... in orbit round a .... called the Sun.

    ii There is a .... in the light from a distant ....

3 Galaxy X has a larger red-shift than galaxy Y.

  a Which galaxy, X or Y, is

    i nearer to us

    ii moving away faster?

  b The light from the Andromeda galaxy is not red-shifted. What does this tell you about Andromeda?

Key points

- The red-shift of a distant galaxy is the shift to longer wavelengths of the light from it because the galaxy is moving away from us.

- The faster a distant galaxy is moving away from us, the greater its red-shift is.

- All the distant galaxies show a red-shift. The further away a distant galaxy is from us, the greater its red-shift is.

- The distant galaxies are all moving away from us because the universe is expanding.

P1 6.5 | The Big Bang ⓚ

Learning objectives

- What is the Big Bang theory of the universe?
- Why is the universe expanding?
- What is cosmic microwave background radiation?
- What evidence is there that the universe was created in a Big Bang?

Figure 1 The Big Bang

??? Did you know ... ?

You can use an analogue TV to detect background microwave radiation very easily – just disconnect your TV aerial. The radiation causes lots of fuzzy spots on the screen.

The universe is expanding, but what is making it expand? The **Big Bang theory** was put forward to explain the expansion. This states that:

- the universe is expanding after exploding suddenly in a Big Bang from a very small initial point
- space, time and matter were created in the Big Bang.

Many scientists disagreed with the Big Bang theory. They put forward an alternative theory, the Steady State theory. The scientists said that the galaxies are being pushed apart. They thought that this is caused by matter entering the universe through 'white holes' (the opposite of black holes).

Which theory is weirder – everything starting from a Big Bang or matter leaking into the universe from outside? Until 1965, most people supported the Steady State theory.

⚙️ How Science Works

Evidence for the Big Bang

Scientists had two conflicting theories about the evolution of the universe: it was in a Steady State or it began at some point in the past with a Big Bang. Both theories could explain why the galaxies are moving apart, so scientists needed to find some way of selecting which theory was correct. They worked out that if the universe began in a Big Bang then there should have been high-energy electromagnetic radiation produced. This radiation would have 'stretched' as the universe expanded and become lower-energy radiation. Experiments were devised to look for this trace energy as extra evidence for the Big Bang model.

It was in 1965 that scientists first detected microwaves coming from every direction in space. The existence of this **cosmic microwave background radiation** can only be explained by the Big Bang theory.

The cosmic microwave background radiation is not as perfectly evenly spread as scientists thought it should be. Their model of the early universe needs to be developed further by gathering evidence and producing theories to explain this 'unevenness' in the early universe.

a How do scientists decide between two conflicting theories?

Cosmic microwave background radiation

- It was created as high-energy gamma radiation just after the Big Bang.
- It has been travelling through space since then.
- As the universe has expanded, it stretched out to longer and longer wavelengths and is now microwave radiation.
- It has been mapped out using microwave detectors on the ground and on satellites.

b What will happen to cosmic microwave background radiation as the universe expands?

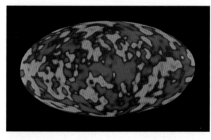

Figure 2 A microwave image of the universe from, the Cosmic Background Explorer satellite

How Science Works

The future of the universe

Will the universe expand forever? Or will the force of gravity between the distant galaxies stop them from moving away from each other? The answer to this question depends on their total mass and how much space they take up – in other words, the density of the universe.

● If the density of the universe is less than a certain amount, it will expand forever. The stars will die out and so will everything else as the universe heads for a Big Yawn!

● If the density of the universe is more than a certain amount, it will stop expanding and go into reverse. Everything will head for a Big Crunch!

Recent observations by astronomers suggest that the distant galaxies are accelerating away from each other. These observations have been checked and confirmed by other astronomers. So astronomers have concluded that the expansion of the universe is accelerating. It could be we're in for a Big Ride followed by a Big Yawn.

The discovery that the distant galaxies are accelerating is puzzling astronomers. Scientists think some unknown source of energy, now called 'dark energy', must be causing this accelerating motion. The only known force on the distant galaxies, the force of gravity, can't be used to explain 'dark energy' as it is an attractive force and so acts against their outward motion away from each other.

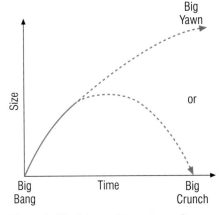

Figure 3 The future of the universe?

c What could you say about the future of the universe if the galaxies were slowing down?

d i An object released above the ground accelerates as it falls. What makes it accelerate?

ii Why are scientists puzzled by the observation that the distant galaxies are accelerating?

Summary questions

1 Copy and complete **a** to **d** using the words below:

created   detected   expanded   stretched

a The universe was .... in an explosion called the Big Bang.
b The universe .... suddenly in and after the Big Bang.
c Microwave radiation from space can be .... from all directions.
d Radiation created just after the Big Bang has been .... by the expansion of the universe and is now microwave radiation.

2 Put the following events A–D in the correct time sequence:
A The distant galaxies were created.
B Cosmic microwave background radiation was first detected.
C The Big Bang happened.
D The expansion of the universe began.

3 a Why do astronomers think that the expansion of the universe is accelerating?
b What would have been the effect on the expansion of the universe if its density had been greater than a certain value?

Key points

● The universe started with the Big Bang, a massive explosion from a very small point.

● The universe has been expanding ever since the Big Bang.

● Cosmic microwave background radiation (CMBR) is electromagnetic radiation created just after the Big Bang.

● CMBR can only be explained by the Big Bang theory.

Summary questions *k*

1 a Place the four different types of electromagnetic waves listed below in order of increasing wavelength.
   A Infrared waves
   B Microwaves
   C Radio waves
   D Gamma rays

b The radio waves from a local radio station have a wavelength of 3.3 metres in air and a frequency of 91 million Hz.
   i Write down the equation that links frequency, wavelength and wave speed.
   ii Calculate the speed of the radio waves in air.

2 In P1 6.1 you will find the typical wavelengths of electromagnetic waves. Give the type of electromagnetic wave for each of the wavelengths given.
   A 0.0005 mm
   B 1 millionth of 1 millionth of 1 mm
   C 10 cm
   D 1000 m

3 Copy and complete **a** and **b** using the words below:

microwave   mobile phone   radio waves   TV

a A .... beam can travel from a ground transmitter to a satellite, but a beam of .... cannot if its frequency is below 30 MHz.

b .... signals and .... signals always come from a local transmitter.

4 Mobile phones send and receive signals using electromagnetic waves near or in the microwave part of the electromagnetic spectrum.

a Name the part of the electromagnetic spectrum which has longer wavelengths than microwaves have.

b Which two parts of the electromagnetic spectrum may be used to send information along optical fibres?

c New mobile phones are tested for radiation safety and given an SAR value before being sold. The SAR is a measure of the energy per second absorbed by the head while the phone is in use. For use in the UK, SAR values must be less than 2.0 W/kg. SAR values for two different mobile phones are given below.
   Phone A   0.2 W/kg
   Phone B   1.0 W/kg
   i What is the main reason why mobile phones are tested for radiation safety?
   ii Which phone, A or B, is safer? Give a reason for your answer.

5 Light from a distant galaxy has a change of wavelength due to the motion of the galaxy.

   a Is this change of wavelength an increase or a decrease?

   b What is the name for this change of wavelength?

   c Which way is the galaxy moving?

   d What would happen to the light it gives out if it were moving in the opposite direction?

6 a Galaxy A is further from us than galaxy B.
   i Which galaxy, A or B, produces light with a greater red-shift?
   ii Galaxy C gives a bigger red-shift than galaxy A. What can we say about the distance to galaxy C compared with galaxy A?

b All the distant galaxies are moving away from each other.
   i What does this tell us about the universe?
   ii What does it tell us about our place in the universe?

7 The diagram shows two galaxies X and Y, which have the same diameter.

a i Which galaxy, X or Y, is further from Earth? Give a reason for your answer.
   ii Which galaxy, X or Y, produces the larger red-shift?

b A third galaxy Z seen from Earth appears to be the same size as X but it has a larger red-shift than X.
   i What can you say about the speed at which Z is moving away from us, compared with the speed at which X is moving away?
   ii What can you deduce about the distance to Z compared with the distance to X? Give a reason for your answer.

Practice questions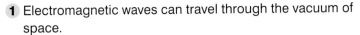

1 Electromagnetic waves can travel through the vacuum of space.

Copy and complete the following sentences using words from the list below. Each word can be used once, more than once or not at all.

energy   frequency   speed   wavelength

All electromagnetic waves travel at the same .... in a vacuum. They do not carry material, but they do carry .... Gamma waves have the greatest .... and the smallest .... (4)

2 Different types of electromagnetic waves have different uses in communications.

Match the type of electromagnetic wave in the list with its use **1** to **4** in the table.

A infrared

B microwaves

C radio waves

D visible light

	These waves are used for
1	producing images in a video camera
2	mobile phone communication
3	television remote controls
4	carrying terrestrial television signals

(4)

3 Microwaves are used for communications. They can be used to send signals to other parts of the world by means of a satellite.

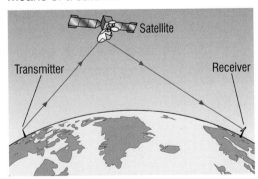

a Give **one** reason why the receiver shown in the diagram can only pick up the signal if a satellite is used. (1)

b Explain why microwaves are used rather than:

   i long wave radio waves (1)

   ii visible light (2)

4 The diagram shows a ray passing through an optical fibre.

a Name **two** types of electromagnetic wave that can travel along an optical fibre. (2)

b Suggest **two** advantages of sending signals along an optical fibre rather than using electrical signals in a metal wire. (2)

5 Scientists have developed a theory about the universe called the 'Big Bang' theory. This theory is supported by evidence. Part of this evidence is the existence of cosmic background microwave radiation.

a What does the 'Big Bang' theory state?

   A The universe began with a massive explosion.

   B The universe will end with a massive explosion.

   C The universe began from a very small initial point.

   D The universe will end at a very small initial point. (1)

b Where does cosmic background microwave radiation come from?

   A people who use microwave ovens to heat food

   B gamma radiation created just after the Big Bang

   C mobile phone transmitters

   D radioactive rocks in the Earth's crust. (1)

c If a scientist finds new evidence that does not support the Big Bang theory what should other scientists do?

   A Change the theory immediately.

   B Check the new evidence to make sure it is reproducible.

   C Ignore the new evidence.

   D Try to discredit the scientist who found the new evidence. (1)

6 Red-shift from distant galaxies provides evidence for the Big Bang theory.

What is meant by red-shift? (2)

7 *In this question you will be assessed on using good English, organising information clearly and using specialist terms where appropriate.*

Explain how red-shift provides evidence for the Big Bang theory. (6)

1   The diagram shows a solar heating panel on the roof of a house.

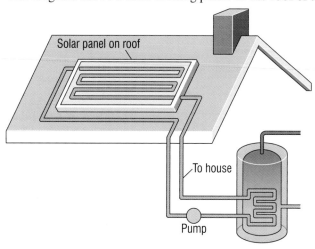

The solar heating panel consists of a flat box backed by a metal plate. The box contains copper pipes filled with a liquid. Liquid pumped through the pipes is heated as it passes through the panel.

Copy and complete the following sentences using words from the list below. Each word can be used once, more than once or not at all.

black   white   transparent   conduction   convection   insulation   radiation

There is a .... cover on top of the panel that allows infrared .... through to heat the metal plate. The metal plate is coloured .... for maximum absorption. There is a sheet of .... under the plate to stop heat loss by .... through the back of the panel.

(5)

2   a   Microwaves are one type of electromagnetic wave.
      i   Which type of electromagnetic wave has a lower frequency than microwaves? (1)
      ii What do all types of electromagnetic wave transfer from one place to another?

(1)

   b   The picture shows a tennis coach using a speed gun to measure how fast the player serves the ball.

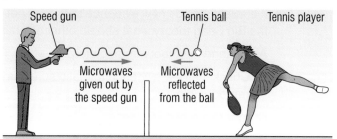

      i   The microwaves transmitted by the speed gun have a frequency of 2.4×10^{10} Hz and travel through the air at 3.0×10^8 m/s.
          Calculate the wavelength in metres of the microwaves emitted from the speed gun.
          Write down the equation you use. Show clearly how you work out your answer and give the unit. (2)
      ii Some of the microwaves transmitted by the speed gun are absorbed by the ball.
          What effect will the absorbed microwaves have on the ball? (2)
      iii Some of the microwaves transmitted by the speed gun are reflected from the moving ball back towards the speed gun. Describe how the wavelength and frequency of the microwaves change as they are reflected from the moving ball.

(2)

AQA, 2009

Study tip

Read through the whole passage first, to get the sense of it, before trying to put the words in.

Study tip

Make sure you have a way of remembering the order of the waves in the electromagnetic spectrum. For example, **g**ood **x**ylophones **u**pset **v**iolins **i**n **m**usical **r**ecitals for **g**amma, **X**-ray, **u**ltraviolet, **v**isible, **i**nfrared, **m**icrowave, **r**adio wave.

3 The diagram shows how electricity is distributed from power stations to consumers.

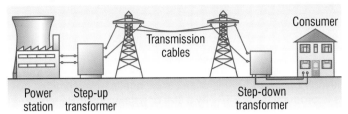

Consumer

Transmission cables

Power station | Step-up transformer | Step-down transformer

a i What name is given to the network of cables and transformers that links power stations to consumers? (1)

ii What does a step-up transformer do? (1)

iii Explain why step-up transformers are used in the electricity distribution system. (2)

b Most of the world's electricity is generated in power stations that burn fossil fuels. State **one** environmental problem that burning fossil fuels produces. (1)

c Electricity can be generated using energy from the wind. A company wants to build a new wind farm. Not everyone thinks that this is a good idea.

i What arguments could the company give to persuade people that a wind farm is a good idea? (2)

ii What reasons may be given by the people who think that wind farms are **not** a good idea? (2)

AQA, 2007

Study tip

There are pros and cons to the use of any source for generating electricity, even the renewable ones. Make sure you know what they are.

4 *In this question you will be assessed on using good English, organising information clearly and using specialist terms where appropriate.*

The diagram shows a vacuum flask. The flask can be used to keep hot liquids hot and cold liquids cold.

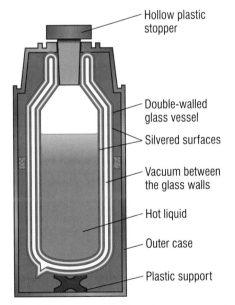

— Hollow plastic stopper

— Double-walled glass vessel

— Silvered surfaces

— Vacuum between the glass walls

— Hot liquid

— Outer case

— Plastic support

Explain how the flask reduces energy transfer by conduction, convection and radiation. (6)

Study tip

Make sure you understand that a vacuum is a completely empty space.

Glossary

A

Absorber A substance that takes in radiation.

Accurate A measurement is considered accurate if it is judged to be close to the true value.

Acid rain Rain that is acidic due to dissolved gases, such as sulfur dioxide, produced by the burning of fossil fuels.

Adaptation Special feature that makes an organism particularly well suited to the environment where it lives.

Adult cell cloning Process in which the nucleus of an adult cell of one animal is fused with an empty egg from another animal. The embryo which results is placed inside the uterus of a third animal to develop.

Agar The nutrient jelly on which many microorganisms are cultured.

Alkali metals The elements in Group 1 of the periodic table, for example, lithium (Li), sodium (Na), potassium (K).

Alkane Saturated hydrocarbon with the general formula C_nH_{2n+2}, for example, methane, ethane and propane.

Alkene Unsaturated hydrocarbon which contains a carbon–carbon double bond. The general formula is C_nH_{2n}, for example, ethene C_2H_4.

Alloy A mixture of metals (and sometimes non-metals). For example, brass is a mixture of copper and zinc.

Aluminium A low density, corrosion-resistant metal used in many alloys, including those used in the aircraft industry.

Amplitude The height of a wave crest or a wave trough of a transverse wave from the rest position.

Angle of incidence Angle between the incident ray and the normal.

Angle of reflection Angle between the reflected ray and the normal.

Anomalous results Results that do not match the pattern seen in the other data collected or are well outside the range of other repeat readings. They should be retested and if necessary discarded.

Antibiotics Drugs that destroy bacteria inside the body without damaging human cells.

Antigen Unique protein on the surface of a cell. They are recognised by the immune system as 'self' or 'non-self'.

Asexual reproduction Reproduction that involves only one individual with no fusing of gametes to produce the offspring. The offspring are identical to the parent.

Atmosphere The relatively thin layer of gases that surround planet Earth.

Atom The smallest part of an element that can still be recognised as that element.

Atomic nucleus Tiny positively charged object composed of protons and neutrons at the centre of every atom.

Atomic number The number of protons (which equals the number of electrons) in an atom. It is sometimes called the proton number.

Auxin A plant hormone that controls the responses of plants to light (phototropism) and to gravity (gravitropism).

B

Bacteria Single-celled microorganisms that can reproduce very rapidly. Many bacteria are useful, for example, gut bacteria and decomposing bacteria, but some cause disease.

Band Part of the radio and microwave spectrum used for communications.

Bar chart A chart with rectangular bars with lengths proportional to the values that they represent. The bars should be of equal width and are usually plotted horizontally or vertically. Also called a bar graph.

Base load Constant amount of electricity generated by power stations.

Big Bang theory The theory that the universe was created in a massive explosion (the Big Bang) and that the universe has been expanding ever since.

Biodegradable Materials that can be broken down by microorganisms.

Biodiesel Fuel for cars made from plant oils.

Biofuel Fuel made from animal or plant products.

Bioleaching Process of extraction of metals from ores using microorganisms.

Biomass Biological material from living or recently living organisms.

Blast furnace The huge reaction vessels used in industry to extract iron from its ore.

Blue-shift Decrease in the wavelength of electromagnetic waves emitted by a star or galaxy due to its motion towards us. The faster the speed of the star or galaxy, the greater the blue-shift is.

Boundary Line along which two substances meet.

C

Calcium carbonate The main compound found in limestone. It is a white solid whose formula is $CaCO_3$.

Calcium hydroxide A white solid made by reacting calcium oxide with water. It is used as a cheap alkali in industry.

Calcium oxide A white solid made by heating limestone strongly, for example, in a lime kiln.

Carbon capture and storage Capture and storage of carbon dioxide produced in fossil fuel power stations. Old gas or oil fields are suitable places to store the carbon dioxide.

Carbon cycle The cycling of carbon through the living and non-living world.

Carbon monoxide A toxic gas whose formula is CO.

Carbon steel Alloy of iron containing controlled, small amounts of carbon.

Carnivore Animal that eats other animals.

Cast iron The impure iron taken directly from a blast furnace.

Catalytic converter Fitted to exhausts of vehicles to reduce pollutants released.

Cement A building material made by heating limestone and clay.

Central nervous system (CNS) The central nervous system is made up of the brain and spinal cord where information is processed.

Charles Darwin The Victorian scientist who developed the theory of evolution by a process of natural selection.

Chemical energy Energy of an object due to chemical reactions in it.

Chromosome Thread-like structure carrying the genetic information found in the nucleus of a cell.

Clone Offspring produced by asexual reproduction which is identical to the parent organism.

Combustion The process of burning.

Competition The process by which living organisms compete with each other for limited resources such as food, light or reproductive partners.

Compost heap A site where garden rubbish and kitchen waste are decomposed by microorganisms.

Compound A substance made when two or more elements are chemically bonded together. For example, water (H_2O) is a compound made from hydrogen and oxygen.

Compression Squeezed together.

Concrete A building material made by mixing cement, sand and aggregate (crushed rock) with water.

Condense Turn from vapour into liquid.

Conduction Transfer of energy from particle to particle in matter.

Conductor Material/object that conducts.

Conservation of energy Energy cannot be created or destroyed.

Contraceptive pill A pill containing female sex hormones which is used to prevent conception.

Control group If an experiment is to determine the effect of changing a single variable, a control is often set up in which the independent variable is not changed, thus enabling a comparison to be made. If the investigation is of the survey type a control group is usually established to serve the same purpose.

Convection Transfer of energy by the bulk movement of a heated fluid.

Convection current The circular motion of matter caused by heating in fluids.

Copper-rich ore Rock that contains a high proportion of a copper compound.

Core The centre of the Earth.

Cosmic microwave background radiation Electromagnetic radiation that has been travelling through space ever since it was created shortly after the Big Bang.

Cost effectiveness How much something gives value for money when purchase, running and other costs are taken into account.

Covalent bond The attraction between two atoms that share one or more pairs of electrons.

Cracking The reaction used in the oil industry to break down large hydrocarbons into smaller, more useful ones. This occurs when the hydrocarbon vapour is either passed over a hot catalyst or mixed with steam and heated to a high temperature.

Crust The outer solid layer of the Earth.

Culture medium A substance containing the nutrients needed for microorganisms to grow.

D

Data Information, either qualitative or quantitative, that have been collected.

Decomposer Microorganism that breaks down waste products and dead bodies.

Denature Change the shape of an enzyme so that it can no longer speed up a reaction.

Density Mass per unit volume of a substance.

Depression A mental illness that involves feelings of great sadness that interfere with everyday life.

Detritus feeder See Decomposer.

Diffraction The spreading of waves when they pass through a gap or around the edges of an obstacle which has a similar size as the wavelength of the waves.

Diffusion Spreading out of particles away from each other.

Direct contact A way of spreading infectious diseases by skin contact between two people.

Directly proportional A relationship that, when drawn on a line graph, shows a positive linear relationship that crosses through the origin.

Displace When one element takes the place of another in a compound. For example, iron + copper sulfate → iron sulfate + copper.

Distillation Separation of a liquid from a mixture by evaporation followed by condensation.

DNA Deoxyribonucleic acid, the material of inheritance.

Doppler effect The change of wavelength (and frequency) of the waves from a moving source due to the motion of the source towards or away from the observer.

Double-blind trial A drug trial in which neither the patient nor the doctor knows if the patient is receiving the new drug or a placebo.

Double bond A covalent bond made by the sharing of two pairs of electrons.

Droplet infection A way of spreading infectious diseases through the tiny droplets full of pathogens, which are expelled from your body when you cough, sneeze or talk.

Drug A chemical which causes changes in the body. Medical drugs cure disease or relieve symptoms. Recreational drugs alter the state of your mind and/or body.

E

E number Number assigned to food additive that has been approved for use in Europe. These are displayed on food packaging.

Echo Reflection of sound that can be heard.

Effective medicine A medicine that cures the disease it is targeting.

Effector organ Muscle and gland that responds to impulses from the nervous system.

Efficiency Useful energy transferred by a device ÷ total energy supplied to the device.

Elastic potential energy Energy stored in an elastic object when work is done to change its shape.

Electrical appliance Machine powered by electricity.

Electrical energy Energy transferred by the movement of electrical charge.

Electricity meter Meter in a home that measures the amount of electrical energy supplied.

Electrolysis The breakdown of a substance containing ions by electricity.

Electromagnetic spectrum A set of radiations that have different wavelengths and frequencies but all travel at the same speed in a vacuum.

Electromagnetic wave Electric and magnetic disturbance that transfers energy from one place to another. The spectrum of electromagnetic waves, in order of increasing wavelength, is as follows: gamma and X-rays, ultraviolet radiation, visible light, infrared radiation, microwaves, radio waves.

Electron A tiny particle with a negative charge. Electrons orbit the nucleus in atoms or ions.

Electronic structure A set of numbers to show the arrangement of electrons in their shells (or energy levels), for example, the electronic structure of a potassium atom is 2, 8, 8, 1.

Element A substance made up of only one type of atom. An element cannot be broken down chemically into any simpler substance.

Emit Give out radiation.

Emitter A substance that gives out radiation.

Emulsifier A substance which helps keep immiscible liquids (for example, oil and water) mixed so that they do not separate out into layers.

Emulsion A mixture of liquids that do not dissolve in each other.

Energy level see Shell.

Energy transfer Movement of energy from one place to another or one form to another.

Enzyme Protein molecule that acts as a biological catalyst.

Epidemic When more cases of an infectious disease are recorded than would normally be expected.

Errors Sometimes called uncertainties.

Error – human Often present in the collection of data, and may be random or systematic. For example, the effect of human reaction time when recording short time intervals with a stopwatch.

Error – random Causes readings to be spread about the true value, due to results varying in an unpredictable way from one measurement to the next. Random errors are present when any measurement is made, and cannot be corrected. The effect of random errors can be reduced by making more measurements and calculating a new mean.

Error – systematic Causes readings to be spread about some value other than the true value, due to results differing from the true value by a consistent amount each time a measurement is made. Sources of systematic error can include the environment, methods of observation or instruments used. Systematic errors cannot be dealt with by simple repeats. If a systematic error is suspected, the data collection should be repeated using a different technique or a different set of equipment, and the results compared.

Error – zero Any indication that a measuring system gives a false reading when the true value of a measured quantity is zero, for example, the needle on an ammeter failing to return to zero when no current flows.

Ethene An alkene with the formula C_2H_4.

Evaporate Turn from liquid into vapour.

Evidence Data which have been shown to be valid.

Evolution The process of slow change in living organisms over long periods of time as those best adapted to survive breed successfully.

Evolutionary relationship Model of the relationships between organisms, often based on DNA evidence, which suggest how long ago they evolved away from each other and how closely related they are in evolutionary terms.

Evolutionary tree Model of the evolutionary relationships between different organisms based on their appearance, and increasingly, on DNA evidence.

Extremophile Organism which lives in environments that are very extreme, for example, very high or very low temperatures, high salt levels or high pressures.

F

Fair test A fair test is one in which only the independent variable has been allowed to affect the dependent variable.

Falling water Water that transfers gravitational potential energy to kinetic energy.

Fermentation The reaction in which the enzymes in yeast turn glucose into ethanol and carbon dioxide.

Flammable Easily ignited and capable of burning rapidly.

Fluid A liquid or a gas.

Food additive A substance added to a food in order to preserve it or to improve its taste, texture or appearance.

Fossil fuel Fuel obtained from long-dead biological material.

Fraction Hydrocarbons with similar boiling points separated from crude oil.

Fractional distillation A way to separate liquids from a mixture of liquids by boiling off the substances at different temperatures, then condensing and collecting the liquids.

Free electron Electron that moves about freely inside a metal and is not held inside an atom.

Frequency The number of wave crests passing a fixed point every second.

FSH Follicle stimulating hormone, a female hormone that stimulates the eggs to mature in the ovaries, and the ovaries to produce hormones, including oestrogen.

G

Gamete Sex cell which has half the chromosome number of ordinary cells.

Gas A state of matter.

Generator A machine that produces a voltage.

Gene A short section of DNA carrying genetic information.

Genetic engineering/modification A technique for changing the genetic information of a cell.

Geothermal energy Energy from hot underground rocks.

Global dimming The reflection of sunlight by tiny solid particles in the air.

Global warming The increasing of the average temperature of the Earth.

Gravitational potential energy Energy of an object due to its position in a gravitational field. Near the Earth's surface, change of g.p.e. (in joules, J) = weight (in newtons, N) × vertical distance moved (in metres, m).

Gravitropism Response of a plant to the force of gravity controlled by auxin.

Greenhouse gas Gases, such as carbon dioxide and methane, which absorb infrared radiated from the Earth, and result in warming up the atmosphere.

Ground heat Geothermal energy that heats buildings directly.

Group All the elements in each column (labelled 1 to 7 and 0) down the periodic table.

H

Hardening The process of reacting plant oils with hydrogen to raise their melting point. This is used to make spreadable margarine.

Hazard A hazard is something (for example, an object, a property of a substance or an activity) that can cause harm.

Herbivore Animal that feeds on plants.

High-alloy steel Expensive alloy of iron mixed with relatively large proportions of other metals, for example, stainless steel which contains nickel and chromium along with the iron.

Homeostasis The maintenance of constant internal body conditions.

Hydration A reaction in which water (H_2O) is chemically added to a compound.

Hydrocarbon A compound containing only hydrogen and carbon.

Hydrogenated oil Oil which has had hydrogen added to reduce the degree of saturation in the hardening process to make margarine.

Hydrophilic The water-loving part of an emulsifier molecule.

Hydrophobic The water-hating hydrocarbon part of an emulsifier molecule.

Hypothesis A proposal intended to explain certain facts or observations.

I

Immune system The body system which recognises and destroys foreign cells or proteins such as invading pathogens.

Immunisation Giving a vaccine that allows immunity to develop without exposure to the disease itself.

Impulse Electrical signal carried along the neurons.

Incomplete combustion When a fuel burns in insufficient oxygen, producing carbon monoxide as a toxic product.

Indicator species Lichens or insects that are particularly sensitive to pollution and so can be used to indicate changes in the environmental pollution levels.

Infectious Capable of causing infection.

Infectious disease Disease which can be passed from one individual to another.

Infrared radiation Electromagnetic waves between visible light and microwaves in the electromagnetic spectrum.

Inheritance of acquired characteristics Jean-Baptiste Lamarck's theory of how evolution took place.

Inherited Passed on from parents to their offspring through genes.

Inoculate To make someone immune to a disease by injecting them with a vaccine which stimulates the immune system to make antibodies against the disease.

Input energy Energy supplied to a machine.

Insulator Material/object that is a poor conductor.

Internal environment The conditions inside the body.

Interval The quantity between readings, for example, a set of 11 readings equally spaced over a distance of 1 m would give an interval of 10 cm.

Ion A charged particle produced by the loss or gain of electrons.

Ionic bond The electrostatic force of attraction between positively and negatively charged ions.

J

Jean-Baptiste Lamarck French biologist who developed a theory of evolution based on the inheritance of acquired characteristics.

Joule (J) The unit of energy.

K

Kidney Organ which filters the blood and removes urea, excess salts and water.

Kilowatt (kW) 1000 watts.

Kilowatt-hour (kW h) Electrical energy supplied to a 1 kW electrical device in 1 hour.

Kinetic energy Energy of a moving object due to its motion; kinetic energy (in joules, J) $= \frac{1}{2} \times$ mass (in kilograms, kg) $\times$ (speed)^2 (in m^2/s^2).

Kingdom The highest group in the classification system, for example, animals, plants.

L

Line graph Used when both variables are continuous. The line should normally be a line of best fit, and may be straight or a smooth curve. (Exceptionally, in some (mainly biological) investigations, the line may be a 'point-to-point' line.)

Limewater The common name for calcium hydroxide solution.

Liquid A state of matter.

Longitudinal wave Wave in which the vibrations are parallel to the direction of energy transfer.

Low-alloy steel Alloy of iron containing small amounts (1–5 per cent) of other metals.

M

Machine A device in which a force applied at a point produces another force at another point.

Malnourished The condition when the body does not get a balanced diet.

Mantle The layer of the Earth between its crust and its core.

Mass The quantity of matter in an object; a measure of the difficulty of changing the motion of an object (in kilograms, kg).

Mass number The number of protons plus neutrons in the nucleus of an atom.

Maximise Make as big as possible.

Mean The arithmetical average of a series of numbers.

Mechanical wave Vibration that travels through a substance.

Menstrual cycle The reproductive cycle in women controlled by hormones.

Metabolic rate The rate at which the reactions of your body take place, particularly cellular respiration.

Microorganism Bacteria, viruses and other organisms that can only be seen using a microscope.

Microwave Part of the electromagnetic spectrum.

Minimise Make as small as possible.

Mixture When some elements or compounds are mixed together and intermingle but do not react together (i.e. no new substance is made). A mixture is *not* a pure substance.

Molecule A group of atoms bonded together, for example, PCl$_5$.

Monitor Observations made over a period of time.

Monomers Small reactive molecules that react together in repeating sequences to form a very large molecule (a polymer).

Mortar A building material used to bind bricks together. It is made by mixing cement and sand with water.

Motor neuron Neuron that carries impulses from the central nervous system to the effector organs.

MRSA Methicillin-resistant *Staphylococcus aureus*. An antibiotic-resistant bacterium.

Mutation A change in the genetic material of an organism.

N

National Grid The network of cables and transformers used to transfer electricity from power stations to consumers (i.e. homes, shops, offices, factories, etc.).

Natural classification system Classification system based on the similarities between different living organisms.

Natural selection The process by which evolution takes place. Organisms produce more offspring than the environment can support so only those which are most suited to their environment – the 'fittest' – will survive to breed and pass on their useful characteristics.

Nerve Bundle of hundreds or even thousands of neurons.

Nervous system See Central nervous system.

Neuron Basic cell of the nervous system which carries minute electrical impulses around the body.

Neutron A dense particle found in the nucleus of an atom. It is electrically neutral, carrying no charge.

Newton (N) The unit of force.

Nitrogen oxides Gaseous pollutants given off from motor vehicles, a cause of acid rain.

Non-renewable Something that cannot be replaced once it is used up.

Normal Straight line through a surface or boundary perpendicular to the surface or boundary.

Nuclear fission The process in which certain nuclei (uranium-235 and plutonium-239) split into two fragments, releasing energy and two or three neutrons as a result.

Nucleus The very small and dense central part of an atom which contains protons and neutrons.

O

Obese Very overweight, with a BMI of over 30.

Oestrogen Female sex hormone which stimulates the lining of the womb to build up in preparation for a pregnancy.

Opinion A belief not backed up by facts or evidence.

Optical fibre Thin glass fibre used to send light signals along.

Oral contraceptive Hormone contraceptive that is taken by mouth.

Ore Rock which contains enough metal to make it economically worthwhile to extract the metal.

Organic waste Waste material from living organisms, for example, garden waste.

Oscillate Move to and fro about a certain position along a line.

Ovaries Female sex organs which contain the eggs and produce sex hormones during the menstrual cycle.

Overweight A person is overweight if their body carries excess fat and their BMI is between 25 and 30.

Ovipositor A pointed tube found in many female insects which is used to lay eggs.

Ovulation The release of a mature egg from the ovary in the middle of the menstrual cycle.

Oxidised A reaction where oxygen is added to a substance (or when electrons are lost from a substance).

P

Pancreas An organ that produces the hormone insulin and many digestive enzymes.

Pandemic When more cases of a disease are recorded than normal in a number of different countries.

Particulate Small solid particle given off from motor vehicles as a result of incomplete combustion of its fuel.

Pathogen Microorganism which causes disease.

Payback time Time taken for something to produce savings to match how much it cost.

Period The stage in the menstrual cycle when the lining of the womb is lost.

Periodic table An arrangement of elements in the order of their atomic numbers, forming groups and periods.

Perpendicular At right angles.

Photosynthesis The process by which plants make food using carbon dioxide, water and light energy.

Phototropism The response of a plant to light, controlled by auxin.

Phytomining The process of extraction of metals from ores using plants.

Pitch The pitch of a sound increases if the frequency of the sound waves increases.

Pituitary gland Small gland in the brain which produces a range of hormones controlling body functions.

Placebo A substance used in clinical trials which does not contain any drug at all.

Plane mirror A flat mirror.

Polymer A substance made from very large molecules made up of many repeating units, for example, poly(ethene).

Polymerisation The reaction of monomers to make a polymer.

Power The energy transformed or transferred per second. The unit of power is the watt (W).

Precise A precise measurement is one in which there is very little spread about the mean value. Precision depends only on the extent of random errors – it gives no indication of how close results are to the true value.

Prediction A forecast or statement about the way something will happen in the future. In science it is not just a simple guess, because it is based on some prior knowledge or on a hypothesis.

Product A substance made as a result of a chemical reaction.

Progesterone Female sex hormone used in the contraceptive pill.

Propene An alkene with the formula C_3H_6.

Proton A tiny positive particle found inside the nucleus of an atom.

Puberty The stage of development when the sexual organs and the body become adult.

Pyramid of biomass A model of the mass of biological material in the organisms at each level of a food chain.

R

Radio wave Longest wavelength of the electromagnetic spectrum.

Range The maximum and minimum values of the independent or dependent variables; important in ensuring that any pattern is detected.

Rarefaction Stretched apart.

Reactant A substance we start with before a chemical reaction takes place.

Reactivity series A list of elements in order of their reactivity. The most reactive element is put at the top of the list.

Real image An image formed where light rays meet.

Receptor Special sensory cell that detects changes in the environment.

Red-shift Increase in the wavelength of electromagnetic waves emitted by a star or galaxy due to its motion away from us. The faster the speed of the star or galaxy, the greater the red-shift is.

Reduction A reaction in which oxygen is removed (or electrons are gained).

Reflector A surface that reflects radiation.

Reflex arc The sense organ, sensory neuron, relay neuron, motor neuron and effector organ which bring about a reflex action.

Reflex Rapid automatic response of the nervous system that does not involve conscious thought.

Refraction The change of direction of a light ray when it passes across a boundary between two transparent substances (including air).

Relationship The link between the variables that were investigated. These relationships may be: causal, i.e. changing x is the reason why y changes; by association, i.e. both x and y change at the same time, but the changes may both be caused by a third variable changing; by chance occurrence.

Renewable Something that is replaced as fast as it is used up.

Renewable energy Energy from sources that never run out including wind energy, wave energy, tidal energy, hydroelectricity, solar energy and geothermal energy.

Repeatable A measurement is repeatable if the original experimenter repeats the investigation using same method and equipment and obtains the same results.

Reproducible A measurement is reproducible if the investigation is repeated by another person, or by using different equipment or techniques, and the same results are obtained.

Resolution This is the smallest change in the quantity being measured (input) of a measuring instrument that gives a perceptible change in the reading.

Resonate When sound vibrations build up in a musical instrument and cause the sound from the instrument to become much louder.

Respiration The process by which food molecules are broken down to release energy for the cells.

Risk The likelihood that a hazard will actually cause harm. We can reduce risk by identifying the hazard and doing something to protect against that hazard.

S

Safe medicine A medicine that does not cause any unreasonable side effects while curing a disease.

Sankey diagram An energy transfer diagram.

Saturated hydrocarbon Describes a hydrocarbon that has single carbon–carbon bonds only (no C=C bonds).

Secreting Releasing chemicals such as hormones or enzymes.

Sense organs Collection of special cells known as receptors which respond to changes in the surroundings (for example, eye, ear).

Sensory neuron Neuron which carries impulses from the sensory organs to the central nervous system.

Sewage treatment plant A site where human waste is broken down using microorganisms.

Sexual reproduction Reproduction which involves the joining (fusion) of male and female gametes producing genetic variety in the offspring.

Shell (or energy level) An area in an atom, around its nucleus, where the electrons are found.

Smart polymer Polymers that change in response to changes in their environment.

Smelting Heating a metal ore in order to extract its metal.

Solar cell Electrical cell that produces a voltage when in sunlight; solar cells are usually connected together in solar cell panels.

Solar energy (light energy) Energy from the Sun or other light source.

Solar heating panel Sealed panel designed to use sunlight to heat water running through it.

Solar power tower Tower surrounded by mirrors that reflect sunlight onto a water tank at the top of the tower.

Solid A state of matter.

Sound A form of mechanical energy.

Species A group of organisms with many features in common which can breed successfully producing fertile offspring.

Specific heat capacity Energy needed by 1 kg of the substance to raise its temperature by 1 °C.

Speed Distance moved ÷ time taken.

Stable medicine A medicine which does not break down under normal conditions.

Stainless steel A chromium–nickel alloy of steel which does not rust.

Start-up time Time taken for a power station to produce electricity after it is switched on.

Statin Drug which lowers the blood cholesterol levels and improves the balance of HDLs to LDL.

Steel An alloy of iron with small amounts of carbon or other metals, such as nickel and chromium, added.

Step-down transformer Used to step the voltage down, for example, from the grid voltage to the mains voltage used in homes and offices.

Step-up transformer Used to step the voltage up, for example, from a power station to the grid voltage.

Steroid (anabolic) Drug that is used illegally by some athletes to build muscles and improve performance.

Stimuli A change in the environment that is detected by sensory receptors.

Stomata Openings in the leaves of plants (particularly the underside) which allow gases to enter and leave the leaf. They are opened and closed by the guard cells.

Sulfur dioxide A toxic gas whose formula is SO_2. It causes acid rain.

Symbol equation A balanced chemical equation showing the formula of each reactant and product in the reaction, for example, $H_2 + Cl_2 \rightarrow 2HCl$

Synapse The gap between neurons where the transmission of information is chemical rather than electrical.

T

Tectonic plates The huge slabs of rock that make up the Earth's crust and top part of its mantle.

Temperature The degree of hotness of a substance.

Temperature difference Difference in temperature between two points.

Territory An area where an animal lives and feeds, which it may mark out or defend against other animals.

Thalidomide A drug that caused deformities in the fetus when given to pregnant women to prevent morning sickness.

Thermal decomposition The breakdown of a compound by heat.

Tide Rise and fall of sea level because of the gravitational pull of the Moon and the Sun.

Tissue culture Using small groups of cells from a plant to make new plants.

Titanium A shiny, corrosion-resistant metal used to make alloys.

Transformer Electrical device used to change an (alternating) voltage. A step-up transformer is used to step the voltage up, for example, from a power station to the grid voltage. A step-down transformer is used to step the voltage down, for example, from the grid voltage to the mains voltage used in homes and offices.

Transition metal These elements are from the central block of the periodic table. They have typical metallic properties and form coloured compounds.

Transverse wave Wave in which the vibrations are perpendicular to the direction of energy transfer.

Trial run Preliminary work that is often done to establish a suitable range or interval for the main investigation.

Turbine A machine that uses steam or hot gas to turn a shaft.

U

Ultrasonic wave Sound wave at frequencies greater than 20 000 Hz which is the upper frequency limit of the human ear.

Unsaturated hydrocarbon A hydrocarbon whose molecules contain at least one carbon–carbon double bond.

Unsaturated oil Plant oil whose molecules contain at least one carbon–carbon double bond.

Urea The chemical produced by the breakdown of amino acids in the liver which is removed by the kidneys.

Urine The liquid produced by the kidneys containing the metabolic waste product urea along with excess water and salts from the body.

Useful energy Energy transferred to where it is wanted in the form it is wanted.

V

Vaccination Introducing small quantities of dead or inactive pathogens into the body to stimulate the white blood cells to produce antibodies that destroy the pathogens. This makes the person immune to future infection.

Vaccine The dead or inactive pathogen material used in vaccination.

Valid Suitability of the investigative procedure to answer the question being asked.

Variable Physical, chemical or biological quantity or characteristic.

Variable – categoric Categoric variables have values that are labels. For example, names of plants or types of material.

Variable – continuous Can have values (called a quantity) that can be given by measurement (for example, light intensity, flow rate, etc.).

Variable – control A variable which may, in addition to the independent variable, affect the outcome of the investigation and therefore has to be kept constant or at least monitored.

Variable – dependent The variable for which the value is measured for each and every change in the independent variable.

Variable – independent The variable for which values are changed or selected by the investigator.

Vegetable oil Oil extracted from plants.

Vibrate Oscillate rapidly (or move to and fro rapidly about a certain position).

Virtual image An image formed where light rays appear to come from.

Virus Microorganism which takes over body cells and reproduces rapidly, causing disease.

Viscosity The resistance of a liquid to flowing or the 'thickness' or resistance of a liquid to pouring.

Visible light The part of the electromagnetic spectrum that can be detected by the human eye.

W

Wasted energy Energy that is not usefully transferred.

Watt (W) The unit of power.

Wavelength The distance from one wave crest to the next wave crest (along the waves).

Wave Disturbance in water.

Wave speed Speed of travel of a wave.

White blood cell Blood cell which is involved in the immune system of the body, engulfing bacteria, making antibodies and making antitoxins.

White light Light that consists of all the colours of the visible spectrum.

Wind Moving air.

Withdrawal symptom The symptom experienced by a drug addict when they do not get the drug to which they are addicted.

Word equation A way of describing what happens in a chemical reaction by showing the names of all reactants and the products they form.

Index